Mechanics of fracture

VOLUME 7

Experimental evaluation of stress concentration and intensity factors

Mechanics of fracture
edited by GEORGE C. SIH

VOLUME 1

Methods of analysis and solutions of crack problems

VOLUME 2

Three-dimensional crack problems

VOLUME 3

Plates and shells with cracks

VOLUME 4

Elastodynamic crack problems

VOLUME 5

Stress analysis of notch problems

VOLUME 6

Cracks in composite materials

VOLUME 7

Experimental evaluation of stress concentration and intensity factors

Mechanics of fracture 7

Experimental evaluation of stress concentration and intensity factors

Useful methods and solutions to experimentalists in fracture mechanics

Edited by

G. C. SIH
Professor of Mechanics and
Director of the Institute of
Fracture and Solid Mechanics

Lehigh University
Bethlehem, Pennsylvania

1981

MARTINUS NIJHOFF PUBLISHERS

THE HAGUE / BOSTON / LONDON

Distributors:

for the United States and Canada

Kluwer Boston, Inc.
190 Old Derby Street
Hingham, MA 02043
USA

for all other countries

Kluwer Academic Publishers Group
Distribution Center
P.O. Box 322
3300 AH Dordrecht
The Netherlands

Library of Congress Catalog Card Number: 81-50356

ISBN 90-247-2558-5 (this volume)
ISBN 90-247-3006-6 (series)

Contents

Editor's preface   IX

Contributing authors   XV

Introductory chapter

Experimental fracture mechanics: strain energy density criterion   *G. C. Sih*   XVII

Chapter 1

Stress concentrations
A. J. Durelli

1.1   Introduction   1
1.2   Advantages and disadvantages of stress analysis methods used to determine stress concentrations   16
1.3   Compilation of results   24
1.4   Geometrically non-linear stress concentrations   43
1.5   Stress concentrations in mixed boundary value problems   67
1.6   Stress concentrations in some specific problems   94
1.7   Stress concentrations in three-dimensional problems   105
1.8   Dynamic stress concentrations   134
1.9   Unconventional approaches to the study of stress concentrations   146
References   155

Chapter 2

Use of photoelasticity in fracture mechanics
C. W. Smith

2.1   Introduction   163
2.2   Analytical foundations for cracked bodies   167
2.3   Experimental considerations   174

2.4   Application of the frozen stress method   177
2.5   Summary and conclusions   186
References   186

Chapter 3

Elastic stress intensity factors evaluated by caustics
P. S. Theocaris

3.1   Introduction   189
3.2   The basic formulas   191
3.3   The equations of caustics   195
3.4   Properties of the caustics at crack tips   201
3.5   The case of birefringent media   218
3.6   The case of anisotropic media   223
3.7   Interacting crack problems   226
3.8   Branched crack problems   229
3.9   Interface crack problems   231
3.10 V-notch problems   234
3.11 Shell problems   238
3.12 Plate problems   240
3.13 Other applications   242
3.14 Discussion   244
3.15 Conclusions   247
References   248

Chapter 4

Three-dimensional photoelasticity: stress distribution around a through thickness crack
G. Villarreal and G. C. Sih

4.1   Introduction   253
4.2   Hartranft-Sih plate theory   254
4.3   Triaxial crack border stress field   259
4.4   Experimental considerations: specimens and materials   262
4.5   Test procedure: frozen stress technique   264
4.6   Comparison of Hartranft-Sih theory with experiments   271
References   279

Chapter 5

Experimental determination of dynamic stress intensity factors by shadow patterns   *J. Beinert and J. F. Kalthoff*

5.1 Introduction   281
5.2 Physical and mathematical principles of the method   282
5.3 Theoretical analysis of the shadow pattern after Manogg: Mode-I-loaded stationary crack   285
5.4 Validity of the analysis for stationary cracks under dynamic loading   296
5.5 The dynamic correction for propagating cracks   297
5.6 Experimental technique and evaluation procedure   308
5.7 Applications   317
5.8 Concluding remarks   328
References   328

Chapter 6

Experimental determination of stress intensity factor by COD measurements   *E. Sommer*

6.1 Introduction   331
6.2 Principle of the interference optical technique   331
6.3 Determination of stress intensity factor from crack opening displacement   335
6.4 Conclusion   346
References   347

Author's Index   349

Subject Index   353

Editor's Preface

Experiments on fracture of materials are made for various purposes. Of primary importance are those through which criteria predicting material failure by deformation and/or fracture are investigated. Since the demands of engineering application always precede the development of theories, there is another kind of experiment where conditions under which a particular material can fail are simulated as closely as possible to the operational situation but in a simplified and standardized form. In this way, many of the parameters corresponding to fracture such as toughness, Charpy values, crack opening distance (COD), etc. are measured. Obviously, a sound knowledge of the physical theories governing material failure is necessary as the quantity of interest can seldom be evaluated in a direct manner. Critical stress intensity factors and critical energy release rates are examples.

Standard test of materials should be distinguished from basic experiments. They are performed to provide routine information on materials responding to certain conditions of loading or environment. The tension test with or without a crack is among one of the most widely used tests. Because they affect the results, with size and shape of the specimen, the rate of loading, temperature and crack configuration are standardized to enable comparison and reproducibility of results. The American Society for Testing Materials (ASTM) provides a great deal of information on recommended procedures and methods of testing. The objective is to standardize specifications for materials and definition of technical terms.

This seventh volume in this series on *Mechanics of Fracture* is concerned with experimental fracture mechanics. The major portion of the book deals with experimental techniques for determining quantities that are fundamental to the formulation of fracture mechanics theories. They include photoelasticity and the frozen stress technique, the method of caustics and COD measurements. These methods enable a coherent tie to be established between theory and experiment so that field data can be interpreted systematically. There is no intention to give a complete coverage of experimental fracture mechanics. Fracture and fatigue testing deserve separate treatment and are purposely excluded from this volume.

The Introductory Chapter emphasizes that fracture mechanics research must necessarily involve a combination of analytical and experimental work. Experimental data become physically meaningful only when they can be fitted into the framework of existing theories. There is a lack of quantitative theory in such areas as ductile fracture and fatigue – even, to some extent, in very simple problems. Therefore, more attention should be focused on the development of theory, for without it experimental fracture mechanics may come to a standstill. It is necessary to advance a criterion that can consistently describe different types of fracture modes as the primary physical parameters are varied.

The discipline of continuum mechanics aims at predicting the global behavior of solids from a knowledge of the properties of the constituents. These properties are assumed to be determinable from conventional test specimens without considering initial defects. This represents a differential of size scale in terms of material damage in comparison with the fracture specimen which accounts for the presence of macrocracks. Predictions are based on knowing the properties of the material in regions near the crack front. The validity of a theory can therefore be judged from its ability to correlate data collected from material and fracture testings. This serves as an independent check.

With an appropriate criterion, basic fracture mechanics can be employed to establish step-by-step procedures for analyzing materials that fail by inelastic deformation and fracture. The critical strain energy density criterion is introduced because it offers a degree of versatility unmatched by other known criteria. It provides a unified approach to establishing a means of weighing between energy density associated with volume change $(\Delta W/\Delta V)_v$ and shape change $(\Delta W/\Delta V)_d$. Their individual contributions are assumed to control yielding and fracture. In fact, the ratio $(\Delta W/\Delta V)_v/(\Delta W/\Delta V)_d$ can be related to fracture toughness divided by yield strength. In this connection, the variation of $(\Delta W/\Delta V)_v/(\Delta W/\Delta V)_d$ across a member and its material properties determine the relative amounts of flat and slant fracture and hence the degree of metal ductility.

Material seperation is inherently an irreversible path dependent process. It is not sufficient to model the degree of irreversibility simply as nonlinear behavior on a load-deformation diagram. Unless the nonlinear behavior is correlated with material damage, no physical meaning can be attached to the data and no new experience can be gained to advance theory. Typically, in metal fracture, a crack first spreads in a stable fashion before rapid growth. During this stage, the crack can change shape and size depending on the local rate of energy dissipation.

A method for generating crack profiles and growth rate is proposed in which the ratio S_j/r_j is constant along the new profiles during the fracture process ($j = 1, 2, \ldots, n$ represents the number of crack growth increments between initiation and termination). This ratio, being equal to $(\Delta W/\Delta V)_c$, is a characteristic material property. Under monotonic loading, each increment of crack growth is taken to correspond to an increment of loading. Similar treatment is also applied to fatigue crack propagation.

Chapter 1 introduces the notion of stress concentration. The stress concentration factor is normally taken as the ratio between the maximum stress in the solid and a reference stress. The latter is obtained from an elementary formula for stress based either on the gross or net area of a cross-section through the plane on which the maximum stress takes place. When reference is made to the net section area in axial loading, the resulting ratio of force and area is referred to as the nominal stress. This stress quantity tends to increase faster than the maximum stress as the net section area is decreased.* it follows that the stress concentration factor decreases with the net section area. The opposite trend prevails when the gross area is used to calculate the reference stress. A clear definition of the particular cross-section over which the reference stress is calculated is needed to avoid confusion.

It is well known that the stress concentration factor is mathematically related to the stress intensity factor. For a narrow ellipse in an infinite body with a remote uniform stress field of magnitude σ, it can be shown that the maximum stress σ_m, at the end of the major axis, is given by $2\sigma\sqrt{a/\rho}$ with ρ being the radius of curvature at the ends of the major axis of length $2a$. Hence, the relation $\sigma\sqrt{a} = (\sigma_m/2)\sqrt{\rho}$ follows immediately. The quantity $\sigma\sqrt{a}$ is the stress intensity factor k_1 whose critical value multiplied by $\sqrt{\pi}$ is the plane strain fracture toughness of material, K_{1c}. Even though K_{1c} is usually obtained from a tensile specimen with a central crack, it is (in linear elastic fracture mechanics) a material constant that should be independent of specimen geometries and loadings.

The various approaches to the study of stress concentrations are reviewed. Solutions for a wide variety of problems of practical interest are also listed. They should provide a wealth of information to those who practice in engineering design. The conversion from stress concentration factors to stress intensity factors can be made in a straight forward manner as already mentioned for the example cited earlier.

---

* This applies to specimens with traction free boundary conditions.

Photoelasticity is a well developed experimental technique for determining the stress distribution in transparent model materials. The application of this method to crack problems was begun approximately two decades ago, and the current state of the art is described in Chapter 2. When a crack is loaded and viewed in a polarized light field, interference bands known as isochromatics or stress fringes are observed. Each fringe represents the locus of points of the same maximum shearing stress, given by the difference of the two principal normal stresses, in the plane normal to the incident light beam. The sum of the two principal stresses is constant along lines known as isopachics. The isochromatic fringe bands form contour-like patterns. Their order is proportional to the maximum shearing stress. Closely spaced narrow bands in the immediate vicinity of the crack tip indicate regions of high stress gradients. The bands become widely separated away from the crack. Once the distribution of the local stresses is measured, the stress intensity factor may be determined for complicated specimen geometries for which an analytical solution may not be possible.

It is possible to apply photoelasticity to crack problems in which their three-dimensional nature must be treated. While conventional photoelasticity is limited to finding stresses only in thin planes, interior stresses may be obtained by applying the frozen-stress technique. A treatment which locks into the specimen the three-dimensional stress pattern is followed by slicing the specimen. Then the interior-stress information can be obtained photoelastically from a series of two-dimensional analyses of each slice. The accuracy depends on the number and thickness of the slices. The freezing technique depends on the diphase behavior of certain polymetric materials which are composed of a network of primary and secondary molecular bonds. Upon increasing the temperature in such polymers, the secondary bonds break down while the stress pattern is maintained by loading and becomes permanently locked in the specimen when the secondary bonds are reformed as the temperature is lowered. This procedure has been applied to evaluate the variation of stresses across a finite thickness plate containing a through crack. The results are presented in Chapter 4.

Another equally useful optical method for finding stresses is the less well known technique of caustics. The fundamental equations and physical meaning of the method are given in Chapter 3. Particular emphasis is placed on evaluating the crack tip stress intensity factors in isotropic and anisotropic media. Several problems dealing with crack interaction, crack branching, interface debonding, etc. are discussed. Applications are also made to the bending of plates and shells.

The caustics method utilizes light generated from a point light source. The stress intensification, say around the crack tip region, causes a reduction of the specimen thickness and refraction index of the material. The light transmitted through the transparent specimen tends to deflect away from the crack tip and cast a shadow image of the crack on a flat screen behind the specimen. The crack tip appears as a dark spot surrounded by a bright light concentration called a caustic. Obviously, the concentration of light decreases with increasing distance from the crack tip because the angular deflection of the light beam decreases. The intensity of the deflected light can be correlated with the crack tip stress field. Since the dimensions of the caustics are quantitatively related to the state of affairs near the crack, the shadow pattern can be used to determine the stress intensity factor. The technique is simple and does not require any sophisticated instrumentation. The same principle can be applied to non-transparent specimens where caustics are produced by light reflecting from the deformed surface.

Chapter 4 presents an experimental verification of the Hartranft-Sih plate theory by application of the frozen-stress photoelastic method. This theory has been used to solve for the three-dimensional stress distribution ahead of the crack in a plate of finite thickness. Unlike conventional plate theories in which the in-plane stress components are arbitrarily assumed to be independent of the transverse variable, the interaction of coupling between the in-plane and transverse variations of the local stress singularity is derived from the condition of plane strain near the crack front except within a thin boundary layer next to the plate surface. The stress solution in this layer is interpolated between the free surface boundary conditions and the requirement of continuity with the interior solution.

The specimen geometry is that of a transverse crack of length $2a$ in a diphase polymeric plate of thickness h. A variety of specimens covering the range from thin to moderately thick plates were examined. The transverse variation of the in-plane stress components was obtained from slices parallel to the plate surfaces. The results are reported in terms of the ratio h/a and agree very well with the theoretical predictions. The slices near the plate surface were made extremely thin in order to monitor the steep stress gradients. The proposed boundary layer thickness equation $\delta = (h/4)/[1 + 4(h/a)]$ gives a useful indication of the region outside of which the two-dimensional solution ceases to be valid. It is in the boundary layer that the stress vary most rapidly, a phenomenon that has not received much attention in the past.

In Chapter 5, the experimental determination of dynamic stress intensity factors is discussed. The method of caustics appears to be particularly advantageous for analyzing dynamic crack propagation because of the simplicity with which measurements can be made on a moving subject. For crack problems attention can be limited to a small area surrounding the tip. Unlike static crack problems in which the stress fields are determined uniquely by external loading and geometric boundary conditions, the dynamic fracture process is complicated by crack motion and reflection of stress waves. Presented in this Chapter are methods for the experimental evaluation of stationary cracks loaded dynamically and moving cracks loaded statically. The results are compared with those obtained under static conditions for several examples.

The sixth and last Chapter in this volume describes a simpler optical interference optical technique that is most suitable for measuring the crack opening displacement (COD) in transparent materials. With the appropriate alignment of the cracked specimen, parallel monochromatic light can be reflected from the opposing crack surfaces. A system of interference fringes is then developed because of the phase difference between the two reflected light waves. The interference fringes correspond to lines of equal distance between two adjacent crack surfaces. Measurements can also be performed on three-dimensional crack configurations. The determination of stress intensity factors from the COD, however, depends on the availability of theoretical results. The lack of reliable stress solutions for three-dimensional crack configurations limits the usefulness of such experimental methods, particularly in obtaining quantities that are not directly measurable.

Before closing this preface, I wish to reiterate the theme of this volume. Experimental fracture mechanics should be distinguished from standard fracture testing in that it is a necessary extension of theoretical analysis. On the other hand, quantitative predictions based on material failure theories can inject new ideas into experimentation. This interplay between theory and experiment is not being widely practiced in fracture mechanics. Unless the objectives of fracture mechanics are more clearly defined among the two groups of workers little progress can be made.

I wish at this point to thank Professor R. J. Hartranft of Lehigh University for helpful discussions concerning the material in the Introductory Chapter. The valuable assistance of my two secretaries, Mrs. Barbara DeLazaro and Mrs. Constance Weaver is also gratefully acknowledged. Their patience and hard work has made the completion of this volume possible.

Lehigh University   G. C. Sih
Bethlehem, Pennsylvania

March, 1981

Contributing authors

J. BEINERT
Institut für Festkörpermechanik der Fraunhofer-Gesellschaft, Freiburg, West Germany

A. J. DURELLI
Mechanical Engineering, University of Maryland, College Park, Maryland

J. F. KALTHOFF
Institut für Festkörpermechanik der Fraunhofer-Gesellschaft, Freiburg, West Germany

G. C. SIH
Institute of Fracture and Solid Mechanics, Lehigh University, Bethlehem, Pennsylvania

C. W. SMITH
Engineering Science and Mechanics, Virginia Polytechnic Institute and State University, Blacksburg, Virginia

E. SOMMER
Fraunhofer-Institut für Werkstoffmechanik, Freiburg, West Germany

P. S. THEOCARIS
Department of Theoretical and Applied Mechanics, The National Technical University of Athens, Athens, Greece

G. VILLARREAL
Department of Civil Engineering, Universidad de Nuevo Leon, Monterrey, Mexico

G. C. Sih

Introductory chapter:

Experimental fracture mechanics: Strain energy density criterion

1. Preliminary remarks

The objective of this Introductory Chapter is to impart a better understanding of the mutual dependence of theory and experiment in fracture mechanics. An experiment cannot be considered successful unless the results are interpreted within the framework of some pre-conceived concept and/or theoretical prediction. Otherwise the test results are of no more value than to predict the results of other tests conducted under exactly the same conditions. Interpreting experimental results requires a sound knowledge of the fracture process and governing physical principles. In addition, theoretical speculations on these processes must be guided by experimental observations. The interplay between theory and experiment is essential to the development of both theories and experiments.

Linear elastic fracture mechanics [1] has not only been receiving world-wide attention from the academic community in recent years but it has also been increasingly employed as a useful engineering approach to optimizing material selection and engineering design. The wide acceptance of this relatively new discipline is largely due to the ease with which the theory can be applied quantitatively for making design decisions to prevent sudden fracture of structural components. As research investigators devoted more effort to ductile fracture, their viewpoints began to diverge; instead of a discipline with unifying concepts and methods, diverse parameters, methods of testing, and jargons emerged. The proliferation of organizations and committees within professional societies had led to the endorsement of a variety of test procedures and specimens. The variety of different specimens now being used in laboratories throughout the world make comparison of results obtained in different laboratories increasingly difficult. In addition, there is a widespread feeling that neither modern electronic computers nor laboratory equipment are being used effectively. Instead of serving to provide insight into the physical fracture process and to establish order within the confusing abundance of data, they seem to have become self-serving and generated an exhaustive amount of numerical results and unsupported data. The close relationship which should exist between methods of analysis and perceptual concepts of the physics of fracture is not being maintained in current research. Analytical work in fracture mechanics is often solely concerned with refinement of computations and fails to grasp the connection of the problem with reality. Experimental work on the other hand often becomes merely the collection of data even after the physical phenomena are better understood. Undoubtedly, these tendencies are responsible for the paucity of progress toward discovering the physical laws that govern material failure. Whenever possible, the separate efforts and multifarious facts should be united by clearly identifying their internal connections.

It is not the intention of this communication to propose a unified theory for fracture mechanics, rather the aim is to call for an increase in the degree of consistency and rigor in research. Too much emphasis cannot be placed on the development of theoretical concepts as an adjunct to experimentation so that different forms of failure may be treated not as unrelated independent events but rather as different aspects of a single, more fundamental phenomenon.

2. Mechanical properties of materials

The study of the mechanical properties of material is concerned with two kinds of quantities: those such as displacement, force, etc., which are directly measurable, and those such as stress, energy, etc., which are defined in terms of measurable quantities by means of mathematical relations. Both types are used to describe the response of materials to external mechanical influences, and are involved in predicting the ability of materials to sustain reversible and irreversible deformations and to resist fracture. The commonly known properties such as modulus of elasticity, yield strength, ultimate strength, etc., have values which measured on a certain scale, depend on rate of loading, size and shape of specimen, environmental conditions such as temperature, moisture, etc.

The material response can vary widely from long-time stable failure by excessive deformation to short-time unstable fracture by sudden separation of material.

Physical hypotheses. To bring some order into this wide range of material behavior, failures should be related to experimentally measurable or mathematically defined quantities through physical hypotheses. Among the more common quantities used as criteria of failure are the maximum normal stress, maximum shear stress, maximum strain, total strain energy density, energy of distortion and octahedral shearing stress. It should be made clear at the start that the validity of a hypothesis cannot be proved mathematically from any relations previously obtained by definition and physical hypothesis. A case in point is the energy release rate quantity defined in linear elastic fracture mechanics to represent the energy dissipated by the unit extension of a crack.*

Mathematics cannot prove or disprove any hypothesis and should be regarded strictly as a tool to obtain verifiable consequences for testing physical hypotheses and through new experiments. The test of any hypothesis of a physical nature rests on the *consistency* of its derived consequences with measurements and its usefulness in forecasting physical events. Passing such a test, the hypothesis is satisfactory and useful.

Continuum scale. All hypotheses advanced in the study of material behavior presuppose that the physical behavior of systems is described using a pre-determined scale. The continuum concept that a material can be repeatedly sub-divided into smaller elements with dimensions approaching zero must, of course, be modified when examining real materials. The element size (volume $\Delta V = \Delta x \Delta y \Delta z$ in Figure 1) must obviously be small enough for numerical accuracy, yet not so small that the continuum model looses its validity. The conflicting needs of numerical analysis and experimentation must be assessed and made compatible with one another. The continuum level, visible to an unaided eye, is normally referred to as the macroscopic scale. Nonhomogeneity of a real material at the microscopic level that prevails within the basic

---

* The calculation of the energy release rate from a stress field involves all of the assumptions used to obtain the stresses. The hypothesis that the crack grows when this equals some critical value is an additional, separate hypothesis.

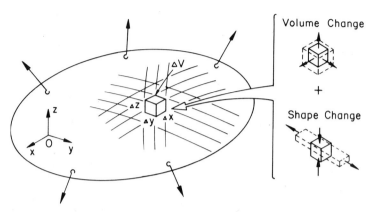

Figure 1. A continuum divided into finite number of macroscopic elements undergoing volume and shape change

element is neglected.* The validity of the continuum approximation is tested through comparison of theory and experiment.

Strain energy density. When the continuum in Figure 1 is subjected to mechanical loading, each element $\Delta x \Delta y \Delta z$ will be deformed. The energy stored per unit of volume ΔV, called the strain energy density function $\Delta W/\Delta V$, will in general vary as a function of x, y, z and t. In terms of the principal stresses, σ_1, σ_2 and σ_3, it is well-known that for linear, isotropic materials,

$$\frac{\Delta W}{\Delta V} = \frac{1}{2E} (\sigma_1^2 + \sigma_2^2 + \sigma_3^2) - \frac{\nu}{E} (\sigma_1\sigma_2 + \sigma_1\sigma_3 + \sigma_2\sigma_3) \tag{1}$$

where ν is Poisson's ratio and E is Young's modulus. Equation (1) may be written as the sum of a part due to change of volume,

$$\left(\frac{\Delta W}{\Delta V}\right)_v = \frac{1-2\nu}{6E} (\sigma_1 + \sigma_2 + \sigma_3)^2 \tag{2}$$

and of a part due to change of shape,

$$\left(\frac{\Delta W}{\Delta V}\right)_d = \frac{1+\nu}{6E} [(\sigma_1 - \sigma_2)^2 + (\sigma_2 - \sigma_3)^2 + (\sigma_3 - \sigma_1)^2] \tag{3}$$

* This simply means that the continuum model usually applied to situations where the microstructure of the material is neglected unless otherwise accounted for.

Hence,

$$\frac{\Delta W}{\Delta V} = \left(\frac{\Delta W}{\Delta V}\right)_v + \left(\frac{\Delta W}{\Delta V}\right)_d \tag{4}$$

The expressions $(\Delta W/\Delta V)_v$ and $(\Delta W/\Delta V)_d$ are useful as their individual contribution when weighed appropriately may be associated with the cause of failure resulting from fracture and yielding.

Test of failure criterion. A typical failure criterion states that failure takes place in a material when some quantity, say $\Delta W/\Delta V$ in equation (1) as proposed by Beltrami and Haigh, reaches certain critical value $(\Delta W/\Delta V)_c$ characteristic of the material. A simple test of the criterion is to compare the results of two simple tests, namely a bar loaded to failure in tension or torsion as shown in Figures 2(a) and 2(b). For a linear elastic material, $\Delta W/\Delta V$ in simple tension can be found from equation (1) by letting $\sigma_1 = \sigma$ and $\sigma_2 = \sigma_3 = 0$. Then

$$\left(\frac{\Delta W}{\Delta V}\right)_c = \frac{1}{2E}\sigma^2 \tag{5}$$

Setting $\sigma_1 = \tau$, $\sigma_2 = -\tau$ and $\sigma_3 = 0$ for case of torsion, equation (1) yields

$$\left(\frac{\Delta W}{\Delta V}\right)_c = \frac{1+\nu}{E}\tau^2 \tag{6}$$

at the instant of failure. If the bars are made of the same material, the

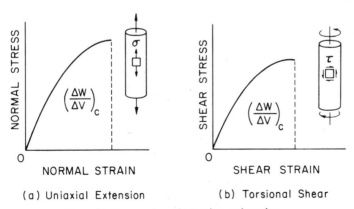

(a) Uniaxial Extension   (b) Torsional Shear

Figure 2. Bar specimen in tension and torsion

normal stress σ in equation (5) and shear stress τ in equation (6) corresponding to failure should be related as

$$\frac{\tau}{\sigma} = \frac{1}{\sqrt{2(1+\nu)}} \tag{6}$$

if the failure criterion is applicable. As ν changes from 0.0 to 0.5, the ratio τ/σ varies in the range 0.707 to 0.577.

The strain energy density criterion is not limited to linear elastic materials but may be applied also to materials with nonlinear behavior as well. The value $(\Delta W/\Delta V)_c$ represents the area under the true stress and strain curve and can be computed directly from experimental data.* Measured values of $(\Delta W/\Delta V)_c$ for a variety of engineering materials can be found in [3, 4].

Specimen size effect. The tensile strength of smooth round bar specimens that are metallurgically and geometrically** similar has been shown to decrease as the bar diameter is increased. This is known as the size effect. The broken surfaces of the specimen show three distinct features of fracture development labelled as I, II and III in Figure 3. Failure by stable crack growth starts from a state of hydrostatic tension

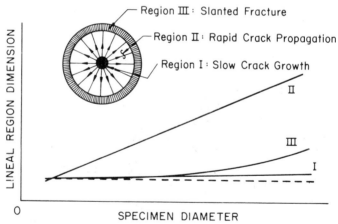

Figure 3. Fracture pattern of metal specimen

* A procedure for determining $(\Delta W/\Delta V)_c$ is presented in an English translation of the Hungarian Standard MSz 4929-76 [2].

** Round bars of different sizes with the same radius to length ratio, say r/L, are said to be geometrically similar. However, even for geometrically similar bars, such quantities as volume to surface ratio will be different for different sizes.

inside the region I. This is followed by rapid or unstable crack propagation denoted by region II. Final separation of the material occurs by slant fracture terminating at the specimen boundary referred to as region III. A plot of the lineal dimensions of these regions versus an order of magnitude increase in the specimen diameter is displayed schematically in Figure 3. Note that region II increases almost directly with specimen size while regions I and III are affected only slightly [5]. The larger specimen would tend to be more brittle because of the larger zone of rapid crack propagation. The lower tensile strength for larger specimens is a direct result of the definition of $\sigma = P/A$ where the rate of increase of the specimen cross-sectional area A exceeds that of the failure load P. The variation of tensile strength with specimen size would be greatly reduced if σ were defined in terms of the area occupied by region I or III instead of the total of all three regions, A. The three regions shown in Figure 3 are typical of aluminum, steel and titanium metal alloys [6] and are relatively independent of the shape of the cross-section.

Size sensitivity of specimens has frequently been attributed to statistical considerations and the weakest-link theory because of inability to understand the fracture development process and to incorporate it into analysis. Since the fracture pattern for many engineering materials is clearly deterministic, the apparent size effect can be resolved by judicious application of a fracture criterion that can methodically account for the physical fracture process. The analysis should distinguish and relate crack initiation, stable and unstable crack growth. These distinct stages of the total fracture process cannot be assumed to coincide nor be treated separately even though their interval of occurrence can be very short for some materials. In fact, the trade-off between specimen size and strength depends intimately on the interaction of the various intervening stages of crack growth. A quantitative assessment of the stability and instability of the fracture process can be made from the strain energy* content of the system whose rate of release** dictates the different stages of fracture. The creation of free surfaces originating from defects is thus an inherent part of the size effect phenomenon associated with strength.

---

* The events of slow and rapid crack growth can be described from the strain energy density criterion, $(\Delta W/\Delta V)_c = S_c/r_c$, where the critical ligament size r_c determines the termination of slow crack growth and the beginning of rapid crack propagation. It is characteristic of the material and cannot be taken as zero. The condition $(\Delta W/\Delta V)_c = $ const. corresponds to the failure of local elements while S_c to the onset of global instability.
** The rate of energy release is determined by conditions around the defect that can either be stress rate or strain rate controlled.

3. Fracture mechanics discipline

Fracture mechanics, like the other branches of mechanics, is concerned with predicting the response of a system to external disturbances. The idealized system is assumed to contain a large aggregate of continuous elements whose mechanical properties can be determined from *material testing*. The specimens and test conditions must be carefully selected to simulate the failure mode of the smaller element as closely as possible. When defects or cracks are present, their significance must be assessed through analysis and *fracture testing* based on certain physical hypotheses. The energy content of specimens for material and fracture testing should be proportioned to control the degree of material damage at different scale levels and monitored as material behavior by the load versus displacement relationships. The connection between observed data and underlying theories must be clearly and consistently established for material damaged at the microscopic and macroscopic scale level. To reiterate, the primary objective of fracture mechanics is to develop procedures so that *data collected on laboratory specimens will be useful in the design of larger size structural components*. The final test of predictive capability lies in expressing results in terms of directly measurable quantities such as allowable load and dimensions of structural components.

Classification of fracture. Several types of fractures in various metal alloys have been observed and classified qualitatively according to the test conditions under which they were created. Two of the most common types of fracture are known as cleavage and shear. They are differentiated by the orientation of the fracture surface with reference to the maximum value of a neighboring stress component. For instance, cleavage fracture would be associated with a surface normal to the largest tensile stress and shear fracture with a surface parallel to the direction of maximum shear stress. This approach can be subjective and lead to ambiguous interpretation. Fundamental difficulties are immediately disclosed when the fracture surface is examined more closely at the microscopic level; an apparent cleavage fracture may consist of minute shear planes and an apparent shear fracture may contain minute clevage planes. This is illustrated by the macrocrack under tension shown in Figure 4. The elastic zone ahead of the macrocrack contains shear planes and the plastic region to the side of the macrocrack contains cleavage planes. Additional confusion arises in attempting to relate brittle fracture to cleavage and ductile fracture to shear. Descrip-

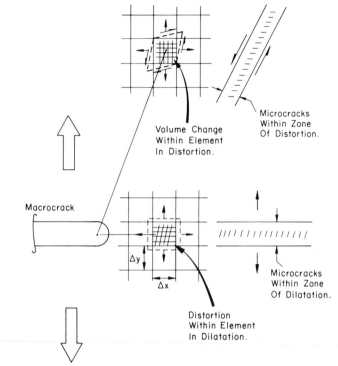

Figure 4. Zones of distortion and dilatation ahead of a macrocrack in tension

tive terminologies based on phenomenological studies alone are not a proper substitute for basic understanding of the fracture of a real material. Although there are no theories at present that can account for the detailed mechanism of fracture, still the problem is not being approached in a more systematic and intelligent manner.

The assumption of material homogeneity is usually invoked in a theory for the determination of stresses and deformations in a solid because of the enormous complexity required to specify the various forms of irregularities in a real material. For this obvious reason, experimental observation would tend to scatter about the orderly behavior predicted by an idealized theory of homogeneous bodies. The mechanical conditions which cause solids to deform permanently or to fracture can be related to shape change (distortion) and volume change (dilatation). In general, a material is subjected to both distortion and dilatation and the corresponding energies can be computed from equations (2) and

(3) depending on the state of stress. Suppose that an element $\Delta x \Delta y$ ahead of the crack in Figure 4 is under the condition

$$\sigma_3 = \nu(\sigma_1 + \sigma_2) \tag{8}$$

known as plane strain. Let σ_1 be directed normal to the crack and σ_2 along the crack. Making use of equation (8), $(\Delta W/\Delta V)_v$ becomes

$$\left(\frac{\Delta W}{\Delta V}\right)_v = \frac{1-2\nu}{E}[\sigma_1 + \sigma_2 + \nu(\sigma_1 + \sigma_2)]^2 \tag{9}$$

and $(\Delta W/\Delta V)_d$ takes the form

$$\left(\frac{\Delta W}{\Delta V}\right)_d = \frac{1+\nu}{3E}[\sigma_1 + \sigma_2 + \nu(\sigma_1 + \sigma_2)]^2 - \frac{1+\nu}{E}[\sigma_1\sigma_2 + \nu(\sigma_1 + \sigma_2)^2] \tag{10}$$

The ratio of equations (9) and (10) can be expressed in terms of σ_1/σ_2 as follows:

$$\frac{\Delta W/\Delta V)_v}{(\Delta W/\Delta V)_d} = \frac{(1-2\nu)(1+\nu)[(\sigma_1/\sigma_2)+1]^2}{2(1+\nu)^2[(\sigma_1/\sigma_2)+1]^2 - 6(\sigma_1/\sigma_2) - 6\nu[(\sigma_1/\sigma_2 + 1)]} \tag{11}$$

A simple calculation for $\nu = 0.3$ shows that $(\Delta W/\Delta V)_v$ is 6.5 times larger than $(\Delta W/\Delta V)_d$ in the case $\sigma_1 = \sigma_2$. This explains why failure is most likely to occur from the crack tip region where the material undergoes a large change in volume. The ratio $(\Delta W/\Delta V)_v/(\Delta W/\Delta V)_d$ decreases rapidly when σ_1/σ_2 increases. The strain energy associated with shape change becomes higher than the strain energy associated with volume change when $\sigma_1/\sigma_2 = 3.36$. Equation (11) is solved numerically for $\nu = 0.3$ and the results are given in Table I.

In general, $(\Delta W/\Delta V)_v$ and $(\Delta W/\Delta V)_d$ are of the same order of magnitude and hence both quantities should be weighed* as a sum according to equation (4). The proportion between $(\Delta W/\Delta V)_v$ and $(\Delta W/\Delta V)_d$

Table I
Values of $(\Delta W/\Delta V)_v/(\Delta W/\Delta V)_d$ for $\nu = 0.3$

σ_1/σ_1	1	2	3	4	5
$(\Delta W/\Delta V)_v/(\Delta W/\Delta V)_d$	6.500	0.839	1.143	2.108	0.696

* The yield criterion of von Mises only accounts for $(\Delta W/\Delta V)_d$ in equation (3) and neglects the energy component due to volume change.

corresponding to a nonuniform state of stress will be different for each element within the continuum and used to explain failure by deformation and fracture. Referring to the macro element $\Delta x \Delta y$ ahead of the crack in Figure 4, the principal stresses σ_1 and σ_2 are equal. Even though $(\Delta W/\Delta V)_v$ is 2.24 times larger than $(\Delta W/\Delta V)_d$, the distortional component of the strain energy density is not negligible and is assumed to be responsible for the creation of slip planes or microcracks within a macroscopic layer where dilatation dominates. Similarly, maximum distortion of the continuum element occurs to the side of the macrocrack. By the same agrument, the dilatational component of the strain energy density is assumed to be associated with the creation of cleavage planes that are perpendicular to the direction of tension. It appears that material damage at the microscopic and macroscopic scale can be connected by referring to the proportion of $(\Delta W/\Delta V)_v$ and $(\Delta W/\Delta V)_d$ within the continuum element. For a macrocrack under tension, the model in Figure 4 suggests macrofracture to coincide with the location where $(\Delta W/\Delta V)_v > (\Delta W/\Delta V)_d$ and macroyielding with the direction where $(\Delta W/\Delta V)_d > (\Delta W/\Delta V)_v$. These locations may be determined by appealing to physical hypotheses.

Yielding and fracture. Since microyielding may attribute to macrofracture and microfracture to macroyielding, the two processes of yielding and fracture are inseparable. They are unique features of material damage and should be treated simultaneously by a single failure criterion, if possible. To this end, the strain energy density function $\Delta W/\Delta V$ will be used in the form

$$\frac{\Delta W}{\Delta V} = \frac{S}{r} \qquad (12)$$

where S is the strain energy density factor[*] [7, 8] and r the radial distance measured from the site of possible failure initiation[**], say the crack tip. The singular dependency $1/r$ is a fundamental character[***] of

[*] The factor S can be computed from the stress intensity factors k_1, k_2 and k_3 only for predicting the initiation of crack propagation in a linear elastic material. The S-criterion is applied through a given stress and strain relation, but is not restricted by it.
[**] The site of failure can be a reentrant corner, inclusion, void, etc., and is not restricted to the crack tip as in the conventional theory of linear fracture mechanics.
[***] Failure criteria based on the amplitude of the asymptotic stresses would tend to be restrictive since the singular character of the stresses depends on the constitutive relation of the material. Moreover, the order of singularity for each stress component may be different as in the case of finite deformation theory.

the Newtonian potential and is independent of the constitutive relation. The response of a particular material is reflected through the factor S. There are three basic hypotheses of the strain energy density criterion:

Hypothesis (1):

The location of fracture coincides with the location of minimum strain energy density $(\Delta W/\Delta V)_{min}$, and yielding with maximum strain energy density, $(\Delta W/\Delta V)_{max}$.

Hypothesis (2):

Failure by stable fracture or yielding occurs when $(\Delta W/\Delta V)_{min}$ or $(\Delta W/\Delta V)_{max}$ reach their respective critical values.

Hypothesis (3):

The amount of incremental growth $r_1, r_2, \ldots, r_j, \ldots, r_c$ is governed by

$$\left(\frac{\Delta W}{\Delta V}\right)_c = \frac{S_1}{r_1} = \frac{S_2}{r_2} = \cdots = \frac{S_j}{r_j} = \cdots = \frac{S_c}{r_c} = \text{const.} \tag{13}$$

There is unstable fracture or yielding when the critical ligament size r_c is reached.

The intent of these proposed hypotheses is to provide a means for correlating the data obtained from material and fracture testing. By this procedure, a means of transferring results from specimen to structure may be gained. Such information is not only valuable to a better understanding of size effect but also opens the door to design in the inelastic range.

Line crack in tension. In order to be more specific, consider a line crack of length $2a$ in a material under uniform tension. Let attention be focused on the variation of the strain energy density $\Delta W/\Delta V$ with the angle θ in Figure 5 for a fixed value of r. The value, r_0, is taken to be sufficiently small that the expression [7]

$$r_0\left(\frac{\Delta W}{\Delta V}\right) = \frac{(1+\nu)}{8E}(3 - 4\nu - \cos\theta)(1 + \cos\theta)\,\sigma^2 a \tag{14}$$

derived from the asymptotic stresses in a linear elastic material is valid. The magnitude of the remote uniform stress is denoted by σ.

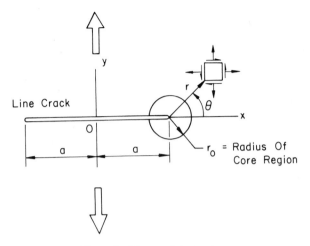

Figure 5. Line crack in tension

The conditions specified by Hypothesis (1) may be satisfied by requiring that $\partial(\Delta W/\Delta V)/\partial\theta = 0$. This gives two solutions for θ. The first solution, $\theta_0 = 0$, corresponds to $(\Delta W/\Delta V)_{min}$ and determines the direction of crack propagation:

$$r_0\left(\frac{\Delta W}{\Delta V}\right)_{min} = \frac{(1 + \nu)(1 - 2\nu)}{2E}\,\sigma^2 a \tag{15}$$

By Hypothesis (2), stable fracture is assumed to occur when $(\Delta W/\Delta V)_{min}$ takes the critical value $(\Delta W/\Delta V)_c$ while Hypothesis (3) assumes that rapid fracture begins when r_0 reaches the critical ligament length, r_c. Equation (15) may be decomposed into the dilatational component

$$r_0\left(\frac{\Delta W}{\Delta V}\right)_v = \frac{2(1 - 2\nu)(1 + \nu)^2}{3E}\,\sigma^2 a \tag{16}$$

and distortional component

$$r_0\left(\frac{\Delta W}{\Delta V}\right)_d = \frac{(1 + \nu)(1 - 2\nu)}{2E}\,\sigma^2 a \tag{17}$$

A comparison of equations (16) and (17) reveals that the strain energy density for volume change in an element along the path of expected crack extension is greater than the strain energy density for shape change. The maximum value of $\Delta W/\Delta V$ corresponds to $\theta_0 = \pm \cos^{-1}(1 - 2\nu)$ and is given by

$$r_0\left(\frac{\Delta W}{\Delta V}\right)_{max} = \frac{(1 + \nu)(1 - \nu)^2}{2E} \sigma^2 a \qquad (18)$$

Based on the definitions of $(\Delta W/\Delta V)_v$ and $(\Delta W/\Delta V)_d$ in equations (2) and (3), equation (18) may be separated into

$$r_0\left(\frac{\Delta W}{\Delta V}\right)_v = \frac{(1 + \nu)^2(1 - \nu)(1 - 2\nu)}{3E} \sigma^2 a \qquad (19)$$

for dilatation and

$$r_0\left(\frac{\Delta W}{\Delta V}\right)_d = \frac{(1 - \nu^2)(1 - \nu + 4\nu^2)}{6E} \sigma^2 a \qquad (20)$$

for distortion. Since $(\Delta W/\Delta V)_d > (\Delta W/\Delta V)_v$ as given by equations (19) and (20) corresponds to yielding in the areas to the sides of the crack at $\theta_0 = \pm \cos^{-1}(1 - 2\nu)$.

Additional light may be shed on yielding and fracture by studying the variation of $(\Delta W/\Delta V)_v/(\Delta W/\Delta V)_d$* in the vicinity of the crack tip. Figure 6 gives a plot of this quantity with Poisson's ratio ν. For an element situated at $\theta_0 = 0$, equations (16) and (17) yield

$$\frac{(\Delta W/\Delta V)_v}{(\Delta W/\Delta V)_d} = \frac{2(1 + \nu)}{1 - 2\nu} \qquad (21)$$

which takes a minimum value of 2.0 at $\nu = 0.0$ and increases rapidly as ν increases. Along the directions of yielding, however, the ratio

$$\frac{(\Delta W/\Delta V)_v}{(\Delta W/\Delta V)_d} = \frac{2(1 + \nu)(1 - 2\nu)}{1 - \nu + 4\nu^2} \qquad (22)$$

acquires a maximum value of 2.0 at $\nu = 0.0$ and decreases to zero at $\nu = 0.5$. The deviation between the curves becomes appreciable as ν is raised beyond 0.2.

For a uniaxial tensile specimen with no crack, it can be shown that

$$\left(\frac{\Delta W}{\Delta V}\right)_v = \frac{1 - 2\nu}{6E} \sigma^2, \left(\frac{\Delta W}{\Delta V}\right)_d = \frac{1 + \nu}{3E} \sigma^2 \qquad (23)$$

---

* The gradient of this ratio is related to the determination of flat and slant fracture [9].

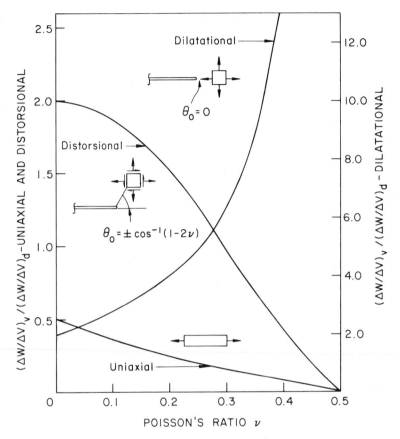

Figure 6. Variations of strain energy density ratio with Poisson's ratio

The ratio

$$\frac{(\Delta W/\Delta V)_v}{(\Delta W/\Delta V)_d} = \frac{1-2\nu}{2(1+\nu)} \tag{24}$$

is also plotted as a function of Poisson's ratio in Figure 6 for comparison purpose. Note that equation (24) is the exact reciprocal of equation (21). Hence, the mechanical behavior of the uniaxial tensile specimen does not appear to represent the state of affairs ahead of the crack. There is the need for improved design in test specimens.

Critical ligament. The circle of radius r_0 in Figure 5 has been referred to

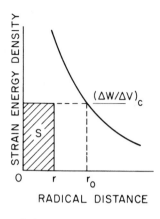
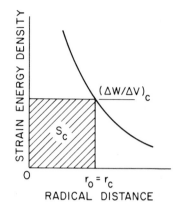

(a) Energy Stored Within   (b) Onset Of Rapid
    Core Region   Crack Growth

Figure 7. Critical strain energy density

as the core region. Depending on the local stress state, r_0 may or may not coincide with the critical ligament, say r_c, that corresponds to the onset of unstable crack extension. For simplicity sake, let $r_0 = r_c$ as illustrated in Figure 7. Material failure is initiated in the core region as soon as the strain energy density level reaches $(\Delta W/\Delta V)_c$, Figure 7(a). However, rapid crack growth will not begin until the area under the $\Delta W/\Delta V$ versus r curve reaches S_c, Figure 7(b). For fracture specimens with load applied perpendicular to the crack, it can be easily shown that

$$S_c = r_c \left(\frac{\Delta W}{\Delta V}\right)_c = \frac{(1+\nu)(1-2\nu)K_{1c}^2}{2\pi E} \tag{25}$$

An approximate estimate of r_c can be made by relating the fracture test result in equation (25) to the tensile yield strength σ_{ys}. The distortional strain energy density reduces to

$$\left(\frac{\Delta W}{\Delta V}\right)_d = \frac{1+\nu}{3E}\,\sigma_{ys}^2 \tag{26}$$

Substituting equations (25) and (26) into (4) and rearranging the result, an expression for r_c is derived:

$$r_c = \frac{3(1-2\nu)}{2\pi}\left(\frac{K_{1c}}{\sigma_{ys}}\right)^2\left[1+\frac{(\Delta W/\Delta V)_v}{(\Delta W/\Delta V)_d}\right]^{-1} \tag{27}$$

Equation (27) shows that r_c depends not only on K_{1c} and σ_{ys} but also on the local stress state through the ratio $(\Delta W/\Delta V)_v/(\Delta W/\Delta V)_d$. Several values of r_c for the 4140 steel with $\nu = 0.3$ are calculated and tabulated in Table II together with other material parameters. A value of $(\Delta W/\Delta V)_v/(\Delta W/\Delta V)_d$ equal to 6.5 is used and is obtained from Figure 6 for $\nu = 0.3$.

The quantity $(\Delta W/\Delta V)_c$ may be obtained independently from the area under the true stress and strain curve or calculated from S_c and r_c by means of equation (12).

Data interpretation. The fundamental physical principles that govern material failure have not been well-established over the years. The value of material and fracture testing is uncertain where physical inter-pretation of data cannot be made possible. More emphasis must be placed on a combination of theory and experimentation. Otherwise, no confidence can be placed on judging the reliability of new results.

A frequent problem of interest is the failure of specimens under combined loading where both normal stress σ and shear stress τ are present. According to the maximum principal stress criterion, the value of σ and τ at failure, when normalized by the ultimate strength σ_u, can be written as

$$\left(\frac{\sigma}{\sigma_u}\right)^2 + \left(\frac{\tau}{\sigma_u}\right)^2 = 1 \tag{28}$$

while the maximum total strain energy density criterion renders

$$\left(\frac{\sigma}{\sigma_u}\right)^2 + 2(1+\nu)\left(\frac{\tau}{\sigma_u}\right)^2 = 1 \tag{29}$$

Table II
Material parameters for 4140 steel

σ_{ys} (ksi)	K_{1c} (ksi$\sqrt{\text{in}}$)	S_c (lb/in)	r_c (in)
220	35	3.379	0.00065
205	52	7.459	0.00164
180	80	17.655	0.00510
165	120	39.724	0.01345

at failure. The choice of selecting equation (28) and (29) depends on the data which may often require fitting an empirical equation of the form

$$A\left(\frac{\sigma}{\sigma_u}\right)^m + B\left(\frac{\tau}{\sigma_u}\right)^n = 1 \tag{30}$$

The parameters A, B, m and n for a particular material are determined from experimental data. It is desirable to consider these parameters as material constants and to attribute apparent changes in material behavior* to the presence of defects or differences in loading and/or specimen geometry.

Internal defect. Suppose that the loading and specimen geometry in a given test are fixed but the material may contain initial defects. The specific case of a cylindrical bar subjected to combined tension and torsion will be considered for evaluating the effect of initial defects on the parameters in equation (30).

Let the cylindrical bar of radius b in Figure 8 contain a flat circular defect of radius a. A force P and torque T are applied such that the bar experiences both a nominal normal stress $\sigma = P/\pi b^2$ and a nominal shear stress $\tau = 2Ta/\pi b^4$. The corresponding stress intensity factors are ** [10]

$$k_1 = \frac{\sigma}{1-(a/b)^2} \sqrt{\frac{ac}{b}} f_1\left(\frac{a}{b}\right) \tag{31a}$$

$$k_3 = \frac{\tau}{1-(a/b)^4} \sqrt{\frac{ac}{b}} f_3\left(\frac{a}{b}\right) \tag{31b}$$

in which the functions $f_1(a/b)$ and $f_3(a/b)$ stand for

$$f_1\left(\frac{a}{b}\right) = \frac{2}{\pi}\left(1 + \frac{1}{2}\frac{a}{b} - \frac{5}{8}\frac{a^2}{b^2}\right) + 0.268\frac{a^3}{b^3} \tag{32a}$$

$$f_3\left(\frac{a}{b}\right) = \frac{4}{\pi}\left(1 + \frac{1}{2}\frac{a}{b} + \frac{3}{8}\frac{a^2}{b^2} + \frac{5}{16}\frac{a^3}{b^3} - \frac{93}{128}\frac{a^4}{b^4} + 0.038\frac{a^5}{b^5}\right) \tag{32b}$$

Referring to [7], the strain energy density factor S for mixed mode crack

---

* Parameters associated with mechanical behavior should be clearly distinguished from those material properties governed by the chemical composition and microstructure.
** The critical value k_{1c} differs from the conventional fracture toughness value K_{1c} by a factor $\sqrt{\pi}$, i.e., $K_{1c} = \sqrt{\pi} k_{1c}$.

extension with $k_2 = 0$ is given by

$$S = a_{11}k_1^2 + a_{33}k_3^2 \tag{33}$$

where the coefficients are given by

$$a_{11} = \frac{1+\nu}{8E}(3 - 4\nu - \cos\theta)(1 + \cos\theta) \tag{34a}$$

$$a_{33} = \frac{1+\nu}{2E} \tag{34b}$$

The angle θ that makes S a minimum is found by taking $\partial S/\partial\theta = 0$. Letting $S_{\min} = S_c$ and using equations (31), it is found that

$$(1 - 2\nu)f_1^2\left(\frac{a}{b}\right)\sigma^2 + \frac{f_3^2(a/b)}{[1 + (a/b)^2]^2}\tau^2 = \frac{2Eb[1 - (a/b)^2]^2 S_c}{ac(1+\nu)} \tag{35}$$

In view of equation (25), the fracture toughness S_c can be determined from a K_{1c} fracture test:

$$K_{1c} = \frac{\sigma_u}{1 - (a_0/b)^2}\sqrt{\frac{\pi a_0 c}{b}}f_1\left(\frac{a_0}{b}\right) \tag{36}$$

in which a_0 is the crack radius at instability. With the aid of equations (25) and (36), equation (35) becomes

$$\frac{f_1^2(a/b)}{f_1^2(a_0/b)}\left(\frac{\sigma}{\sigma_u}\right)^2 + \frac{[1 + (a/b)^2]^{-2}}{1 - 2\nu}\frac{f_3^2(a/b)}{f_3^2(a_0/b)}\left(\frac{\tau}{\sigma_u}\right)^2 = \frac{a_0 c_0[1 - (a/b)^2]^2}{ac[1 - (a_0/b)^2]^2} \tag{37}$$

Equation (37) may be rearranged and cast into the form shown by equation (30) with $m = n = 2$. The parameter A is a complicated function of defect size and cylinder geometry and does not depend on material constants. Poisson's ratio ν is the only material constant in the expression for B.

A graphical display of τ/σ_u versus σ/σ_u is given by Figure 8 for $a_0 = 0.016$ in., $b = 2.0$ in. and a material with $\nu = 1/3$. The three solid curves labelled $a/b = 0.008$, 0.012 and 0.018 are calculated from equation (37) and they are plotted next to the dotted curves found from equations (28) and (29) with no consideration given to the presence of initial defects. The relation between τ/σ_u and σ/σ_u is seen to depend sensitively on the size of defects. Therefore defects may contribute significantly to data scattering if they are not accounted for in the analysis.

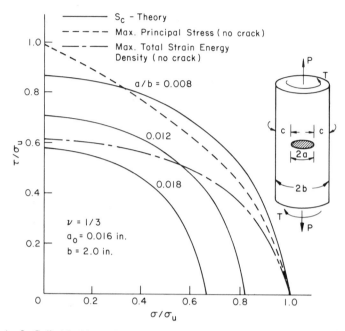

Figure 8. Cylindrical bar with an internal crack subjected to tension and torsion

Surface defect. A cylindrical bar with a surface defect in the form of a circumferential crack may be treated by the same procedure. The configuration of the cracked bar is shown in Figure 9. Under combined tension and torsion, there arise the stress intensity factors [10]

$$k_1 = \sigma \left(\frac{b}{c}\right)^2 \sqrt{\frac{ac}{b}} \, g_1\left(\frac{c}{b}\right) \tag{38a}$$

$$k_3 = \tau \left(\frac{b}{c}\right)^3 \sqrt{\frac{ac}{b}} \, g_3\left(\frac{c}{b}\right) \tag{38b}$$

The functions $g_1(c/b)$ and $g_3(c/b)$ take the forms

$$g_1\left(\frac{c}{b}\right) = \frac{1}{2} \left[1 + \frac{1}{2}\frac{c}{b} + \frac{3}{8}\frac{c^2}{b^2} - 0.363 \frac{c^3}{b^3} + 0.731 \frac{c^4}{b^4} \right] \tag{39a}$$

$$g_3\left(\frac{c}{b}\right) = \frac{3}{8} \left[1 + \frac{1}{2}\frac{c}{b} + \frac{3}{8}\frac{c^2}{b^2} + \frac{5}{16}\frac{c^3}{b^3} + \frac{25}{128}\frac{c^4}{b^4} + 0.208 \frac{c^5}{b^5} \right] \tag{39b}$$

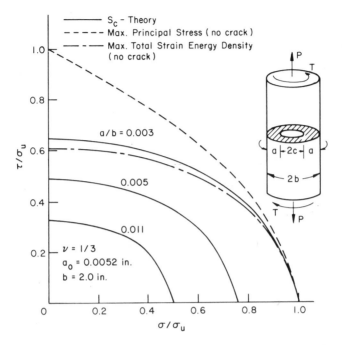

Figure 9. Cylindrical bar with an external crack subjected to tension and torsion

As in the previous example, the S factor obtained from equation (33) by means of equations (38) may be minimized with respect to θ in the coefficients a_{ij} ($i, j = 1, 3$). The resulting expression is

$$(1 - 2\nu) g_1^2\left(\frac{c}{b}\right) \sigma^2 + \left(\frac{b}{c}\right)^2 g_3^2 \left(\frac{c}{b}\right) \tau^2 = \frac{2Eac^3 S_c}{b^3(1 + \nu)} \tag{40}$$

With K_{1c} known as

$$K_{1c} = \sigma_u \left(\frac{b}{c_0}\right)^2 \sqrt{\frac{\pi a_0 c_0}{b}} \, g_1\left(\frac{c_0}{b}\right) \tag{41}$$

S_c in equation (40) may be eliminated using equation (25) to yield the failure envelope,

$$\frac{g_1^2(c/b)}{g_1^2(c_0/b)} \left(\frac{\sigma}{\sigma_u}\right)^2 + \frac{(b/c)^2}{1 - 2\nu} \frac{g_3^2(c/b)}{g_3^2(c_0/b)} \left(\frac{\tau}{\sigma_u}\right)^2 = \frac{a_0 c^3}{a c_0^3} \tag{42}$$

Equation (42) is evaluated numerically for $a_0 = 0.0052$ in, $b = 2.0$ in and $\nu = 1/3$. The results are displayed in Figure 9 for $a/b = 0.003$, 0.005 and 0.011. The curves in Figure 9 for a surface defect are quite different from those in Figure 8 for an internal defect. Both the size and type of defect can affect the combination of σ and τ leading to fracture. This may explain why conventional failure criteria cannot consistently correlate fracture data.

4. Incremental crack growth

Crack growth as a continuous motion is thinkable only in mathematical terms. It is not a precise description because of the discrete character of nature. Physically speaking, what is perceived as a continuous motion is actually a series of ordered states of rest. The interval between these states is not necessarily constant. Depending on the rate of energy release, the increment of length in each successive stage of crack growth can either increase or decrease. The former may lead to unstable crack propagation and the latter to crack arrest. Another commonly observed phenomenon of ductile fracture is the transition from stable to unstable cracking. The quantitative assessment of such behavior requires not only a knowledge of the availability of energy for creation of free surface but also a criterion that determines the rate of incremental crack growth and shape of crack profiles which are needed for experimental verification.

For illustration, the cracking behavior of the two specimens in Figure 10 will be considered. When a crack is completely engulfed in a uniform

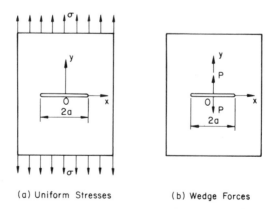

(a) Uniform Stresses   (b) Wedge Forces

Figure 10. Cracked specimens for (a) unstable crack growth and (b) stable crack growth

stress field as shown in Figure 10(a), the energy available to create new free surfaces increases as the crack expands. For the case of a pair of wedge forces applied to the crack surfaces in Figure 10(b), cracking once initiated tends to slow down because the energy available to drive the crack decreases.

Monotonically increasing crack growth. Let the crack motion of the specimen in Figure 11(a) be regarded as a series of finite segments r_1, r_2, etc., of crack growth. The strain energy density factor S is proportional to $\sigma_c^2 a$, where σ_c is the critical stress level* maintained after crack initiation. Hence, the strain energy density factors S_1, S_2, etc., will increase with each increment of growth r_1, r_2, etc., i.e.,

$$S_1 < S_2 < \cdots < S_j < \cdots < S_c \tag{43a}$$

$$r_1 < r_2 < \cdots < r_j < \cdots < r_c \tag{43b}$$

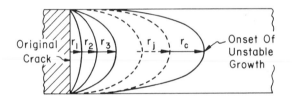

(a) Unstable Growth

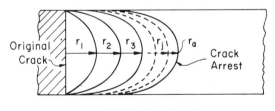

(b) Stable Growth

Figure 11. Incremental crack growth leading to (a) instability and (b) crack arrest

* An overload is inherent in the process of fracture initiation where the crack tip tends to rest against a stronger grain which causes an abrupt increase in energy demand. This is followed by a sudden excess of energy available for release when the local material inhomogeneity is overcome.

It follows from the growth criterion in equation (13) that

$$\frac{a}{r_1} = \frac{a + r_1}{r_2} = \frac{a + r_1 + r_2}{r_3} = \cdots = \text{const.} \qquad (44)$$

If crack initiation starts when $r_1 = r_0$, then a recursion relation for incremental crack advancement can be obtained:

$$r_{n+1} = \left(1 + \frac{r_0}{a}\right) r_n, \quad n \geq 1 \qquad (45)$$

Each consecutive segment of growth is seen to increase and unstable fracture occurs when r_j reaches the critical ligament r_c. In reality, the crack front acquires a curvature across the specimen thickness with the maximum extension taking place in the mid-plane as illustrated in Figures 11. The growth increments $r_1, r_2, \ldots, r_j, \ldots, r_c$ in equation (45) correspond to the maximum values on each crack profile, Figure 11(a).

Monotonically decreasing crack growth. Refer to the crack loaded by wedge forces in Figure 10(b). The crack begins to move by segments r_1, r_2, etc., with decreasing strain energy density S_1, S_2, etc., because S is inversely proportional to the half crack length a, i.e., $S \sim P_c^2/a$ with P_c being the critical load. The conditions

$$S_1 > S_2 > \cdots > S_j > \cdots > S_a \qquad (46a)$$

$$r_1 > r_2 > \cdots > r_j > \cdots > r_a \qquad (46b)$$

imply that the fracture process leads to eventual crack arrest as r_j approaches r_a. Since $(\Delta W / \Delta V)_c = \text{const.}$ during the crack growth process, equation (13) requires that

$$\frac{1}{ar_1} = \frac{1}{(a + r_1)r_2} = \frac{1}{(a + r_1 + r_2)r_3} = \cdots = \text{const.} \qquad (47)$$

which for $r_1 = r_0$ gives the expression

$$r_{n+1} = \frac{r_n}{1 + (r_n/r_0)(r_n/a)}, \qquad n \geq 1 \qquad (48)$$

Figure 11(b) shows the curved crack profiles through the specimen thickness and the decreasing increments of maximum crack growth r_1, $r_2, \ldots, r_j, \ldots, r_a$ at the mid-plane.

Development of crack profile. It is useful to construct a procedure for developing crack profiles so that experiments can be devised to check the validity of the theory. In three dimensions, all the elements located at the prospective sites of the crack profile at a given instance are assumed to satisfy the condition of constant strain energy density $(\Delta W/\Delta V)_c$ as stated by equation (13). The growth condition can be written more precisely as

$$\left(\frac{\Delta W}{\Delta V}\right)_c = \frac{S_1^{(i)}}{r_1^{(i)}} = \frac{S_2^{(i)}}{r_2^{(i)}} = \cdots = \frac{S_j^{(i)}}{r_j^{(i)}} = \cdots = \text{const.} \tag{49}$$

where $i = 1, 2, \ldots, m$ represents the number of points or elements on a crack profile and $j = 1, 2, \ldots, n$ represents the number of crack profiles required to simulate the fracture process.

The procedure for applying the constant strain energy density criterion can be illustrated by referring to a crack located in a tensile plate specimen as shown in Figure 12(a). Because of symmetry, consider only one-half of the plate thickness and a typical crack profile constructed by six points or elements labelled as 1, 2,..., 6. The precise locations of these elements $r_1, r_2, \ldots, r_6$ are fixed by the condition of $(\Delta W/\Delta V)_c =$ const. The variations of $\Delta W/\Delta V$ with the distance measured from the crack front may be found analytically or experimentally. Figure 12(b) shows the $\Delta W/\Delta V$ versus r relation for each of the six elements in

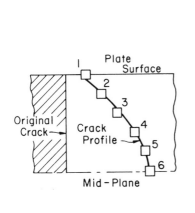

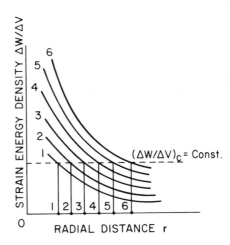

(a) Curved Crack Profile.   (b) Strain Energy Density Variations

Figure 12. Crack growth development

Figure 12(a). The intersections of the line $(\Delta W/\Delta V)_c = $ const. for a given material with the curves determines the values of $r_1, r_2, \ldots, r_6$ that form the crack profile. This step by step procedure can be repeated such that a series of crack profiles are developed to describe the total fracture process from initiation to termination.

Transition from stable to unstable fracture. Instantaneous crack instability is a mathematical abstraction that has no physical counterpart. All fracture processes may be regarded as transitions from stable* to unstable crack propagation. In particular, the commonly observed phenomenon of ductile fracture in structural components is characterized by slow and stable cracking prior to rapid crack propagation. This is normally associated with a nonlinear response on the load versus deformation curve and is attributed to non self-similar crack growth and material deformation beyond the elastic limit. These two effects occur simultaneously under load. There are no known experimental methods that can separate them and measure their individual contributions. It is inaccurate and incorrect to simply identify the nonlinear behavior of metal alloys with plasticity. The presence of a slowly growing crack also contributes to nonlinearity. Nonlinearity is controlled by the rate of loading, specimen size and geometry, crack growth pattern and material properties and is not a well-understood subject for metals. It depends on the scale level, size of physical domain, and time interval on the basis of which the response is recorded. Yielding of the grains in a polycrystal does not necessarily correspond to yielding of polycrystalline specimens measured by a tensile test. Comparison of physically recorded results on the different scale levels should be made with the utmost care. Quantities such as microstress and macrostress, which are frequently interchanged in the literature, must be distinguished in the analysis.

Deformation and fracture are unique in that they occur in an orderly fashion dictated by the laws of nature. Therefore, it is only logical that man-made measurements should be identified with physical events. Otherwise, the results are neither meaningful nor helpful for gaining new experience. This is particularly true in conventional material and fracture tests relative to the initiation of yielding and fracture. It is not clear how these quantities could be used to predict the performance of the materials tested when operated under conditions that are different from those in the laboratory. It is fundamental knowledge that a nonlinear process is path dependent and so must be identified with material

---

* The details of stable fracture initiation require a consideration of material damage at the microscopic level.

damage at each stage of loading on the load-deformation curve. Such capability must be established theoretically and verified experimentally to resolve the size effect* problem.

It is well-known that the deformation and fracture pattern in a material can be controlled by the rate of loading, specimen size and temperature. However, it would be overwhelming to discuss failure modes corresponding to all possible combinations of the variables. To focus ideas, only the specimen thickness h will be altered. The specimen size effect is customarily exhibited by a plot of the critical stress σ_c times the square root of the half crack length a and π, i.e., $\sigma_c \sqrt{\pi a}$,** against the plate thickness h, Figure 13(a). Only when $\sigma_c \sqrt{\pi a}$ ceases to change for h sufficiently large is the quantity $K_{1c} = \sigma_c \sqrt{\pi a}$ a measure of the fracture toughness of the material. In this case, the excess available energy produces fracture surfaces that are predominantly flat. As h is decreased, $\sigma_c \sqrt{\pi a}$ looses its physical meaning and should no longer be referred as fracture toughness or stress intensity factor. The fact that ductile fracture is accompanied by an increase in the load carrying capacity of the specimen should not be interpreted as a change in

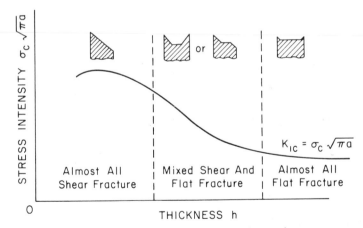

Figure 13. Size effect for a cracked tensile specimen

* The conventional approach of measuring a so-called "fracture toughness" parameter for each different specimen thickness is not sound and does not provide any predictive capacity. It only confirms that the measured parameter cannot be adequately used as a material constant.

** Because the stress intensity factor concept is restricted to flat fracture, the measurement of K_{1c} requires the specimen to be overly thick so that a large amount of energy can be stored and released to initiate the onset of rapid crack propagation.

fracture toughness. First of all, yielding affects only the material behavior on a load deformation diagram and does not constitute a change in material failure properties. On physical grounds, if the fracture toughness is interpreted as characterising the resistance of the material to fracture, it should be a constant with or without yielding. Because K_{1c} is specimen size sensitive, it is more consistent to regard it simply as a measure of incipient rapid fracture rather than an inherent material parameter. Plastic deformation or yielding reduces the amount of energy available to cause unstable fracture. Ductile fracture can only be explained and resolved when the interaction between deformation and fracture are properly accounted for. The nature of this path dependent process must be understood before experiments can be designed. Reference can be made to [11, 12] for a more detailed discussion on this subject.

An inherent characteristic of ductile fracture is that both flat and slant fracture surfaces are produced. The strain energy density criterion [7, 8] has been shown in many previous problems to be well-suited for analyzing mixed mode fracture where the path of crack extension is not known as *a priori*. It is informative to visualize the deformation and fracture history in a tensile specimen with thickness h, Figure 10(a). A crack of length $2a$ with a straight edge lies in a plane normal to the applied stress σ. Figure 14(a) shows the crack profiles during the stage of slow and stable growth while the material near the plate surfaces is deformed beyond the yield point. The local plastic deformation constrains crack growth and delays the start of

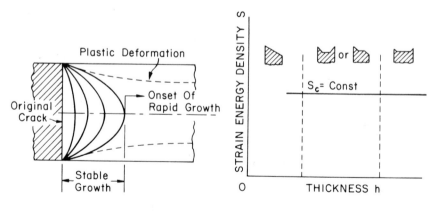

(a) Transition Of Stable To   (b) Constant Fracture Toughness
Unstable Fracture

Figure 14. Ductile fracture in monotonically loaded tensile specimen

unstable crack propagation. This problem has been analyzed [13] for the ductile fracture of plates with different thicknesses. The critical strain energy density factor or fracture toughness S_c was held constant, Figure 14(b), for several plate thickness to crack length ratio. As the plate thickness is continuously decreased, deformation becomes the dominant mode of failure and S_c fails to be applicable. This is why the constant S_c line has not been extended beyond the value of h below which fracture instability no longer occurs.

The transition from stable to unstable fracture is controlled by the critical ligament size r_c and fracture toughness measured by S_c which is proportional to $(\Delta W/\Delta V)_c$. Figures 15 give a comparison of $(\Delta W/\Delta V)_c$, S_c and critical load for two different materials A and B. Material A is said to be tougher than material B, Figure 15(a), if

$$\left(\frac{\Delta W}{\Delta V}\right)_A > \left(\frac{\Delta W}{\Delta V}\right)_B, \quad \left(\frac{\Delta W}{\Delta V}\right)_A r_A > \left(\frac{\Delta W}{\Delta V}\right)_B r_B \quad (50)$$

The tougher material may also sustain a higher critical load as shown in Figure 15(b).

The above procedure gives a complete description of deformation* and

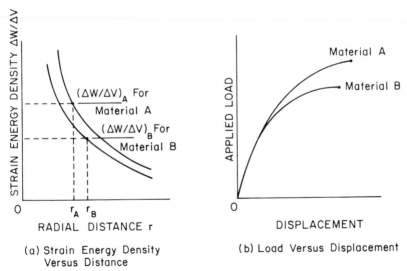

(a) Strain Energy Density   (b) Load Versus Displacement
    Versus Distance

Figure 15. Relationship between toughness and critical load for two materials

---

* In a continuum analysis, microscopic damage of the material enters into the problem through plastic deformation [13].

fracture, Figure 14(a), as the material is loaded incrementally. Each point
on the load-deformation curve, Figure 15(b), can be identified with material
damage such that the critical load corresponds to satisfying the condition
$S_c = r_c(\Delta W/\Delta V)_c$. Allowable load and net section size, which are the basic
information needed in engineering design, can thus be found for structural
members undergoing ductile fracture. The method and analytical results in
[13] serve as a guide to experimentally identify each point of the
load-deformation diagram with material damage. Such analytical capability when verified by experiments will place increased confidence in
translating laboratory data for use in the design of larger size structural
members.

5. Fatigue crack growth: a path dependent process

The behavior of metals in fatigue is a complex process which is not yet
properly understood despite the enormous amount of data generated by
testing. Admittedly, much has been done to examine fracture test
specimens at the microscopic level by identifying cracks at the earliest
possible stage and following their subsequent growth. Many of the
results have unavoidably been couched in terms of microscopic entities
and their interactions: e.g., dislocation, vacancies, grain inhomogeneity,
inclusions, voids, etc. Although these studies offer a detailed description
of the mechanisms fatigue failure, little insight will be gained unless the
results are tied together by theory.

Fatigue is normally separated into initiation* and propagation. The
necessity for addressing these two processes** individually arises
because the theory is not able to bridge the gap between physical events
observed at the microscopic and macroscopic scale level. Unless efforts
are made to relate macroscopic variables with microscopic entities, the
fatigue process will not be completely understood.

No attempt will be made to explain all qualitative observations, for
this would lead to no useful results. The objective is to develop a
procedure within the framework of continuum mechanics such that
experimental data can be interpreted with order and consistency. As in
the case of monotonic loading, there is a need to infuse fatigue crack
growth data with results obtained from materials testing.

---

* The mechanism of fatigue initiation has been attributed to the dislocation process of
extrusion and intrusion and will not be discussed in this communication.
** Fatigue initiation and propagation are intimately related and cannot be justified as two
separate processes on the basis of difficulties arising from theoretical treatment.

Fatigue specimen. Unlike monotonically loaded tensile specimens, the load amplitude is maintained at a level such that the specimen will not break on the first peak load but will eventually break after many cycles of loading. The implication is that damage is being accumulated in the material. A stress amplitude versus number of cycles to failure diagram is often obtained for determining the endurance limit* of a particular material. The loading usually has constant amplitude and frequency while the mean stress level can be changed to determine its effect on the fatigue life

A basic assumption of continuum mechanics is that the mechanical properties of a material can be measured from a test specimen, Figure 16(a). This notion applies to cyclic loading as well where the material now experiences reversal of stress and strain forming hysteresis loops are shown in Figure 16(b). The object is to predict the global response of a solid by assuming that the material properties of all the local elements within the solid can be determined from material testing. As mentioned previously for materials subjected to monotonic loading, there are difficulties associated with this procedure because of the well-known size effect. The uniaxial test specimen may not accurately simulate the failure behavior ahead of a crack. This was discussed in terms of the strain energy density ratio $(\Delta W/\Delta V)_v/(\Delta W/\Delta V)_d$ which is a more sensitive measure of material failure due to deformation and fracture. The fundamental problem of specimen size effect must also overcome in fatigue.

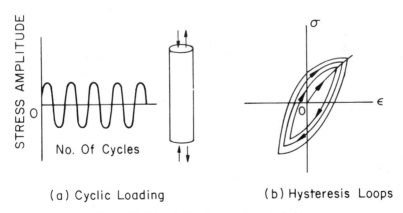

(a) Cyclic Loading   (b) Hysteresis Loops

Figure 16. Material testing specimen in fatigue

* The endurance limit is defined as the stress amplitude corresponding to that portion of the curve which no longer changes with the number of cycle of loading. Not all metals possess this property.

Cracked specimen. Fracture testing considers a larger size specimen with an initial macrocrack. The case of a transverse crack* in a plate has been used extensively for collecting crack propagation data. The standard cyclic loading has constant amplitude and frequency, Figure 17. The uniaxial stress amplitude and number of cycles applied to the specimen at distances remote from the crack shall be denoted by Σ and N, respectively. An element of material ahead of the crack, however, is in a multiaxial stress state where the amplitude and frequency response of a typical stress component, say σ, may, in general, be different from the input. This is illustrated in Figure 18(a) with n being the number of

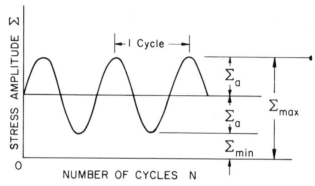

Figure 17. Cyclic loading applied to a crack specimen

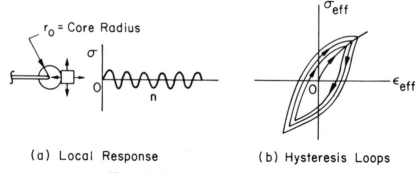

(a) Local Response   (b) Hysteresis Loops

Figure 18. Response of element ahead of crack

* When the material deforms plastically, this specimen can no longer be considered as two-dimensional as the most important aspects of material damage occur in the plate thickness direction.

cycles experienced by the local element. The influence of phase shift between the applied load spectrum (Σ, N) and material response (σ, n) should not be overlooked*. A knowledge of the cyclic response on the local element is pertinent to testing the fatigue properties of metals. Fatigue, as a path dependent process, depends on the load history. Hence, the applied cyclic loads on the test specimen in Figure 16(a) cannot be chosen at will.

Damage accumulation. Under fatigue, the material microstructure is damaged before any failure can be detected or described as a process of macrocrack propagation. It is in this sense that material damage is said to be accumulated**. A convenient measure of this accumulation can be expressed in terms of the effective*** stress σ_{eff} and strain ϵ_{eff} from which the hysteresis energy density, $\Delta W/\Delta V$, can be computed, Figure 18(b).

Suppose that an estimate of the additional hysteresis energy density accumulation in the material element ahead of the crack can be made for each cycle, then a saturation point will eventually be reached at which the element is broken and the macrocrack advances. The amount of growth will depend on the rate of energy release which cannot be arbitrarily assumed. Let the total damage be represented as an average,

$$\sum_{j=1}^{n} \left(\frac{\Delta W}{\Delta V}\right)_j = \left(\frac{\Delta W}{\Delta V}\right)_{ave} \Delta n \qquad (51)$$

where n is the number of cycles for the element to break. In equation (51), an average hysteresis energy density $(\Delta W/\Delta V)_{ave}$ is defined in terms of the interval number of cycles Δn. For simplicity, assume that the energy density datum line corresponds to zero. A fatigue crack growth hypothesis can be stated:

---

* The problem of an elastic plate with an angle crack subjected to cyclic uniaxial loading has been solved [14]. The stress intensity factors k_1 and k_2 are found to depend on the frequency of loading.

** The terminology "damage accumulation" should not imply a direct correspondence between fatigue life and number of load cycles. An occasional high positive overload is known to increase the fatigue life of cracked specimens. Nonlinear behavior of materials is not always intuitively obvious.

*** The effective stress and strain can be defined in terms of the principal stresses σ_1, σ_2 and σ_3 and the principal strains ϵ_1, ϵ_2 and ϵ_3, i.e.,

$$\epsilon_{eff} = \tfrac{1}{3}[(\sigma_1 - \sigma_2)^2 + (\sigma_2 - \sigma_3)^2 + (\sigma_3 - \sigma_1)^2]$$

$$\epsilon_{eff} = \tfrac{1}{3}[(\epsilon_1 - \epsilon_2)^2 + (\epsilon_2 - \epsilon_3)^2 + (\epsilon_3 - \epsilon_1)^2]$$

Crack growth is assumed to occur when the total hysteresis energy density $(\Delta W/\Delta V)_{ave}\Delta n$ *reaches a critical value that is characteristic of the material.*

The above assumption can be expressed as

$$\left(\frac{\Delta W}{\Delta V}\right)_{ave}\Delta n = A \tag{52}$$

in which A is a material constant. With the help of equation (12), equation (52) can be put into the form

$$\left(\frac{\Delta W}{\Delta V}\right)_{ave} = \frac{\Delta S}{\Delta r} = \frac{A}{\Delta n} \tag{53}$$

If the load is applied symmetrically with respect to the crack plane, $\Delta r = \Delta a$ and an expression for $\Delta a/\Delta N$ is obtained:

$$\frac{\Delta a}{\Delta N}\left(\frac{\Delta N}{\Delta n}\right) = C(\Delta S) \tag{54}$$

Neither the constant $C = 1/A$ nor $\Delta N/\Delta n$ can be set to unity unless demonstrated by analysis or experimentation. The quantity ΔS is the change of strain energy density factor for a given increment of crack growth and must be computed for a material that dissipates energy through irreversible deformation and fracture. Unfortunately, there does not exist a suitable constitutive relation* that can realistically describe the response of metals undergoing fatigue. The theory of continuum plasticity leaves much to be desired since the yield conditions have yet to be established on a firm foundation. One of the major shortcomings of the conventional elastic-plastic crack analyses is that the yield criterion is not consistent with the criterion giverning the propagation of the macrocrack. The nature of physics suggests that a single criterion should govern material damage at both the microscopic and macriscopic level.

Fracture surface appearance. Fatigue cracks grow under the conditions of a sustained mean stress $(\Sigma_{max} + \Sigma_{min})/2$ with a superimposed cyclic

* Hopefully, nonlinear analyses can be performed to establish the range of validity of fatigue predictions based on results obtained from the linear theory of elasticity which is used to compute the change of stress intensity factors ΔK in many crack growth models. The conditions under which the linear analysis yields a good approximation should be clearly understood and established.

stress of amplitude Σ_a, Figure 17. The loads may also vary in amplitude depending on the service conditions. For many engineering structures, cracks are initiated very early so that most of the fatigue life is taken up with its growth.*

It has been observed in aluminum sheet specimens [15] that fatigue cracks tend to advance more rapidly in the mid-plane than at the surface. Figure 19 shows that fracture surfaces produced in fatigue have two distinct features, one appearing dull and the other bright. The dull regions are produced by sudden bursts of crack growth. This is dominated by the release of dilatational energy ahead of the crack and occurs relatively fast. It tends to spread more in the interior while the material at the trailing edge near the free surface is being deformed. In the bright textured areas, growth is controlled by distortion in the material near the surface** while much less growth (denoted in x_m in Figure 19) occurs at the mid-plane. This is because most of the energy is momentarily stored in the material near the surface layer and less in the central portion. When sufficient crack growth has taken place at the surface, say x_s, to allow an increase in energy at the center, another sudden burst of growth occurs in the interior. As growth continues, the crack length at the center exceeds than at the surface by increasing amounts until the specimen is completely broken.

Crack growth rate. The intermittent behavior of fatigue cracks with periodic changes from slow growth to quick bursts indicates that the

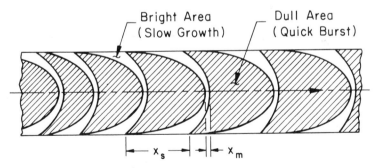

Figure 19. Fatigue fracture growth pattern

* The proportion of crack initiation and propagation life of a polycrystalline metal will depend on the microstructure of the material and surface treatment which can prolong the time before crack initiation.
** Fatigue crack growth has been related to the number of striations observed from fractured specimen surfaces.

shape of the crack front may be the controlling factor. The two stage
propagation process is basically the same as ductile fracture under
monotonic loading characterized by slow stable crack growth prior to
rapid fracture. The method for developing the crack profiles in Figure 19
follows the same procedure described earlier for ductile fracture.

The crack growth condition will be assumed to follow a state of
critical strain energy density accumulation. The hysteresis energy den-
sity $(\Delta W/\Delta V)_j$, $j = 1, 2, \ldots, n$, ahead of the crack for each cycle of loading
is counted and an increment of crack growth is assumed to occur when
the critical condition in equation (52) is fulfilled. The quantity ΔS in
equation (54) for each increment of crack growth is found from

$$\Delta S = S_{j+k} - S_j \tag{55}$$

which represents the area underneath the strain energy density versus
radial distance curve, Figure 20. The corresponding segment of crack
growth is Δr or Δa as the load is cycled from N_j to N_{j+k} with k being the
number of cycles corresponding to an increment of growth Δa. The
difference between the two curves in Figure 20 represents the amount of
irreversibility within the system. An elasticity calculation will not exhibit
such a character since the curves for N_j and N_{j+k} would be identical. This
process is repeated and a new ΔS is calculated for each increment of
crack growth. A crack growth relationship between a and N can then be
formed with the initial crack length $2a_0$ being specified and the critical

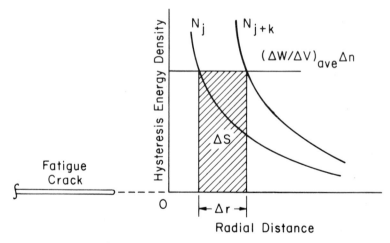

Figure 20. An increment of crack growth

crack length $2a_c$ determined from the fracture toughness value K_{1c}. However, the crack path between a_0 and a_c will depend on the rate of crack growth and/or the spectrum of cyclic loading. The path dependency nature of the fatigue crack growth process can be better understood by considering the following two cases:

Case (1):

Determine the number of cycles for each constant increment of crack growth.

Case (2):

Determine the non-uniform rate of crack growth increment for a fixed number of load cycles.

The conditions under which Cases (1) and (2) yield different or similar crack path can be examined for a variety of load spectrums and specimen geometries.

Material microstructure. Having discussed the interaction between cyclic loading and specimen geometry, a few remarks on the influence of material microstructure is in order. Needless to say, much uncertainty is encountered when attempting to analyze the conversion of available energy to irreversible material damage due to microcracks in the grains, grain boundary rumpling, creation of voids, etc. A more manageable approach would be to account for the microscopic entities through a crack tip radius of curvature which varies as the fatigue crack grows and encounters microstructure inhomogeneity. The change in the crack tip radius can be related to the grain size, local modulus and characteristic dimensions representing microdefects or inclusions. Special attention should be given to quantities that are sensitive to changes in scale such as microstress and macrostress.

Experimental data for materials with different microstructures can be collected with conventional and crack growth fatigue specimens from which the constants A, C and $\Delta a/\Delta N$ can be obtained. Once ΔS is found, $\Delta N/\Delta n = C(\Delta S)/(\Delta a/\Delta N)$ becomes known and can give an indication of load transfer to cause damage in the material near the crack tip region and will reflect the influence of material microstructure. The success depends on a reliable determination of ΔS by theory or experiment which is by no means straightforward.

6. Concluding remarks

An effort has been made to define the objective of experimental fracture mechanics within the framework of continuum mechanics. Without a clear concept as to how the several types of fractures in metals can be explained and described categorically by theory, there is nothing else that remains other than to perform tests and study the conditions of fracture on phenomenological grounds. Such an approach has limited use in comparing different material behavior under the same operating conditions but does not establish predictive capability for estimating the remaining life of structural members containing defects or cracks.

Non-destructive testing methods have demonstrated the capability for locating and describing the size and shape of defects in structures. This information is incomplete unless a fracture criterion can be applied to determine the critical combinations of defect size, geometry of the structure and loading. Otherwise, no reliable inspection procedures could be developed. There is a definite need to generalize and refine the theory of fracture mechanics beyond the conventional approach of fracture testing. The discipline has the potential to resolve the fundamental problem of scaling or size effect in the field of solid mechanics.

This communication is intentionally confined to the basic aspects of deformation and fracture in metals leaving out the abundance of test data and fracture observations which often tend to prejudice the development of theory. Attention is focused on applying the strain energy density criterion, mainly because it offers a degree of quantitativeness and qualitativeness never before achieved from any single criterion. In particular, the $1/r$ dependence of the strain energy density $\Delta W/\Delta V$ offers a unique characterization of the crack tip field that is completely independent of the constitutive relations. The factor S, defined as $r(\Delta W/\Delta V)$ in equation (12), represents the local energy release for a segment of crack growth r. It can be related to the tearing energy [16] commonly used in measuring the rate at which energy is released by extension of a crack in rubber. Another salient feature of the criterion is that it can be most easily integrated with stress analysis for determining crack profiles and growth in three dimensions. Such capability is a prerequisite for analyzing ductile and fatigue fracture that are triaxial in nature even though the loading may be confined to a plane. The step-by-step procedures described in this chapter offer guidance to analytical and experimental work in fracture mechanics. Particular emphasis should be given to the measurement of crack front profiles in terms of parameters that control the rates of crack growth.

It can be concluded that the basic concepts of fracture mechanics are sound. Much of what remains to be done concerns with finding the appropriate constitutive equations to account for the influence of material damage. This implies that changes in material behavior due to physical damage and alteration in microstructure and/or chemical composition should be distinguished in terms of measurable quantities. The varying degree of irreversibility needs to be quantified.

References

[1] *Linear Fracture Mechanics*, edited by G. C. Sih, R. P. Wei and F. Erdogan, Envo Publishing Co., Inc., Lehigh Valley, Pennsylvania (1976).

[2] *The Determination of Absorbed Specific Fracture Energy*, Hungarian Standard MSz 4929-76 (English Translation), Institute of Fracture and Solid Mechanics Publication, Lehigh University (1978).

[3] Gillemot, L. F., *Criterion of crack initiation and spreading, Journal of Engineering Fracture Mechanics*, 8, pp. 239–253 (1976).

[4] Czoboly, E., Havas, I. and Gillemot, F., The absorbed specific energy till fracture as a measure of the toughness of metals, *Proceedings of International Symposium on Absorbed Specific Energy and/or Strain Energy Density criterion*, edited by G. C. Sih, E. Czoboly and F. Gillemot, Sijthoff and Noordhoff International publishers, Alphen aan den Rijn, pp. 107–130 (1981).

[5] Sih, G. C., Fracture toughness concept, *American Society of Testing Materials*, STP 605, pp. 3–15 (1976).

[6] DeSisto, T. S., Carr, F. L. and Larson, F. R., *Proceedings of American Society for Testing and Materials*, 63, pp. 768–779 (1963).

[7] Sih, G. C., A special theory of crack propagation: methods of analysis and solutions of crack problems, *Mechanics of Fracture I*, edited by G. C. Sih, Noordhoff International Publishing, Leyden, pp. 21–45 (1973).

[8] Sih, G. C., A three-dimensional stress energy density factor theory of crack propagation: three-dimensional crack problems, *Mechanics of Fracture II*, edited by G. C. Sih, Noordhoff International Publishing, Leyden, pp. 15–53 (1975).

[9] Sih, G. C., The analytical aspects of macrofracture mechanics, *Proceedings of International Conference on Analytical and Experimental Fracture Mechanics*, edited by G. C. Sih and M. Mirabile, Sijthoff and Noordhoff International Publishers, Alphen aan den Rijn, pp. 3–15 (1981).

[10] Sih, G. C., *Handbook of Stress Intensity Factors*, Institute of Fracture and Solid Mechanics Publication, Lehigh University, Bethlehem, Pennsylvania (1973).

[11] Sih, G. C., Elastic-plastic fracture mechanics, *Proceedings of International Conference on Prospects of Fracture Mechanics*, edited by G. C. Sih, H. C. van Elst and D. Broek, Noordhoff International Publishing, Leyden, pp. 613–621 (1974).

[12] Sih, G. C., Mechanics of ductile fracture, *Proceedings of International Conference in Fracture Mechanics and Technology*, edited by G. C. Sih and C. L. Chow, Sijthoff and Noordhoff International Publishers, Alphen aan den Rijn, pp. 767–784 (1977).

[13] Sih, G. C., Mechanics of crack growth: geometrical size effect in fracture, *Proceedings of International Conference on Fracture Mechanics in Engineering Application*,

edited by G. C. Sih and S. R. Valluri, Sijthoff and Noordhoff International Publishers, Alphen aan Den Rijn, pp. 3–29 (1979).

[14] Hartranft, J. R. and Sih, G. C., Application of the strain energy density fracture criterion to dynamic crack problems, *Proceedings of International Conference on Prospects of Fracture Mechanics*, edited by G. C. Sih, H. C. van Elst and D. Broek, Noordhoff International Publishing, Leyden, pp. 281–297 (1974).

[15] Forsyth, P. J. E. and Ryder, D. A., Some results of examination of aluminum alloy specimen fracture surfaces, *Metallurgia*, 63, No. 377, pp. 117–124 (1961).

[16] Lake, G. J. and Thomas, A. G., Mechanics of fracture of rubber-like materials, *Proceedings of International Symposium on Absorbed Specific Energy and/or Strain Energy Density Criterion*, edited by G. C. Sih, E. Czoboly and F. Gillemot, Sijthoff and Noordhoff International Publishers, Alphen aan den Rijn, pp. 223–242 (1981).

A. J. Durelli

1 | *Stress concentrations*

1.1 Introduction

Stress is defined as a limiting process. It takes place at a mathematical point. It is a tensor. One of the components of the stress may have the same value at two neighboring points. In that case these points belong to the same contour or locus (for instance, all points on the same isochromatic line have the same value of maximum shear stress, all points on the same isobar have the same value of normal stress). In general, stress components change from point-to-point.

A uniform state of stress is usually illustrated by a straight column subjected to uniaxial load provided this load is applied also in a uniform manner at the ends of the column. A case like this is seldom if ever found in practice. In general, the state of stress changes from point-to-point and if any of the components of stress (or the resultants) are represented by a family of loci, peaks and valleys of the values appear. The stress is therefore distributed in a non-uniform manner and the not always precisely defined concept of stress concentration is associated with peaks, and gradients around the peaks, of the stress components.

The previous comments apply to an ideally continuous matter. They apply with even more reason if the material is made of discrete particles: inclusions, grains, crystals, atoms, etc. In general, the state of stress will not be uniform at the interface between discrete particles.

These considerations apply equally well to strain, strain components and strain concentrations. If the concept of stress concentration is more popular than the concept of strain concentration it is due to the fact that laws of failure are more commonly related to stress, and that historically, engineers found the computation of loads, and the stresses derived from loads easier to handle than the measurement of displacements, and the determination of their derivatives, which are necessary to obtain strains.

Stress Concentration Factors. The stress concentration factor is a means to evaluate quantitatively the stress concentration. Another means which is less commonly used but also has importance in fracture is the gradient of the stress.

The stress concentration factor can be defined as the ratio between the peak (or maximum) stress in the body and some other stress (or stress-like quantity) taken as reference. The reference stress is arbitrary. When the stress concentration is associated with the disturbance produced in the stress field by the presence of an inside boundary, the reference stress can be computed using elementary equations applied to either the gross or the net area of a cross-section through the point at which the peak stress takes place. When the stress concentration is associated with a change in the outside geometry of the body, the reference stress can be computed using elementary equations applied to a cross-section through the point at which the peak stress takes place, or alternatively the equations can be applied to a cross-section at some remote region of the body. In most cases it is convenient to give stress concentration factors taking as reference the stress computed over net areas rather than gross areas. Besides being intuitively preferable, this method avoids that the factors approach infinity when the net area is small Figure 1.1.

In some mixed boundary-value problems of the restrained shrinkage type (see Section 1.5) it is convenient to define the stress concentration factor as the ratio between the maximum normal, or shear, stress and the product of the free shrinkage of the body multiplied by the modulus of elasticity of the body.

In dynamic problems it is usually convenient to define the stress concentration factor as the ratio of a component of the stress at a point at which its value is maximum, to the same component at the same point of the body built without the geometric discontinuity. The body without the discontinuity may be subjected to the same dynamic load as the body with it, or to a static system of forces equivalent to the dynamic load. The stress concentration is then a function of time and may occur at different points, for different times.

Stress concentration factors are defined by Roark and Young [1] as the ratio of the true maximum stress to the stress calculated by the ordinary formulae of mechanics (flexure formula, torsion formula, etc.). Peterson [2] defines the factor in a similar way as the ratio of the maximum value of the stress to the nominal value obtained using linear distribution over the cross-section, and so does Griffel [3]. All these

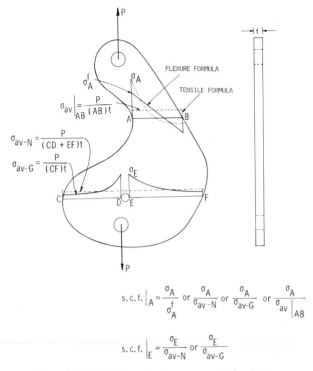

$$\text{s. c. f.}\bigg|_A = \frac{\sigma_A}{\sigma_A^f} \text{ or } \frac{\sigma_A}{\sigma_{av-N}} \text{ or } \frac{\sigma_A}{\sigma_{av-G}} \text{ or } \frac{\sigma_A}{\sigma_{av}}\bigg|_{AB}$$

$$\text{s. c. f.}\bigg|_E = \frac{\sigma_E}{\sigma_{av-N}} \text{ or } \frac{\sigma_E}{\sigma_{av-G}}$$

Figure 1.1. Definitions of stress concentration factors.

definitions are convenient in many cases but not applicable in others, as those instances in which stresses associated with pressure (either on the surface of a shell or plate, or on the inside surface of hole) are given as related to the amount of that pressure. Peterson [2] in Figure 126 uses this convention which is inconsistent with his definition. It seems more reasonable to state that the factor takes as reference any quantity with the dimensions of a stress, as long as that reference is defined.

The lack of consistency in the use of the particular cross-section over which the stress is averaged for reference, may also introduce confusion. Tetelman and McEvily [4] define the elastic stress concentration factor as the ratio of the local stress near the discontinuity to the nominal stress acting across the gross cross-section, (p. 16). Few lines later the statement is made however that the maximum stress concentration factor decreases from 3 (for the case of the circular hole) as the ratio of the hole size to plate size increases. Actually, when the gross area is taken as reference, the stress concentration factor increases toward infinity. It decreases when the net area is taken as reference [6] (p. 2.7).

Stress concentration factors are not always defined as the ratio of a peak stress to another stress taken as reference. Sometimes they are defined as the ratio of any stress in the perturbated region to a reference stress. See for instance Tetelman and McEvily [4] Equation (1.13) and Savin [5], figures IV-5 and IV-7. It seems less confusing to call 'stress ratio' the value of any stress referred to a convenient reference or nominal stress, and call "stress concentration factor" the ratio of the maximum stress referred to that reference stress. Both stress ratios and stress concentration factors are pure numbers.

Physical significance of stress concentrations. The theory of elasticity is based on the assumption of continuity of matter, indefinitely divisible, each particle retaining its original properties during the division.

The physical world, however, is more complicated than that idealized model. Materials are made of structural elements which may be crystals or molecules, both of which are combinations of atoms. At the dimensional level of these structural elements, matter is not uniform or homogeneous. Its properties change from point to point. They may also change with direction at a point. If a circular hole in an infinite homogeneous, isotropic plate, subjected to uniaxial load, develops a complicated stress field (Kirsch problem), it is easy to understand the difficulties which would be encountered in the rigorous solution of the stress distribution present in a material made of such a mosaic of elements. The following considerations may help in the understanding of the scope of the application of the theory of elasticity and of the stress concentrations factors obtained using its basic assumptions. They are obviously important in the study of problems related to fracture mechanics since in fracture mechanics the study of crack propagation and of stress distributions around cracks, are basic, and the order of the dimension of the crack tips may be the same as the order of the structural elements of matter.

Heterogeneity and discontinuity Figures 1.2 to 1.8 show a series of photographs of materials in which the range of dimensions of the structural elements varies from about one inch to about 10^{-7} inches. At one extreme of the range, the structural elements are visible to the bare eye, (in the case of some civil engineering structures like concrete

Figure 1.2. Actual grains fragmented from a titanium alloy ingot. XI 1/2 (W. Rostoker).

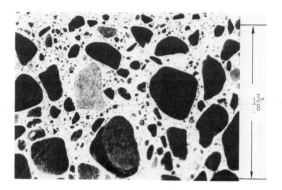

Figure 1.3. Structure of a concrete made of portland cement, water, sand and stones (Bureau of Reclamation).

dams, the structural elements may have dimensions of the order of feet), at the other end of the range, powerful microscopes are necessary to see the structural elements.

An ideal parallelepiped is introduced in the theory of elasticity to obtain the basic equations. It is assumed to be sufficiently small for the forces acting on its faces to be essentially uniform and represented in the first approximation by single average loads acting in the center of their areas of application. It is obvious, however, that this wording is relative and that the parallelepiped will be small or large by comparison.

0.1 in.

Figure 1.4. Structure of a mortar made of cement, water and sand. Many voids are present. (Portland Cement Association).

The grain size of 1020 steel is of the order of 10^{-4} inches, the grain size of 1045 steel is of the order of 10^{-5} inches.

In Figure 1.9 the cross-section of the structural elements of the solid propellant shown in Fig. 1.5 has been outlined. It will be assumed that the elements have, perpendicularly to the paper, dimensions of the same order as the ones shown in the paper. Uniform forces have been applied to the sides (about 0.05 in.) of the parallelepiped. The normal stress σ_x along a vertical cross-section will likely have a complicated distribution as shown, which will depend on the relative rigidity of the elements and of the embedding matrix. Because of equilibrium, however, the average has to be equal to the value of the stress applied to the side of the parallelepiped. A smaller parallelepiped (sides about 0.001 in.) as the one located in position 1 will be sufficiently small for the forces acting on its

0.01 in.

Figure 1.5. Structure of a rocket solid propellant. X73 (K. Bills).

faces to be represented by a single average acting in the center. If the smaller parallelepiped is at position 2, it will still be too large for this approximation to be sufficiently accurate. A parallelepiped one tenth that size may be necessary.

Consider now the tungsten grains of Figure 1.7, which have been outlined in Figure 1.10. If a small parallelepiped of this material (sides about 0.015 in.) is taken from a loaded body, the distribution of the stresses on the sides of the parallelepiped may look as shown in Figure 1.10. The distribution is complicated and will depend on the relative rigidities of the structural elements and of the embedding matrix; but it should be noticed that these stresses are statically equivalent to the applied uniform stresses. The theory of elasticity developments would apply to this parallelepiped if the analysis is limited to the average stress over the face of the parallelepiped, the scatter about the average is neglected, and any other discontinuity present is of much larger size than the size of the elements. In the figure, only σ_y has been represen-

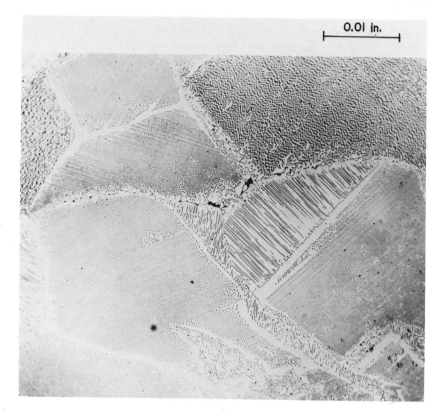

Figure 1.6. Structure of chromium–chromium carbide eutectic. X100 (W. Rostoker).

ted, for clarity; but the same reasoning applies to σ_x and τ_{xy}, and to the other three stress components.

So far the illustrations have been schematic. There are methods, however, which in some cases can illustrate the previous considerations, experimentally and with relatively good accuracy. Figure 1.11 shows the isochromatic pattern obtained from a photoelastic coating applied to a concrete block 3 in. $\times$ 3 in. $\times$ 3 in. This concrete is a mixture of sand, cement, water and gravel. The coating is a thin film of epoxy that follows faithfully at the interface the deformations of the surface of the block of concrete. The block was subjected to a uniform pressure on its vertical faces. If the assumptions of the theory were satisfied, the total surface would be uniformly light or uniformly dark. The fringes shown in the figure are loci of points of equal maximum shear strain

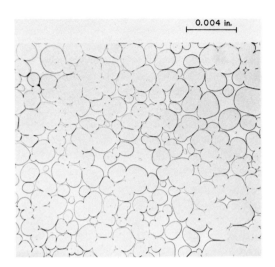

Figure 1.7. Duplex structure of tungsten grains in a nickel–iron envelope matrix. X250 (W. Rostoker).

neglecting the influence on the birefringence, of gradients of stresses in the direction of the light. The points indicated by letters correspond to the maxima of shear strain. The distribution is obviously extremely complicated.

The illustrations can also be obtained from transparent models photoelastically analyzed. Figure 1.12 shows the maximum shear stress distribution in a matrix shrunk around circular inserts much more rigid than the matrix. The loading condition is equivalent to a uniform biaxial-loading of biaxiality equal to unity. Figure 1.13 shows the distribution for the case of inclusions of different sizes, rigidities and shapes. Since the linear theory of the continuum applies to bodies irrespective of actual size, it can be visualized that the size of the inserts is of the order of a fraction of a thousandth of an inch. It can also be imagined that the inserts are not necessarily circular, nor regularly spaced, and that they may exhibit different rigidities. This would illustrate the stress distribution present, at that scale, in a material like the one shown Figures 1.5 and 1.7.

A circular hole in a heterogeneous medium. Suppose that a small hole (diameter about 0.001 in.) is drilled in a material like the chromium-chromium carbide represented in Figure 1.6 and suppose a unidimen-

Figure 1.8. Platinum crystal hemisphere—Radius about 1900 Å. (Magnification: 530,000 times.) Each dot is a platinum atom. The circles are crystal facets. Erwin S. Mueller (No. 14254).

sional load is applied as indicated in Figure 1.14 If the hole is at position 1, the stress distribution will likely not be too for from the one predicted by Kirsch, unless anisotropy is very pronounced. Suppose, however, that the hole is slightly larger and located at position 2. It would be practically impossible to predict the stress distribution using theoretical means.

Suppose finally that the hole diameter is appreciably larger than the structural element size (about 10 times). This case is illustrated in Figure 1.15. With some scatter, the theoretical curve may represent approximately the actual stress distribution.

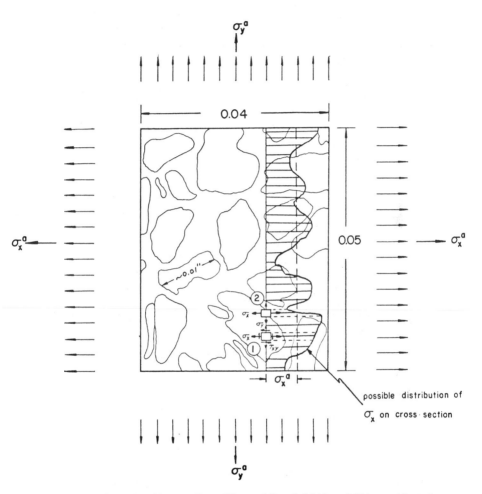

Figure 1.9. Portion of solid propellant (Figure 1.5), of 0.04 in. × 0.05 in., subjected to uniform loads on its sides. An ideal parallelepiped of 0.001 in. at position 1 will have forces about uniformly distributed on its sides. The same size parallelepiped at position 2 will not.

A. J. Durelli

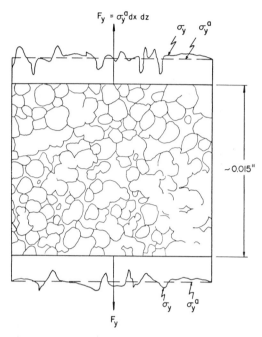

Figure 1.10. Structural elements of the tungsten shown in Figure 1.7. A parallelepiped of about 0.015 in. side, taken from the body, may have on its sides stresses distributed as shown. The theory of elasticity developments may still hold if applied to the average stress.

Concentrated loads. Frequently, in theoretical developments, loads are assumed to be concentrated on a point. An example is Boussinesq's problem of the concentrated load acting normal to the edge of a thin plate of large expanse. The elasticity solution gives an infinite stress at the point of contact.

Physically there is no such thing as an infinite stress, just like there is no such thing as a stress without a deformation, since no material is infinitely rigid. Any load transmitting member must have finite cross-section. And any material to which a narrow loading member is applied will show a deformed area under the load. Even for ideally homogeneous and isotropic materials no infinite stress would appear at the point of application of the concentrated load.

Sometimes a theoretical solution shows an infinite stress because the body exhibits a sudden change in geometry, as a reentrant corner. This is also physically impossible. First, no corner has a zero radius. Any corner observed under a sufficiently large magnification exhibits a finite

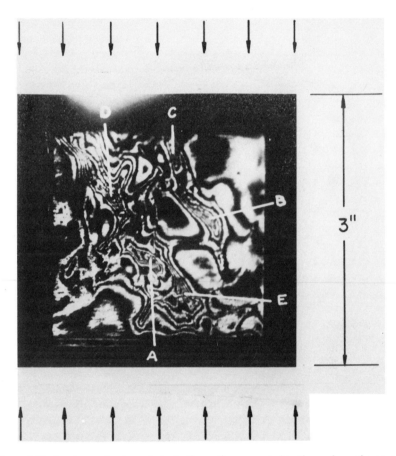

Figure 1.11. Isochromatics in a photoelastic coating cemented to the surface of a concrete block, subjected to uniform pressure on its top and bottom surfaces. Identified points exhibit zero shear. (G. U. Oppel).

Figure 1.12. Isochromatic fringe pattern in Hysol 8705 with seven circular inclusions.

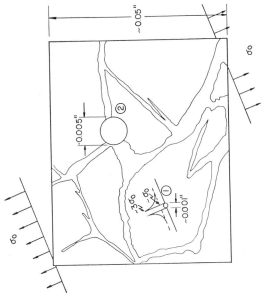

Figure 1.14. Small holes in a material like the chromium–chromium carbide eutectic (Figure 1.6). Stress distribution in hole at position 1 can be determined approximately. There is no available solution for stresses around hole at position 2.

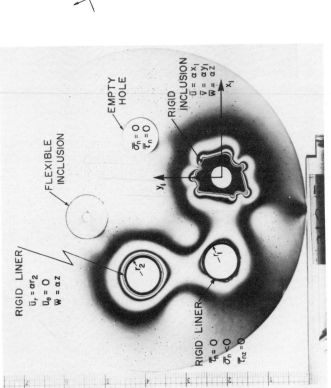

Figure 1.13. Isochromatics associated with restrained shrinkage around several inclusions of different kinds allowing free motion between inclusions.

15

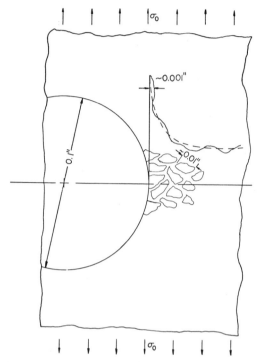

Figure 1.15. Possible stress distribution in the neighborhood of a relatively large hole in a plate of a material with relatively small structural elements.

radius. Second, the high stress associated with the sharp corner either produces a yielding in the material redistributing the stresses or distributes itself over some larger area of a finite size element. A third associated phenomenon is that even in theoretically isotropic, homogeneous and continuous materials, as soon as a stress acts on the boundary strains appear which deform the boundary and alter its radius of curvature.

1.2 Advantages and disadvantages of stress analysis methods used to determine stress concentrations

Many stress analysis methods are available to determine the stress distribution in loaded bodies of complicated shapes. Some of these methods use mathematical approaches based usually on the assumptions of the theory of elasticity. Among these some are numerical methods replacing differential equations with difference equations or replacing the continuum by a system of finite elements.

Other methods are experimental and consist in measuring physical properties, or changes in physical properties, which can be related to stresses, strains, or displacements. Among these methods the most commonly used are two- and three-dimensional, static and dynamic photoelasticity, brittle coatings, photoelastic coatings, electrical strain gages using ohmic resistance or capacitance or inductance properties, mechanical and optical strain gages, grids, moiré, holography, speckle, etc.

The advantages and disadvantages of these methods to study stress and strain distributions in different kinds of bodies subjected to different kinds of loading conditions have been considered in previous publications [7, 8]. Here the emphasis will be put on the particular subject of the determination of stress and strain concentration factors using the above mentioned methods.

The fact that stress concentrations occur as peak values in a field and that the gradients around these peaks are usually steep, present particular problems in the evaluation of the strains and stresses and make certain methods more adaptable than others for the study of the concentrations.

Some mathematical approaches. Four theoretical approaches are frequently considered: (1) collocation (also called 'boundary collocation' or point 'matching'), (2) conformal mapping, (3) finite differences, and (4) finite elements.

The first two methods are, in general, combinations of analytical and numerical procedures, while the last two are essentially numerical.

When the collocation method is used it is simple to satisfy the boundary conditions. Once a series solution is established for the governing differential equation, the method requires that the boundary conditions be satisfied only at discrete points along the boundary [9]. The accuracy and the convergence of the solution obtained using this method are in general very much influenced by the choice of the collocation points.

In problems of stress concentration such as plates with holes or notches, the method is capable of yielding highly accurate results if the chosen function also satisfies exactly the boundary conditions around the discontinuity, as is the case in problems with circular or elliptical holes or semicircular notches. But in situations where this is not possible, the method would determine only approximate values of stress concentration factors, depending on the number and location of collocation points. The rate of convergence of the method can be improved when a least square criterion is added to the technique [10].

For problems governed by Airy's biharmonic equation, Mushkelishvili [11] has developed the conformal mapping method to obtain the exact solution if the mapping function is known (usually in polynomial form). In some cases the function can be approximated in a fairly simple fashion, but in general one must make use of an extensive numerical procedure to obtain it. In the case of a simply connected domain, finding the mapping function reduces to the solution of an integral equation [12–14] and for doubly-connected domains (the case of some rocket motor cross sections) a system of coupled integral equations must be solved [15].

The mapping function must be very accurate in the neighborhood of points where stress concentrations occur. When this method is used it is frequently difficult to reproduce the geometry of sharp corners. This is one of the significant limitations of the method.

Finite-difference methods reduce continuous systems to equivalent lumped-parameter systems. Instead of obtaining a continuous solution of the differential equation, the differential equation is replaced by a finite-difference equation and the approximate values of the solution are found at isolated points. The procedure leads to a system of n simultaneous linear algebraic equations when n discrete points are selected. A common way of solving this system of simultaneous equations is to use iteration or relaxation methods. When large stress gradients are present in the field, numerical instability and errors may take place. These methods have become more practical since the advent of digital computers.

The method that has taken the greatest advantage of the new development of digital computers is the finite-element method. When this method is used the continuous structure is replaced by a large number of small elements and equilibrium and compatability are applied to the whole of each of them. The method has found applications in many fields of mechanics and can be used when bodies have complicated boundaries and inhomogeneous anisotropic nonelastic properties. Smaller-size elements can be used where gradients are high and storage capacity can be made very large, but there is a practical limit to the number of elements in the matrix to be used. Some investigators [16] propose hybrid techniques in which the local concentration is separated from the regular field, permitting an appreciable reduction in the number of elements. This technique has been introduced in some standard programs [17].

Depending on the type of problem to be solved, finite-difference methods may be more or less efficient than finite-element methods. Considerations on the subject have been made by Bushnell [18] who calls attention to some of the advantages of finite-difference energy methods. Key and Krieg [19] also compare both approaches and point out situations in which the finite-difference method may give better results. The introduction of arbitrary meshes in the finite-difference methods by Perrone [20] makes the method in many cases more competitive with finite-elements methods. General considerations on finite-element methods from a historical perspective and also in comparison with finite-difference methods have been presented by Zienkiewicz [21] and by Oden [22].

Photoelasticity. Measurements can be taken of changes in indices of refraction, most frequently changes in the difference between indices of refraction (relative retardation), and relate them to some components of the stress or strain tensor. Usually photoelastic determinations yield directly the maximum shear in the plane of observation. The method can be very sensitive and very precise and is self-sufficient for the determination of the magnitude and directions of maximum shear stresses, but requires supplementary information to determine the individual principal stresses. For the common cases of elasto-static small deformations at room temperature the calibration is easy. However, if the calibration has to be conducted for elevated temperatures, if the loads are applied rapidly or the deformations are large, or irreversible, then the calibration can be very complicated.

Photoelastic coatings can be applied to prototype surfaces of complicated shape. In general, they lack sufficient sensitivity, their maximum response being about one fringe when applied to structural steel subjected to elastic strains. Determination of individual principal stresses is complicated and lacks precision.

It is frequently forgotten that a low shear in the plane of observation may be associated at the same point with a large shear in a plane perpendicular to the plane of observation. In plane stress problems the isochromatic observation is frequently limited to free boundaries and the analysis for a stress concentration factor is then complete. However when the photoelastic coating is applied to a structural member of complicated shape stress concentrations may exist that are not detected.

Photoelasticity is very well-suited to the determination of stress concentration factors in two- and three-dimensional problems (the points of interest are usually subjected to one-dimensional or two-dimensional states of stress). Photoelastic coatings, because of thickness effects, may give appreciable errors.

Grids. Grid methods are not commonly used. They lack sensitivity for the determination of small strains, but they are very well suited for the determination of large strains in rubber-like materials, in plastics, or in metals undergoing plastic deformation. Grids can be applied to the surface of practically any prototype, but like moiré, are more often used on models.

Grids have the advantage of giving a direct geometric representation of the body. A limitation is the level of error associated with the measurement of the geometric transformation. The precision of the determinations will depend on the sharpness of the marks, the recording method (photograph, etc.) and the precision of the measuring instrument. (Indices of precision are usually associated with the standard deviation of the repeated measurements). The location of each point of the grid is determined by means of two readings on the instrument scale, the one corresponding to the zero point (or origin) and the one corresponding to the point of interest. Displacement components are obtained from the difference in position of points before and after deformation and so will have a somewhat lower degree of precision than the one of the values of the coordinates of the points.

Strains or strain-like quantities are obtained by further treatment of the measurements in the form $l_f - l_i/l_i$, $\Delta l/l$, $\Delta u/\Delta x$, $\partial u/\partial x$, etc. operations which further reduce the degree of precision. For a given grid line sharpness and measuring system, the error in the strain depends on the magnitude of Δl. This change in length of a segment can be increased by increasing the strain or increasing the base length. The base length l, however, has to be relatively short since the determined strain is assumed to be uniform over its length. A typical base length is $l = 0.4$ in. and a typical standard deviation is about 0.001 in. Thus

$$\varepsilon = \frac{\Delta l + 0.001}{0.4 + 0.001} \tag{1.1}$$

For a given Δl it is seen that the strain will have a standard deviation of the order of 0.002. The overall curves presented will be more accurate due to the averaging effect of drawing a smooth curve through the points.

A disadvantage of grids is the laborious point-by-point analysis. An attempt to reduce the time required to evaluate grid patterns has been described in [23].

Grids can be drawn or printed representing any desired coordinate system (polar or elliptical for instance) which may be difficult using other methods. In certain cases this may be an important advantage.

The main limitation of grids methods for the determination of stress concentrations is the relatively large base length over which the strain is averaged [7]: The accuracy, however, can be appreciably improved if the displacement curve is drawn, as suggested by Fischer [24, 25], and a graphical differentiation of the curve is conducted, giving a better estimate of the maximum strain.

Moiré. Moiré directly yields components of displacement without the need for an intermediate physical property. To obtain strains the displacement data must be differentiated.

The strain tensor at every point is then completely determined. The use of moiré on prototypes for the solution of two-dimensional static-elastic problems, despite the techniques of fringe multiplication available today, is limited in general in its application by the 40 *l*pmm maximum density of the available gratings. Without multiplication the maximum response of moiré is about one fringe per inch when applied to a structural steel subjected to elastic strains.

Multiplication and differentiation techniques are becoming more practical. Up to now, in-plane moiré in general cannot be applied to curved surfaces except when they are developable, or the radius of curvature is large. Projected gratings and shadow-moiré are, however, very practical to determine deflections in plates and shells.

Moiré-of-moiré patterns are approximate in the sense that they represent the average of the partial derivative over the shifted interval [7]. When the shift takes place normally to a boundary, the boundary value of the partial derivative has to be estimated by extrapolation from the recorded pattern. Otherwise it should be obtained by graphical differentiation of the displacement field. More detailed discussion of these points can be found in [26].

The standard deviation of the moiré grating pitch is probably less than 0.00001 and so the standard deviation in strain for a shift of 0.03 will be the order of

$$\varepsilon = \frac{0.00001}{0.03} \simeq 0.0003 \qquad (1.2)$$

The moiré grating is so dense that the fringes are the average of a large number of grating line intersections. This averaging effect increases the precision beyond the one indicated above.

The same comment made about the base length for stress concentration determinations using grids applies also to moiré, but less critically, because the length of measurement is much shorter.

Brittle coatings. Principal stress trajectories can be obtained easily and with good precision using brittle coatings. The determination of the value of the principal stresses is subjected to larger errors and requires more skill in the operator. Even for this purpose, the method is very practical when the temperature is about room temperature and does not change abruptly. It can be applied to the actual prototypes, or components, without requiring the manufacture of a model [7]. The upper limit on strains is about 0.01. The method is less practical when several loading conditions are applied to the same body since a new coating will be necessary for each loading condition. (Photoelastic coatings and strain gages are not subjected to this limitation.) The sensitivity of brittle coatings is of the order of 500×10^{-6}, but with refrigerating techniques, and some sacrifice of precision, it can be as low as 100×10^{-6}.

For the solution of dynamic problems the coating should be properly calibrated. The techniques of application to dynamic problems are appreciably more complicated than for static work, and one of the difficulties to be overcome is the 'closing-up' of cracks at stress levels lower than the ones at which cracks close under static load. In respect to photoelastic coatings, brittle coatings have the important advantage of being much thinner.

The main limitation of brittle coatings to determine stress concentrations is the step-wise method used to apply loads. The value of the stress producing the failure of the coating is associated with a load that falls within the interval between the last load that does not produce a crack, and the first load producing the crack. The estimate can be improved by drawing a curve representing stresses, through the point of maximum stress, but the location of this point may also be subjected to error [25].

Holography. Holography [27] yields whole-field displacement information but it is easier to obtain the out-of-plane displacements than the in-plane displacement. This constitutes a serious limitation. In addition, the need for high stability and darkness during recording (when using a cw laser) also limit its applicability. A great advantage is the very high sensitivity of the method and that measurements can be made on diffusely reflecting prototypes requiring no special preparation. It has found so far few applications in the determination of stress concentration factors, but it has found a wide field of application as an inspection of materials methods since flaws in materials produce an appreciable disturbance in the hologram.

Speckle. Methods based on laser speckle are a fairly recent development. Techniques have been developed to measure out-of-plane deflection and slopes, and in-plane displacements and strains. As of this time the technique for in-plane strains is not as practical as the others [28].

In general speckle methods, like holography, have the advantages of being noncontacting and requiring no surface preparation, which makes easier the application to prototypes. Depending on which technique is being used, speckle can yield either point-by-point or whole-field information. Some techniques also have the advantage of being able to vary the sensitivity over a large range. In contrast to holography (with which its shares many features), there is no problem with fringe localization and thus, evaluation of specklegrams is a much less ambiguous and simpler process. The main drawback of the speckle techniques is the generally poor quality of the fringe patterns. In addition, similar to holography, the out-of-plane measurements are easier and more precise than in-plane measurements. As is the case of holography it has so far been seldom used for stress concentration determination.

Strain gages. Electrical resistance strain gages are very sensitive, easy to apply and frequently have very small base lengths. They can be used on models or on prototypes. One of their main advantages is their ability to respond correctly to high-frequency strains since their rise time is of the order of 0.1×10^{-6} second [29]. They have the following limitations: (1) the temperature should usually be lower than 400°F and relatively steady; (2) the time of observation should usually be less than a few days; (3) the strain should not be larger than about 0.02; and (4) changes in moisture may affect the output. Special gages do, however, exist which can accommodate up to 20 percent strain and, independently, 800°F. The output can be recorded several convenient ways [7].

Mechanical strain gages (of the Huggenberger type, or using dial gages) have longer base lengths, cannot be used for dynamic applications, and their output has to be recorded visually. However, they are not subjected to the time, temperature, and humidity limitations of electrical resistance strain gages. Capacitance strain gages are better suited for determinations at elevated temperatures and are more stable as function of time, but their base length is usually longer than the base length of resistance type gages.

All these gages give point-by-point and unidirectional information. When 'rosettes' are used two or three of them are combined. Gages are very practical when the location of the point of stress concentration is known. The information obtained is an average over the base length of the gage and should therefore be relatively small. This is particularly true when the determination of the stress concentration factor is desired.

Interferometry. Interferometric techniques can be applied to a large class of problems. In some cases they provide all the necessary data while in others they are used in conjunction with other techniques.

In general, interferometric techniques have the advantage of high sensitivity, being able to measure on the order of the wavelength of light. The main drawbacks are that generally they require high stability, are difficult to zero, and are affected by rigid body motions. In addition, they require high-quality model surfaces for both reflection and transmission applications. They have been used few times to determine stress concentrations.

1.3   Compilation of results

The object of this section is to list and describe, and in some cases critically analyze, the most important handbooks published with compilations of stress concentration factors. It was thought useful to include the equations obtained by Neuber, for a large number of geometries and boundary conditions, since they deal with many of the most important cases of interest in fracture, and they are not easily available in English.

The circular and elliptical holes.   Two problems are of basic importance in engineering design and in fracture prevention and analysis. They are the problems of the stress distribution around a circular hole and the problem of the stress distribution around an elliptical hole, in infinite plates. The solution to these problems have been obtained by Kirsch [30] and by Inglis [31] respectively. These solutions are important sometimes by themselves, sometimes by comparison with the solution of similar problems. And of course the solution for the circular hole is a particular case of the solution for the elliptical hole.

The solution of the problem of stress distribution around a circular empty hole in a thin plate of infinite width subjected to unidimensional uniformity distributed load is:

$$\sigma_r = \frac{\sigma_{aG}}{2}\left(1-\frac{a^2}{r^2}\right) + \frac{\sigma_{aG}}{2}\left(1-\frac{4a^2}{r^2}+\frac{3a^4}{r^4}\right)\cos 2\theta \qquad (1.3a)$$

$$\sigma_\theta = \frac{\sigma_{aG}}{2}\left(1+\frac{a^2}{r^2}\right) - \frac{\sigma_{aG}}{2}\left(1+\frac{3a^4}{r^4}\right)\cos 2\theta \qquad (1.3b)$$

$$\tau_{r\theta} = -\frac{\sigma_{aG}}{2}\left(1+\frac{2a^2}{r^2}-\frac{3a^4}{r^4}\right)\sin 2\theta \qquad (1.3c)$$

$$\sigma_z = \tau_{zx} = \tau_{xy} = 0 \qquad (1.3d)$$

At the edge of the hole:

$$\sigma_\theta = \sigma_{aG}(1-2\cos 2\theta); \quad \sigma_r = \tau_{r\theta} = 0 \qquad (1.4)$$

For the cross-section through the center and perpendicular to the x axis, along which the load is applied:

$$\sigma_\theta = \frac{\sigma_{aG}}{2}\left(2+\frac{a^2}{r^2}+\frac{3a^4}{r^4}\right) \qquad (1.5)$$

A detailed representation of loci of stress components can be found in [32].

Consider the case of a circular hole in a biaxial stress field. The solution may be obtained by superposition. Calling $\sigma^{(1)}$ and $\sigma^{(2)}$ the two perpendicular uniform fields, $k = (\sigma_1/\sigma_2)$ and $c = (a/r)$ the maximum shear stresses (isochromatics) are given by:

$$\sigma_1 - \sigma_2 = \sigma^{(2)}[(k-1)^2(9c^8 - 12c^6 - 2c^4 + 4c^2 + 1)\sin 2\phi +$$
$$+ c^4(1+k)^2 + (1-k)^2(9c^8 - 12c^6 + 10c^4 - 4c^2 + 1)\cos^2 2\phi +$$
$$+ c^2(1-k^2)(2 + 6c^4 - 4c^2)\cos 2\phi]^{1/2} \qquad (1.6)$$

at the edge of the hole:

$$(\sigma_1 - \sigma_2)_{c=1} = \sigma^{(2)}[(1+k) + 2(1-k)\cos 2\phi] \qquad (1.7)$$

and the sum of the principal stresses anywhere in the field is given by:

$$\sigma_1 + \sigma_2 = \sigma^{(2)}[(1+k) + 2c^2(1-k)\cos 2\phi] \tag{1.8}$$

The individual values of $\sigma^{(1)}$ and $\sigma^{(2)}$ can be obtained by successive addition and subtraction of the isochromatics and isopachics.

For the case of the elliptical hole Inglis [31] used elliptical coordinates α and β such that:

$$x = C \cosh \alpha \cos \beta \tag{1.9a}$$

$$y = C \sinh \alpha \sin \beta \tag{1.9b}$$

and for $\alpha = $ constant:

$$\frac{x^2}{C^2 \cosh^2 \alpha} + \frac{y^2}{C^2 \sinh^2 \alpha} = 1 \tag{1.10}$$

and for $\beta = $ constant

$$\frac{x^2}{C^2 \cos^2 \beta} - \frac{y^2}{C^2 \sin^2 \beta} = 1 \tag{1.11}$$

The three components of the stresses at the generic point are given in the form of infinite series, rather cumbersome to compute, but the particular case of the edge of the hole, which is in general the one of greatest interest, is relatively simple. It has been evaluated for the case of combined loading $\sigma^{(1)}$ and $\sigma^{(2)}$, perpendicular to each other. Calling 2a and 2b the lengths of the major and minor axes of the ellipse; α_0 the elliptical coordinate which is constant along the edge of the hole; ϕ the angle between the 2a axis and the directions of loading $\sigma^{(1)}$; and $k = \sigma^{(1)}/\sigma^{(2)}$ the stresses at the boundary of the elliptical hole are given by:

$$\sigma_\alpha = \tau = 0 \tag{1.12a}$$

$$\sigma_\beta = \frac{(\sigma^{(1)} + \sigma^{(2)})\sinh 2\alpha_0 + (\sigma^{(1)} - \sigma^{(2)})[\cos 2\phi - e^{2\alpha_0}\cos 2(\phi - \beta)]}{\cosh 2\alpha_0 - \cos 2\beta} \tag{1.12b}$$

When the directions of loading (1) and (2) coincide with the axes b and *a*, respectively, of the ellipse then $\phi = \pi/2$ and

$$\sigma_\beta = \frac{(\sigma^{(1)} + \sigma^{(2)}) \sinh 2\alpha_0 + (\sigma^{(1)} - \sigma^{(2)})[e\, 2\alpha_0 \cos 2\beta - 1]}{\cosh 2\alpha_0 - \cos 2\beta} \qquad (1.13)$$

At the ends of the major (point *B*) and minor (point *A*) axes of the ellipse the stresses are:

$$\sigma_A = \sigma^{(2)}\left(1 + 2\frac{b}{a}\right) - \sigma^{(1)} \qquad (1.14a)$$

$$\sigma_B = \sigma^{(1)}\left(1 + 2\frac{a}{b}\right) - \sigma^{(2)} \qquad (1.14b)$$

Since at the apex of the major axis $2a$ of the ellipse, the radius of curvature is $\rho = b^2/a$, the above expression for the stress concentration factor can be given in terms of ρ. For the case of uniaxial loading, p:

$$\sigma_{max} = p\left[1 + 2\left(\frac{a}{\rho}\right)^{1/2}\right] \qquad (1.15)$$

The stress concentration factors for a typical elliptical hole $a/b = 2$, and for a circular hole, subjected to biaxial loading conditions, are shown in Figure 1.16. The solution for the elliptical hole has been found independently by Kolosov [33].

Isida [34] gave a great deal of attention to the stress concentrations present at the edge of elliptical holes. He obtained solutions in the form of power series for the case of finite strips. Results of an experimental systematic study [35] on finite strips are shown in Figures 1.17 to 1.19.

All these developments refer to very small, or 'infinitesimal' strains. When the strains are 'finite,' sufficiently large to change appreciably the shape of the hole, the results may be considerably different from the ones recorded above. The analysis for finite strains is reported in Section 1.4.

Coker and Filon's 'Treatise on Photo-elasticity'. This classic work [36] not only on photoelasticity, but on many aspects of the theoretical treatment of elasticity problems, contains the first systematic treatment of stress concentrations around holes, fillets, notches, cracks and numerous machine components. The integration of the theoretical and

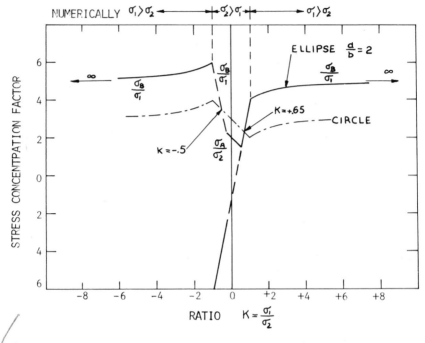

Figure 1.16. Stress concentration factor for an elliptical hole $a/b = 2$, and a circular hole, in an infinite plate, subjected to biaxial loading.

the experimental developments and their applications to engineering problems is unique.

Problems like the determination of stresses in eccentric hollow cylinders, or in cylinders with outer contour of square form (both under pressure) and problems like the determination of stresses in cutting tools, in spline shafts and in nuts and bolts have not only historic interest but permanent value. The same should be said of the study of concentrations in circular and elliptical chain links. The chapter on holes and cracks includes the treatment of eyebars and is fundamental.

Lehr's 'Distribution of Stresses in Elements of Construction'. Lehr's book [37] was probably the first of all the treatises dealing with the experimental analysis of strain and including all the information available at the time (1934), on stress concentrations. These concentrations are defined as the ratio of the peak stresses to the 'nominal' stresses, which are the ones obtained using the elementary formulae. Six main

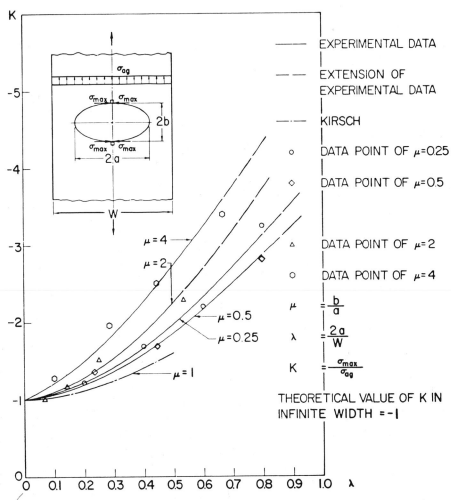

Figure 1.17. Stress concentration factors (K) for the points under maximum compression in a finite width plate with an elliptical hole.

A. J. Durelli

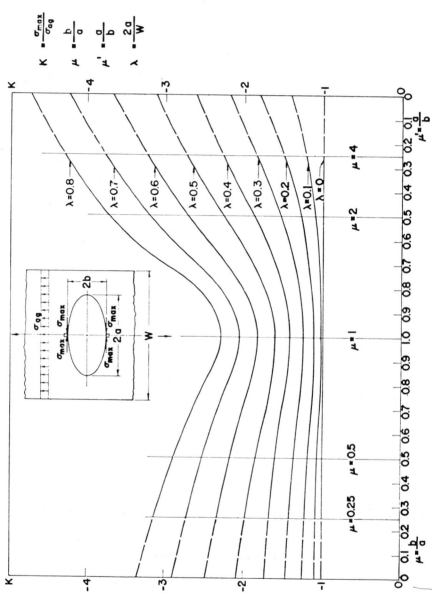

Figure 1.18. Stress concentration factors (K) for the points under maximum compression using $\lambda = 2a/W$ as a parameter.

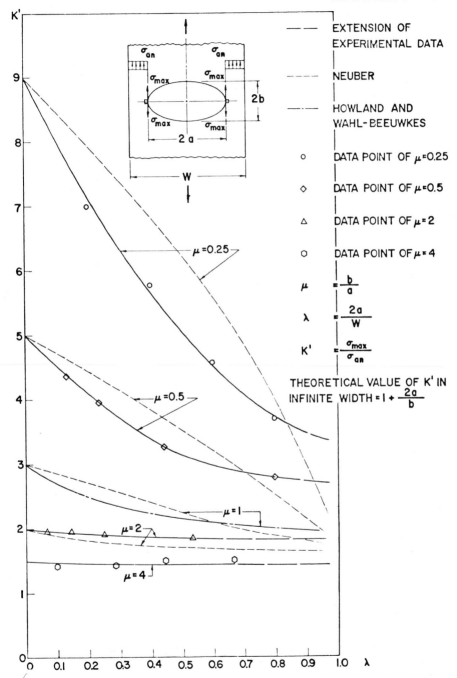

Figure 1.19. Stress concentration factors (K') for the points under maximum tension in a finite plate with an elliptical hole.

31

groups of geometries are considered: empty holes, holes with pins, shafts with grooves, frames, fillets and keyholes. The information is presented parametrically and the author evaluated the available results using his judgment to interpolate and extrapolate results obtained by other investigators. Results are also included for several machine components like crankshafts, gears and pressure vessels.

Neuber's approach. A book [38] of fundamental importance was published by Neuber in 1937. The determination of stress concentrations was simplified appreciably by: (1) reducing the number of cases to a few typical ones, (2) deciding that the curvature at the base of the notch is of primary importance while the effect of the flank angle at the ends of the notch is of secondary importance, (3) solving for the two limiting cases of shallow and deep notches, and (4) obtaining results from interpolation between the results corresponding to the two limiting cases. Other simplifying assumptions are introduced in the development of particular cases. Neuber calls his approach 'principles for exact calculation' but it obviously is only a useful approximation. Final results are summarized in a nomograph that permits the solution of a very large number of stress concentration problems in plates, rods and tubes, with shallow or with deep notches, subjected to axial load, to torsion, or to bending. The equations developed for the limiting cases (and which are not easily available in English literature) are the following:

(a) Deep external notches on both sides of a bar subjected to axial load (bars referred to in items (a) to (i) have rectangular cross-section)

$$\alpha_1 = \frac{\alpha_{max}}{p} = \frac{2\left(\frac{a}{\rho} + 1\right)\sqrt{\frac{a}{\rho}}}{\left(\frac{a}{\rho} + 1\right) \text{arctg}\sqrt{\frac{a}{\rho}} + \sqrt{\frac{a}{\rho}}} \qquad (1.16)$$

calling ρ the radius of curvature at the base of the notch, a half the width of the narrowest cross-section and p the average stress over the neck of thickness d, which is given by:

$$p = \frac{P}{2ad} \qquad (1.17)$$

(b) Deep external notches on both sides of a bar subjected to pure bending:

$$\alpha_2 = \frac{\sigma_{max}}{p} = \frac{4\frac{a}{\rho}\sqrt{\frac{a}{\rho}}}{3\left[\sqrt{\frac{a}{\rho}} + \left(\frac{a}{\rho} - 1\right)\arctan\sqrt{\frac{a}{\rho}}\right]} \tag{1.18}$$

where $p = \dfrac{3M}{2da^2}$.

(c) Deep external notches on both sides of a bar subjected to pure shear:

$$\frac{\tau_{max}}{p} = \frac{\frac{a}{\rho}\sqrt{\frac{a}{\rho}}}{\sqrt{\frac{a}{\rho} + 1}\left[\left(\frac{a}{\rho} + 1\right)\arctan\sqrt{\frac{a}{\rho}} - \sqrt{\frac{a}{\rho}}\right]} \tag{1.19a}$$

$$\frac{\sigma_{max}}{p} = \frac{\frac{a}{\rho}\sqrt{\frac{a}{\rho} + 1}}{\left(\frac{a}{\rho} + 1\right)\arctan\sqrt{\frac{a}{\rho}} - \sqrt{\frac{a}{\rho}}} \tag{1.19b}$$

where $p = V/2ad$ and V is the shearing force.

(d) Deep external notches on one side of a bar subjected to axial load:

$$\frac{\sigma_{max}}{p} = \frac{\alpha_1 - 2C}{1 - \dfrac{C}{\sqrt{\dfrac{a}{\rho} + 1}}} \tag{1.20}$$

where

$$C = \frac{\alpha_1 - \sqrt{\dfrac{a}{\rho} + 1}}{\dfrac{4}{3\alpha_2}\sqrt{\dfrac{a}{\rho} + 1} - 1} \tag{1.21}$$

(e) Deep external notch on one side of a bar subjected to bending:

$$\frac{\sigma_{max}}{p} = \frac{2\left(\dfrac{a}{\rho} + 1\right) - \alpha_1\sqrt{\dfrac{a}{\rho} + 1}}{\dfrac{4}{\alpha_2}\left(\dfrac{a}{\rho} + 1\right) - 3\alpha_1} \tag{1.22}$$

(f) Deep external notch on one side of a bar subjected to shear: same equations as in (e) but with

$$p = \frac{V}{ad} \tag{1.23}$$

(g) Shallow external notch on one side, or shallow external notches on two sides, of a bar subjected to axial load:

$$\frac{\sigma_{max}}{p} = 3\sqrt{\frac{t}{2\rho}} - 1 + \frac{4}{2 + \sqrt{\frac{t}{2\rho}}} \tag{1.24}$$

where

$$p = \frac{P}{bd} \quad \text{or} \quad p = \frac{P}{2bd} \tag{1.25}$$

depending on whether there is notch on one side or both sides.
(h) Shallow external notch on one side, or shallow external notches on two sides, of a bar subjected to bending:
Same equations as for (g) with

$$p = \frac{6M}{b^2d} \quad \text{and} \quad p = \frac{3M}{2b^2d} \tag{1.26}$$

for one side notch or two sides notches respectively.
(i) Shallow external notch on one side, or shallow external notches on two sides, of a bar subjected to shear:

$$\tau_{max} < 1.5p. \tag{1.27}$$

where

$$p = \frac{V}{bd} \quad \text{or} \quad p = \frac{V}{2bd} \tag{1.28}$$

for one side or two sides notches respectively.
(j) Deep external circumferential notch in a round bar subjected to

axial load

$$\frac{\sigma_{max}}{p} = \frac{1}{N}\left[\frac{a}{\rho}\sqrt{\frac{a}{\rho}+1} + \left(0.5 + \frac{1}{m}\right)\frac{a}{\rho} + \left(1 + \frac{1}{m}\right)\left(\sqrt{\frac{a}{\rho}+1} + 1\right)\right] \qquad (1.29)$$

where

$$N = \frac{a}{\rho} + \frac{2}{m}\sqrt{\frac{a}{\rho}+1} + 2; \quad p = \frac{P}{\pi a^2} \qquad (1.30)$$

$1/m$ Poisson's ratio and a the radius at the neck.

(k) Deep external circumferential notch in a round bar subjected to bending:

$$\frac{\sigma_{max}}{p} = \frac{1}{N}\frac{3}{4}\left(\sqrt{\frac{a}{\rho}+1} + 1\right)\left[3\frac{a}{\rho} - \left(1 - \frac{2}{m}\right)\sqrt{\frac{a}{\rho}+1} + 4 + \frac{1}{m}\right] \qquad (1.31)$$

where

$$p = \frac{4M}{\pi a^3} \qquad (1.32)$$

and

$$N = 3\left(\frac{a}{\rho}+1\right) + \left(1 + \frac{4}{m}\right)\sqrt{\frac{a}{\rho}+1} + \frac{1 + \dfrac{1}{m}}{1 + \sqrt{\dfrac{a}{\rho}+1}} \qquad (1.33)$$

(l) Deep external circumferential notch in a round bar subjected to shear:

$$\frac{\sigma_{max}}{p} = \frac{\left(\frac{a}{\rho}+1\right)\left(\frac{a}{\rho} + 2\sqrt{\frac{a}{\rho}+1} + 2\right)\left(\frac{a}{\rho} + 2 + \frac{2}{m}\right)}{2\sqrt{\frac{a}{\rho}}\left(\sqrt{\frac{a}{\rho}+1} + \frac{1}{m}\right)\left(\frac{a}{\rho}+2\right)^{3/2}} \qquad (1.34)$$

where

$$p = \frac{V}{\pi a^2} \qquad (1.35)$$

(m) Deep external circumferential notch in a round bar subjected to torsion:

$$\frac{\tau_{max}}{p} = \frac{3\left(1 + \sqrt{\frac{a}{\rho} + 1}\right)^2}{4\left(1 + 2\sqrt{\frac{a}{\rho} + 1}\right)}$$

(1.36)

where

$$p = \frac{2M}{\pi a^3}$$

(1.37)

(n) Shallow circumferential internal notch in a round bar subjected to axial load:

$$\frac{\sigma_{max}}{p} = \frac{1}{N}\left\{2\left(\frac{t}{\rho}\right)^2 - \left(1.5 - \frac{1}{m}\right)\frac{t}{\rho} + 1 - \frac{1}{m} + \left[-\left(1.5 + \frac{1}{m}\right)\frac{t}{\rho} + \frac{1}{m}\right]\frac{t}{\rho}c\right\}$$

(1.38)

where

$$c = \frac{\operatorname{arctg}\sqrt{\frac{t}{\rho} - 1}}{\sqrt{\frac{t}{\rho} - 1}}, \quad \text{if } \frac{t}{\rho} > 1$$

(1.39a)

$$c = \frac{\ln\left(1 + \sqrt{1 - \frac{t}{\rho}}\right) - \frac{1}{2}\ln\frac{t}{\rho}}{\sqrt{1 - \frac{t}{\rho}}}, \quad \text{if } \frac{t}{\rho} < 1$$

(1.39b)

and

$$N = \frac{t}{\rho} + 1 - \frac{1}{m} + \left(\frac{t}{\rho} - 2 + \frac{2}{m}\right)\frac{t}{\rho}c - \left(1 + \frac{1}{m}\right)\left(\frac{t}{\rho}\right)c^2$$

(1.40)

and where t is the depth of the notch, b the radius of the bar, and

$$p = \frac{P}{\pi b^2}$$

(1.41)

(o) Shallow circumferential internal notch in a round bar subjected to bending:

$$\frac{\sigma_{max}}{p} = \frac{1}{N}\left\{g[g(10t^2 - 2\alpha) - 36t^2 + 16 - 4\alpha] - 16t^2\right\}$$ (1.42)

where

$$g = \frac{\dfrac{t}{\rho}}{\left(\dfrac{t}{\rho} - 1\right)^2}\left[3\frac{t}{\rho}c - 2\frac{t}{\rho} - 1\right]$$ (1.43)

(p) Shallow circumferential internal notch in a round bar subjected to shear:

$$\frac{\sigma_{max}}{p} = \frac{1}{N}2\sqrt{\frac{t}{\rho} - 1}$$ (1.44)

where

$$p = \frac{3m + 2}{2m + 2}\frac{V}{\pi b^2}$$ (1.45a)

$$N = \left(\frac{t}{\rho} - 1\right)^{-3/2}\left\{\left[\left(2 - \frac{1}{m}\right)\frac{t}{\rho} + 1 + \frac{1}{m}\right]\left[\frac{t}{\rho}c - 1\right] - 2\frac{t}{\rho} + 2\right\}$$ (1.45b)

and c is given by equations (1.39a) and (1.39b).

(q) Shallow circumferential internal notch in a round bar subjected to torsion:

$$\frac{\tau_{max}}{p} = \frac{2\left(\dfrac{t}{\rho} - 1\right)^2}{3\left(\dfrac{t}{\rho}\right)^2 c - 5\dfrac{t}{\rho} + 2}$$ (1.46)

where

$$p = \frac{2Mt}{\pi b^4}$$ (1.47)

(r) Circular and elliptical hole in a very wide bar, subjected to axial load:

$$\frac{\sigma_{max}}{p} = 1 + 2\sqrt{\frac{t}{\rho}} = 1 + 2\frac{t}{s} \qquad (1.48)$$

t being half the axis of the ellipse, in the direction perpendicular to the applied load, and s half the axis in the direction parallel to the applied load.

(s) Circular and elliptical hole in a very wide bar, subjected to bending

$$\frac{\sigma_{max}}{p} = 1 + \sqrt{\frac{t}{\rho}} \qquad (1.49)$$

(t) Circular and elliptical hole in a very wide bar, subjected to pure shear:

$$\frac{t_{max}}{p} = 3\frac{\left(\frac{t/\rho + 2}{3}\right)^{3/2} - \sqrt{\frac{t}{\rho}}}{\left(\sqrt{\frac{t}{\rho}} - 1\right)^2} \qquad (1.50a)$$

$$\frac{\sigma_{max}}{p} = \frac{3}{2}\left(\sqrt{\frac{t}{\rho}} + 2 + \frac{1}{\sqrt{\frac{t}{\rho}}}\right) \qquad (1.50b)$$

This monumental work of great imagination showing the unusual mastery that the author has of the theory of elasticity equations ends with a sheet in which three nomographs summarize the whole book. All stress concentrations factors can be computed with two or three motions of the pencil. This is a unique accomplishment in the literature of stress concentrations. In the second edition an extension was made to account also for the influence of nonlinear stress-strain relationships. Experimental work, conducted mainly after the publication of the book, shows in some cases appreciable differences with the values obtained from Neuber's nomographs (see for instance Figure 1.19). These differences are frequently due to the approximate method Neuber uses to interpolate between results corresponding to shallow and deep notches. Although they should be a warning to the scientist using the nomographs, the differences do not detract from the overall value of Neuber's attempt at giving a unified presentation of the field.

Roark and Young's 'Formulas for Stress and Strain'. Roark and Young's book [39] on formulae for stresses and strains is organized like a regular treatise on advanced strength of materials, but with little description of principles and methods, and great emphasis on the presentation of problems already solved. Large number of tables contain the problems described schematically. Next to them are listed the equations necessary to solve them and all the necessary auxiliary data. Frequently the most important values for design are given directly. In many of the tables coefficients to solve the problem parametrically are included. The components analyzed are beams (straight and curved), bars under torsion, plates, columns, shells and pressure vessels. Analysis of contact stresses, buckling, dynamic and thermal stresses are also included.

A chapter at the end of the book deals specifically with stress concentrations. The material is presented for 23 geometric shapes in the same format of the rest of the book. Results are not given in graphical form as in Peterson's handbook, nor in the form of more or less general equations to be frequently worked out, as in Savin's book, but by means of simple algebraic equations, with reference to the sources which report mostly experimental work.

The matter of the definition of stress concentrations should be mentioned again. Some stress concentration factors are listed in the body of the book, which are not listed in the table dealing with stress concentration factors. An instance is the large stress present in rectangular beams of great depth, when subjected to uniform loads, or to concentrated loads. Stresses twenty times larger than those obtained using elementary formulae may appear. The fact that the boundaries are straight does not seem good enough reason not to call this situation a concentration of stress. Another example is the case of contact stresses.

Lipson, Noll and Clock's 'Handbook on Stress and Strength'. In this book [40], 45 charts have been assembled giving in graphical form the stress concentration factors for holes, fillets and grooves in solid and hollow circular shafts subjected to tension, bending and torsion. The information was obtained from theoretically and experimentally obtained data. The main emphasis of the book however is in the presentation of almost 100 charts for so-called 'fatigue' stress concentration factors K_f, corresponding to annealed and to quenched steel. Most of these have been derived from the 'theoretical' stress concentration factor K_t and values of notch sensitivity q obtained averaging experimental data obtained by several investigators, and using the expression:

$$K_f = 1 + q(K_t - 1). \tag{1.51}$$

The book has been written for designers and many of the results given are only a first approximation to the solution.

Savin's 'Stress Distribution Around Holes'. Although G. N. Savin's book [41] does not deal with discontinuities at outside boundaries, it is a unique treatise on stress distributions around inside discontinuities. The treatment of all kinds of holes is dealt in an exhaustive manner. It includes any number of holes of any shape, in linear and nonlinear materials, subjected to small or large deformations, and also when the materials behave plastically or viscoelastically. It also includes analysis of cases in which the materials with the holes are anisotropic and when the equations of equilibrium include moments. Most of the problems have been solved in an approximate manner using the method which consists in finding two analytical functions of a complex variable, the complex potentials of Kolosov and Muskhelishvili and of Cauchy-type integrals. In some cases the method of perturbation of the form of the boundaries has been used.

A great amount of information has been assembled in this book, each chapter of which contains a very large number of bibliographic references. The main limitations seem to be the approximate character of the solutions obtained, the fact that by conformal mapping not always the desired geometry is obtained and that the mapping function must be very accurate in the neighborhood of points where stress concentrations occur. References to experimental work are much less numerous than the references to theoretical contributions and do not seem to be up-to-date. It may be added that by contrast with engineering design handbooks, the graphic results are difficult to use because of the small size of the reproduction, or because some of the necessary information is not directly available. In some cases sketches in the figures are misleading like the ones representing infinite plates with holes as if they were finite plates. On the other hand, the number of equations assembled to approach mathematically the solution of stress distributions around holes is enormous.

Foeppl and Sonntag's 'Tables for Strength of Materials'. This book [42] presents the subject in a manner similar to the one used by Roark.

In a schematized way most problems in strength of materials are dealt with profusion of tables and graphs, and equations ready to be used by the designer. Chapters deal with bending of straight and curved bars, torsion, two-dimensional problems, three-dimensional problems, plates and shells. Stress concentrations are studied with more detail than by Roark and follow to a large extent Neuber's approach.

The problems of thick curved beams of variable height, and stress concentrations associated with protuberances are included.

The lack of reference to the work of Coker and Filon, and the very few references to experimental results are surprising.

Peterson's 'Stress Concentration Factors'. Whereas Savin's treatise is the result of the theoretical work of a school of investigators using a few particular methods, Peterson's handbook [43] is the compilation of the results obtained by hundreds of investigators using any available methods. And whereas Savin most of the time details the methods used and the step-by-step procedure to obtain the results, Peterson compiles the results with little or no reference to the methodology. Few of Savin's graphs have been prepared with enough precision to be used directly for measurements. All of Peterson's graphs have been prepared with this purpose in mind, and most of them present the information parametric-ally. Although conformal mapping methods are approximate methods, Savin's work emphasizes theoretical precision. In Peterson's handbook the emphasis is on usefulness to designers and judgment is expected in the application of the information to give answers to designers' problems.

The scope of Peterson's handbook is larger than the scope of Savin's treatise. It includes notches and grooves, shoulder fillets, holes in two and in three-dimensional bodies, and an appreciable number of structural components: shafts with key seats, splined shafts, gear teeth, shrink-fitted members, bolts and nuts, springs, crankshafts, hooks, box sections, rotating disks, rings, pressure vessels, etc. Considerations on fatigue are numerous, reflecting the experience of the author on this subject, but viscolasticity, plasticity and large deformations are not considered.

Strangely enough only two minor references to Savin's work can be found in Peterson's handbook and little advantage seems to have been taken from the sizable amount of information accumulated by Savin.

Griffel's 'Handbotk for Formulas for Stress and Strain'. The book [44] assembles the usual formulae for the bending of straight and curved beams, the bending of plates, the design of pressure vessels and the vibration frequencies of plates. A large part of the book deals with properties related to the geometry of cross-sections of components. A chapter is included dealing with stress concentration factors in beams and plates with holes. Several of the results are taken from Savin's book. Here again sketches are misleading showing finite plates with very large holes as if they were infinite plates. More serious, however, are statements like: 'the data can certainly be applied with assurance to cases where the hole diameter is three-fifths the total width of the plate....' (p. 246). For the case of the circular hole the stress concentration factor in the finite plate is 5.2 when referred to the gross area, and 2.1 when referred to the net area. The value given in the graph is 3. This and similar approximations do not seem acceptable.

Nisitani's approach. Using a socalled body force method Nisitani has determined a large number of stress concentration factors for fillets and notches in two- and three-dimensional problems. Some have been included in Peterson's handbook. Others, with particular emphasis on solutions for crack problems have been published in [45]. These include elliptical holes and notches with cracks at the apexes and the interaction between a crack and an elliptical notch. The results are given graphically and in tables and the relation between stress concentration factors and stress intensity factors is frequently pointed out.

Bibliographic sources. With the exception of Neuber's, the nine handbooks mentioned above have large numbers of bibliographic references. Savin's has more than 700, Coker more than 400, Peterson about 380, and Roark about 360. Another important source of references is one of the books by Heywood [46] with more than 1000 references.

Two reviews of the literature should be mentioned, a survey by Hogan with particular interest in holes and pressure vessels [47] with about 500 references, and another, very general, by Neuber and Hahn [48], with 250 references. Also related to the problem is a review by Sternberg [49] and a comparative evaluation of experimental methods by Durelli [50]. Sternberg confined his survey to the 'analytical treatment of three-dimensional stress concentrations within the linear equilibrium theory of homogeneous and isotropic, elastic media.' The papers surveyed used 'exact' methods to solve the homogeneous displacement equations of equilibrium in terms of stress functions. The problems include spherical, spheroidal and elliptical cavities and inclusions, cracks and notches, with 57 references.

1.4 Geometrically non-linear stress concentrations

Few studies have been conducted to determine the influence of non-linearities on stress concentration factors. Important contributions have been made by Neuber [51] who for certain cases of loading investigated theoretically the influence of material non-linearities, and by Savin [41] following Green and Adkins, who studied geometric non-linearities in circular and elliptical holes in finite plates. Some results obtained experimentally, using linear materials, after the original shape changes appreciably, will be reported here. The subject is certainly relevant to the study of cracks since their shape will likely deform appreciably before fracture.

Geometric non-linearity. Stress concentration factors obtained when deformations are infinitesimal, in materials behaving linearly, are applicable to any material behaving linearly, and subjected to any level of deformation, provided this level is within the proportional limit of the material. When geometric non-linearities are involved the stress concentration factor changes with the deformation level, and factors are applicable only to the particular level for which they have been obtained.

The definition of stress concentration also may change. In some instances, it has been found convenient to use:

$$\frac{\bar{\sigma}_{max}}{E} \qquad (1.52)$$

where $\bar{\sigma}$ is a 'natural' stress defined as the integral of instantaneous, or incremental, stresses, each of them computed using the area corresponding to the particular incremental load level:

$$\bar{\sigma} = \lim_{\Delta P \to 0} \sum \frac{\text{instantaneous change in load}}{\text{instantaneous area}} (\Delta P) \qquad (1.53a)$$

$$\bar{\sigma} = \int_{m=i}^{m=f} \frac{\Delta P}{A_m} \qquad (1.53b)$$

where m is used to represent the increment. P is the load on the area A, and ΔP is the increase in load in the area A_m.

Use is also made of the concept of natural strain:

$$\bar{\varepsilon} = \ln \frac{1}{1 - \varepsilon^E} \qquad (1.54)$$

and of the two classic definitions of strain; the Lagrangian, or conventional engineering strain:

$$\varepsilon^L = \frac{\text{total change in length}}{\text{initial length}} = \frac{l_f - l_i}{l_i} \tag{1.55}$$

and the Eulerian strain:

$$\varepsilon^E = \frac{\text{total change in length}}{\text{final length}} = \frac{l_f - l_i}{l_f} \tag{1.56}$$

Besides the 'natural' stress definition mentioned above, the two classic definitions of stress, Lagrangian or engineering:

$$\sigma^L = \frac{P}{A_i} \tag{1.57}$$

referred to the undeformed area, and Eulerian:

$$\sigma^E = \frac{P}{A_f} \tag{1.58}$$

referred to the deformed area are used:
   The materials used for all the experimental determinations reported in this section are several varieties of polyurethane rubbers, all of them behaving linearly and elastically to high levels of strain, of order of 80% in the natural representation.

Stress concentration around elliptical holes in finite plates subjected to uniform load producing finite strains. In Section 1.3 results of the analysis conducted using the infinitesimal theory were reported. Here the case of the elliptical holes in plates subjected to loads producing finite strains will be considered [52].
   Figure 1.20 shows the change in the shape of the boundary of an elliptical hole $a_i/b_i = 2.0$ as the load is applied, for different ratios of the major axis of the hole to the width of the plate. Subscript i stands for 'initial' values as subscript f for 'final' values.
   Figure 1.21 shows the stress distribution at the boundary of the hole for a very large plate: $\eta = a_i/w = 0.112$ where w is the width of the plate. Figure 1.22 shows the corresponding distribution for the case of a

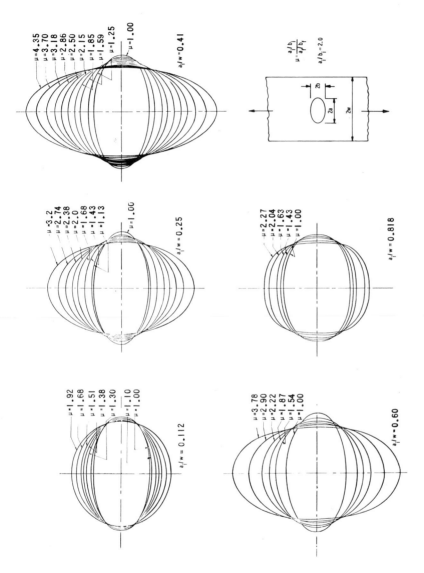

Figure 1.20. Metamorphosis of an elliptical hole in a plate subjected to deformation.

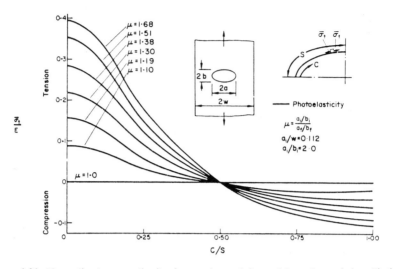

Figure 1.21. Normalized stress distribution at the undeformed boundary of the elliptical hole in a large plate ($\eta = 0.112$) for a linear material subjected to several levels of in-plane axial load.

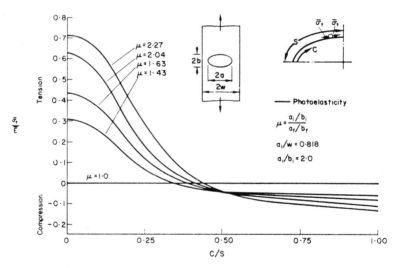

Figure 1.22. Normalized stress distribution at the undeformed boundary of the elliptic hole in a finite plate ($\eta = 0.818$) for a linear material subjected to several levels of in-plane axial load.

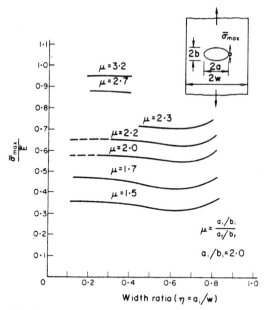

Figure 1.23. Normalized stress ratios as a function of width for a linear material subjected to in-plane axial loading.

narrow plate ($\eta = 0.818$) and Figure 1.23 shows the maximum stresses, parametrically, for all the cases investigated. The results indicate that the maximum stress is primarily a function of the shape of the hole in its deformed state. The investigation was conducted using photoelasticity, grids and moiré effects on a polyurethane rubber plate.

In the previous figures the normalization was obtained by dividing stresses by the modulus of elasticity. This is a correct but unconventional way. The following figures [53] use the more common method of dividing the stresses by some average stress, which requires knowledge of the load. The first method indicates the relationship of the stress to the deformed shape, the second emphasizes the stress-load relationship. Figure 1.24 shows the stress concentration factor, using the load as parameter, Figure 1.25 the change in shape and Figure 1.26 the stress concentrations factor using the hole shape as parameter.

Stress and strain concentrations in spheres diametrically loaded and subjected to finite strains. The values of the stresses and strains, at the center of a solid sphere, as the level of deformations increases, are of

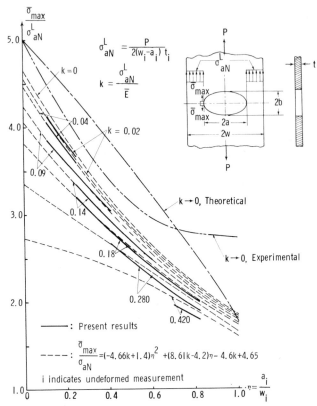

Figure 1.24. Stress concentration factor at the boundary of an elliptic hole in a plate of linear material, subjected to several levels of in-plane axial load, using normalized load as a parameter.

general interest [54] (Figure 1.27). They have been obtained using embedded gratings and moiré effects from a material with a strain energy function

$$W = C_1(I_1 - 3) + C_2 \ln\left(\frac{I_2}{3}\right) \tag{1.59}$$

where $C_1 = 17$ psi and $C_2 = 25$ psi.

The Eulerian vertical stresses on the horizontal axis of the sphere when subjected to several load levels of diametral vertical compression are shown in Figure 1.28. Further results can be found in [54].

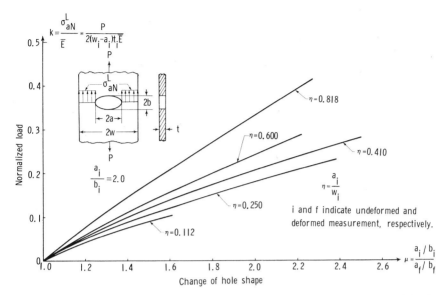

Figure 1.25. Change of shape of an elliptic hole in a finite plate subjected to several levels of in-plane axial load as a function of normalized load.

When the sphere is hollow ($OD/ID = 2$) the strain concentration factor, as shown in Figure 1.29 is amazingly close to constant, as function of the amount of load, despite very large changes in shape [55]. Figure 1.30 gives the complete distribution of principal strains on the inner boundary. Further results can be found in [57].

Stress and strain concentrations in a disk subjected to diametral load producing finite strains. It is important to know, mainly for problems related to fracture, the change in the position and value of the stress concentration factors in a disk, subjected to diametral compression, when deformations are large, but the material behaves in a linear manner [56].

The most striking feature of the influence of the level of deformation on the stress concentration factor, is that the location of the maximum in-plane shear, which takes place at, or near the point of load application when deformations are small, moves in toward the center of the disk, very appreciably Figure 1.31. At the same time the values of the maximum normal stresses decrease, also appreciably. The information is given in Figure 1.32 where $\bar{E}$ = Young's modulus based on true stress

50   *A. J. Durelli*

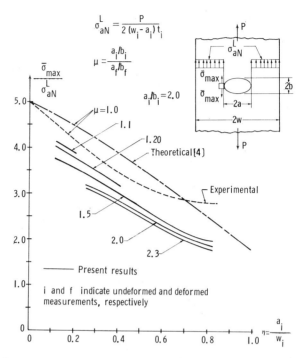

Figure 1.26. Stress concentration factor at the boundary of an elliptic hole in a plate of linear material, subjected to several levels of in-plane axial load, using change of the hole shape as a parameter.

versus natural strain, t_0 = thickness of undeformed disk, r_0 = radius of undeformed disk; $\bar{\sigma}_x$ and $\bar{\sigma}_y$ = Lagrangian true stresses, $\bar{\varepsilon}_x$, $\bar{\varepsilon}_y$, $\bar{\varepsilon}_z$ = Cartesian natural strains and n = photoelastic fringe order which gives the maximum natural shear stress.

The stress results shown are applicable to any disk, of any size, of any linear material. The strains are given for a material with a Poisson's ratio $\nu = 0.48$.

Stress and strain concentrations in a circular ring subjected to diametral compression producing finite strains. Using grids, moiré and photo-elasticity and a polyurethane rubber as model material, strains and stresses have been determined in a ring subjected to eight levels of deformations [57, 58]. The ratio of the outside diameter to the inside

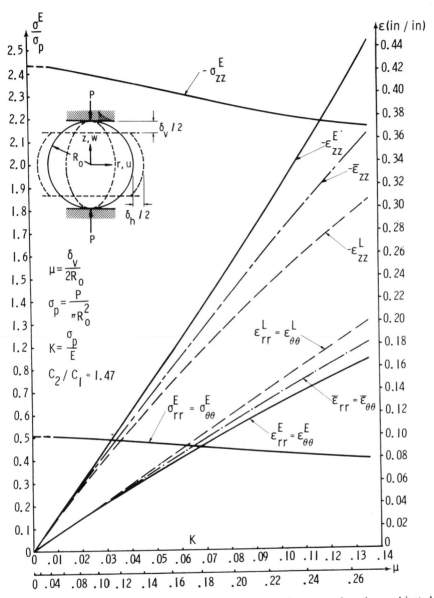

Figure 1.27. Stresses and strains, as a function of load, at the center of a sphere subjected vertically to several levels of diameteral compression.

52   A. J. Durelli

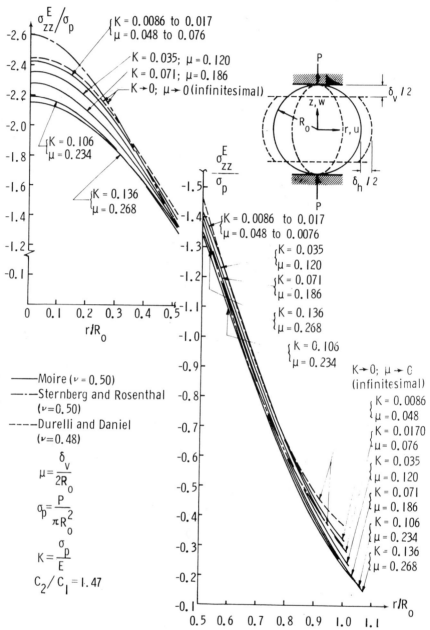

Figure 1.28. Eulerian vertical stress on the horizontal axis of a sphere subjected vertically
to several load levels of diametral compression.

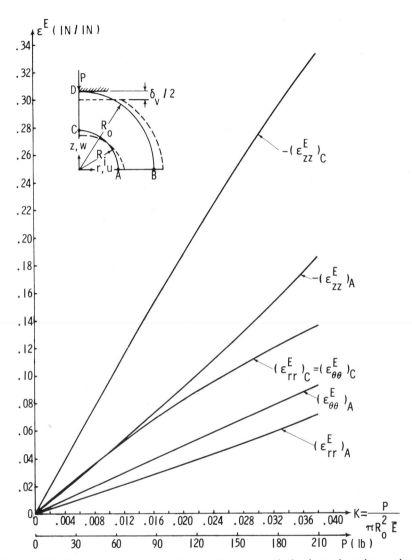

Figure 1.29. Strains at the intersection of the axes and the inner boundary sphere $(OD/ID) = 2$ subjected vertically to several load levels of diametral compression.

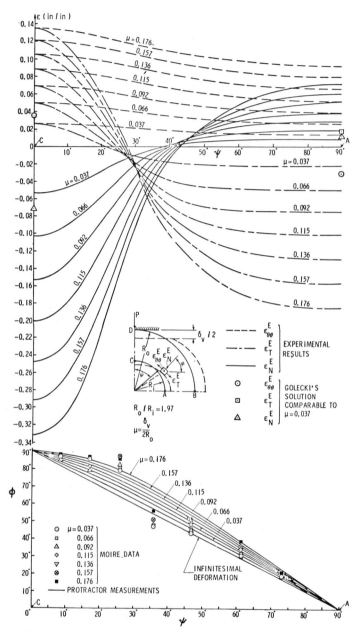

Figure 1.30. Principal strains $\varepsilon_{\theta\theta}^{E}$, ε_{N}^{E} and ε_{T}^{E} and their orientation on the inner boundary of a hollow sphere subjected vertically to seven levels of diametral compression in lagrangian description.

54

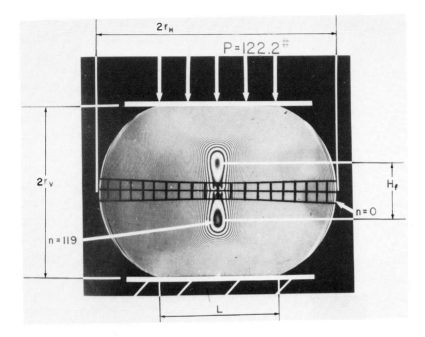

Figure 1.31. Isochromatics and grids in a polyurethane disk (solithane 113) subjected to diametral loading ($P = 122.2$ lbs., $K = 0.155$).

diameter of the ring is approximately 1.5., and the level of deformation is given by the parameter $\lambda = \delta_v/OD$, δ_v being the change in length of the loaded diameter. Some of the more significant results are shown in the figures. Figures 1.33 and 1.34 show the distorted grids and the Lagrangian strains tangential to the inner boundary, Figure 1.35 the Eulerian horizontal strains on the vertical axis, Figure 1.36 the Eulerian horizontal strains in the horizontal axis, Figure 1.37 the Eulerian vertical strains on the horizontal axis and Figure 1.38 the Eulerian principal strains and their orientations on the inner boundary. Typical record of the physical evidence is shown in Figure 1.39.

Linear and non-linear elastic and plastic stress concentration in a plate with a large circular hole loaded axially in its plane. The stress concentration at the edge of the circular hole, in a finite plate loaded axially

A. J. Durelli

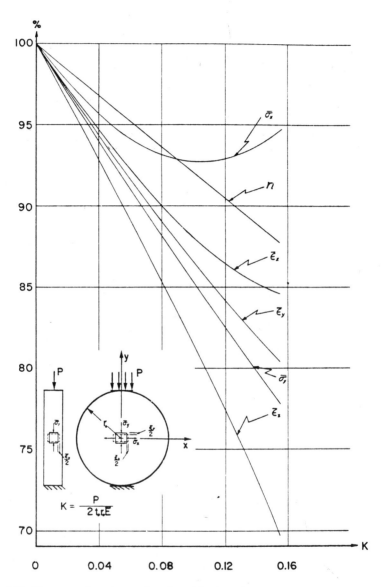

Figure 1.32. Stresses and strains, as function of load, at the center of a disk diametrically
loaded, given as percent of the values obtained from the small strain theory.

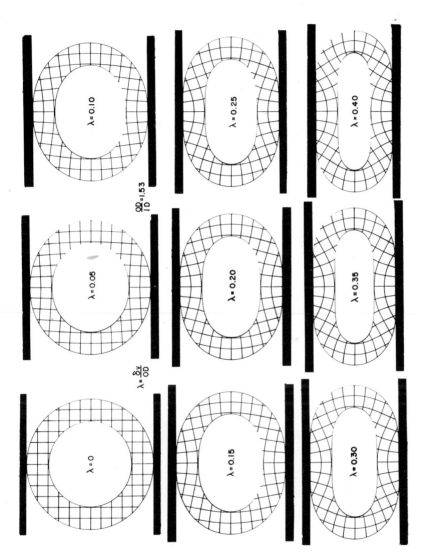

Figure 1.33. Cartesian grid on a circular ring subjected to several levels of diametral compression.

58

A. J. Durelli

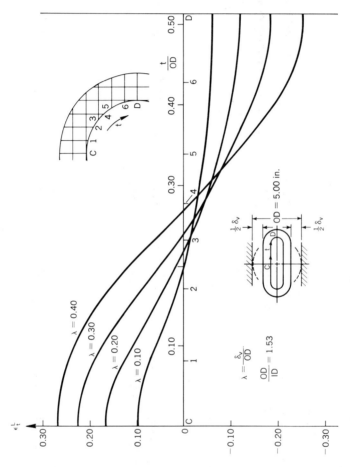

Figure 1.34. Lagrangian strain ϵ_t^L tangential to the inner boundary of a circular ring subjected to several levels of diametral compression.

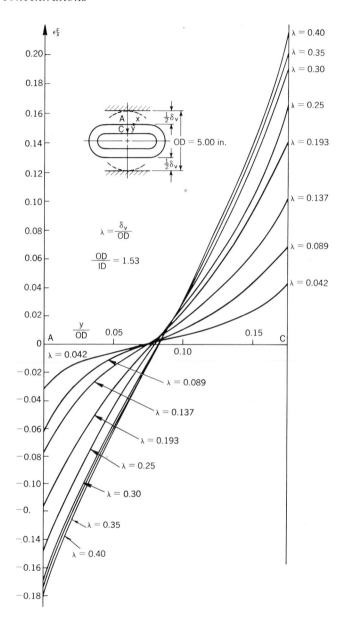

Figure 1.35. Eulerian horizontal strain ε_x^E on the vertical axis of a circular ring subjected to several levels of diametral compression.

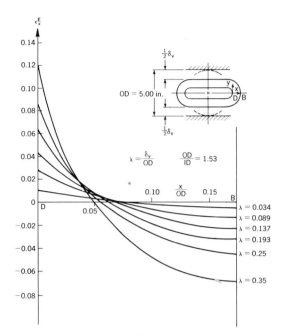

Figure 1.36. Eulerian horizontal strains ε_x^E on the horizontal axis of a circular ring subjected to several levels of diametral compression.

in its plane, is given in Figure 1.40 as function of D/W. As the hole becomes larger and it approaches the edge of the plate, large deformations may take place and the phenomenon is non-linear. Inconsistencies between results obtained by several investigators are probably due to the fact that their determinations were obtained at different load levels.

The matter was investigated [59] using the geometry shown in Figure 1.41 and the two materials the behavior of which is shown in Figure 1.42.

The strain concentration factor is given as a function of the level of deformation, expressed by ε_{av}, across the ligament or net area (Figure 1.43). The results obtained are particularly interesting in two respects: (1) the strain concentration factor varies from about 2 to 1, whether there is or not material non-linearity, and (2) there seems to be little variation between the strain concentration factor in the aluminum and the rubber, until the whole width of the aluminum ligament reaches the yield point. At this load, the strain concentration factor in the aluminum very quickly drops to unity.

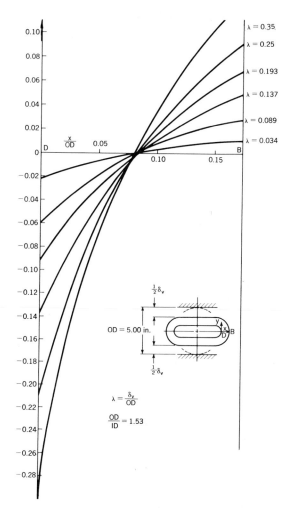

Figure 1.37. Eulerian vertical strain ε_y^E on the horizontal axis of a circular ring subjected to several levels of diametral compression.

62

A. J. Durelli

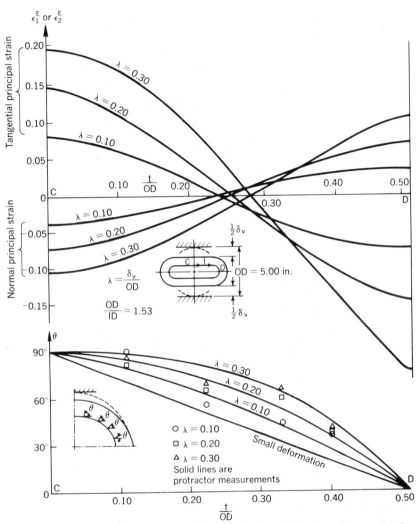

Figure 1.38. Principal eulerian strains ε_1^E and ε_2^E and their orientation on the inner boundary of a circular ring subjected to several levels of diametral compression.

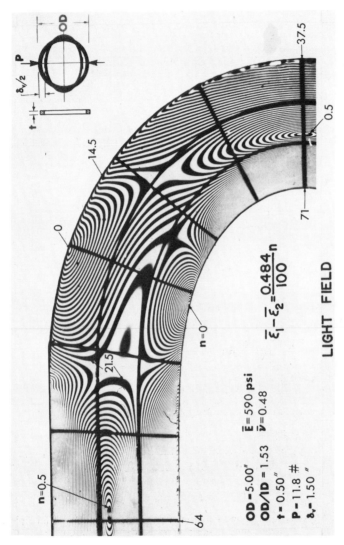

Figure 1.39. Isochromatics in a circular ring subjected to large diametral compression (obtained photoelastically).

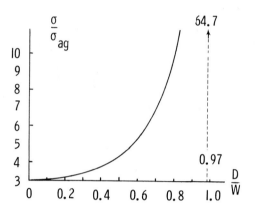

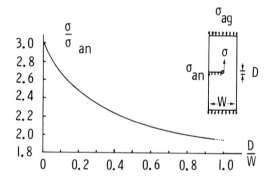

Figure 1.40. Stress concentration factors in an in-plane axially loaded finite plate with a circular hole, as function of the size of the hole.

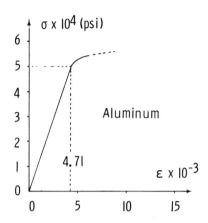

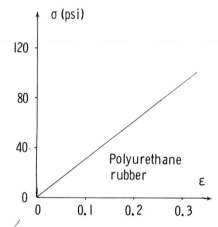

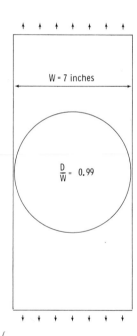

√ Figure 1.41. Geometry of the in-plane axially loaded plate with a big circular hole.

√Figure 1.42. Mechanical properties of the materials used to determine the non-linear s.c.f.

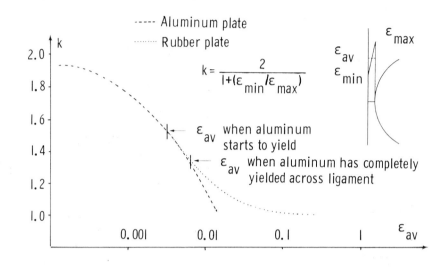

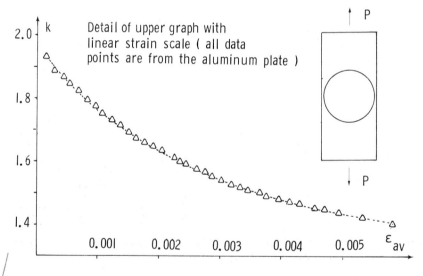

Figure 1.43. Stress concentration in a plate with a big hole, axially loaded in its plane, as a function of the level of deformation.

1.5 Stress concentrations in mixed boundary value problems

Most of the stress concentration factors listed in catalogs and easily available in the literature correspond to first boundary value problems and deal with notches and fillets. In this section some results will be shown of studies conducted to analyze concentrations in mixed-boundary value problems, and in particular at or near corners. The loading is the 'restrained shrinkage' produced by the difference in the free contraction or expansion of two bodies bonded to each other. Several thermal problems fall in this category. Restrained shrinkage can also develop in the process of curing or by absorption (or loss) of moisture. Applications are numerous in solid propellant rocket grains, composite materials, weldments, etc. Only the results obtained for two-dimensional problems are summarized in this section. Results corresponding to three-dimensional problems can be found in Section 1.7.

Stress concentration on the bonded interface of a strip with different end configurations. The long edge of the strip is bonded to a rigid bar. Loading is produced by the restrained shrinkage of the strip. The short ends of the strip have fillets of different configurations [60]. An illustration of the principal stresses at the interface is given in dimensionless manner in Figure 1.44 referred to α, free shrinkage of the strip, and E its modulus of elasticity. The stress concentration factor is shown in Figure 1.45.

Stress concentrations at variously shaped corners in rectangular bars bonded on two long edges, and subjected to restrained shrinkage. This problem arises when flat bars are bonded, cemented or welded on their top and bottom surfaces to other bodies of different mechanical or thermal characteristics. The free shrinkage of the bar is called α. The restraint in the longitudinal and the transverse direction of the bar could be the same in all directions, or have two perpendicular components α and $k\alpha$, where k a number smaller than unity is the ratio between the transverse and the longitudinal restraint. Restraint may be in only one direction, the direction of the length of the plate ($k = -\nu$, Poisson's ratio). By subtraction of one from the other of these loading conditions a variety of other biaxial conditions can be obtained [62, 63], Figure 1.46. The geometry of the corners investigated is shown in Figure 1.47.

The distributions of maximum shear stresses, and the distribution of

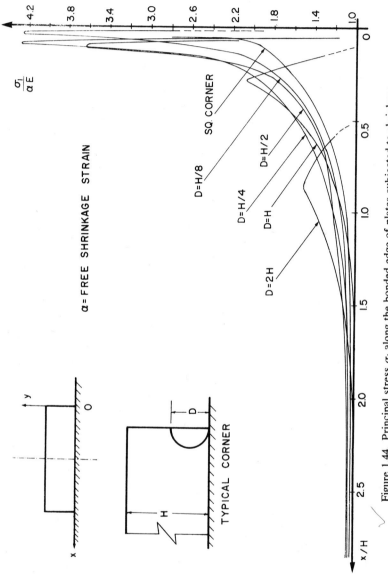

Figure 1.44. Principal stress σ_1 along the bonded edge of plates subjected to shrinkage.

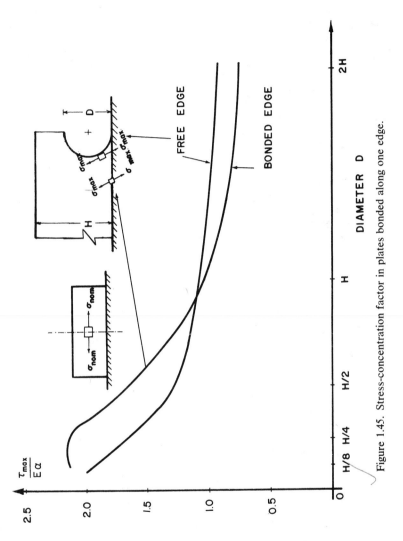

Figure 1.45. Stress-concentration factor in plates bonded along one edge.

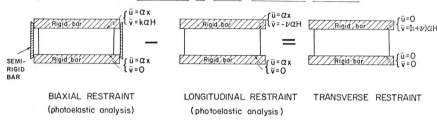

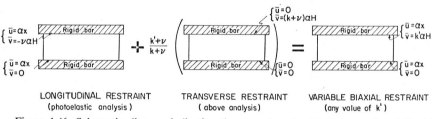

Figure 1.46. Schematic diagram indicating the procedure to obtain any ratio of biaxial restrained shrinkage.

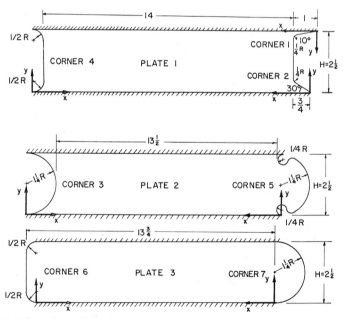

Figure 1.47. Dimensions of the three models used to analyze the strains associated with 7 different geometries of junctions of bonded and free boundaries.

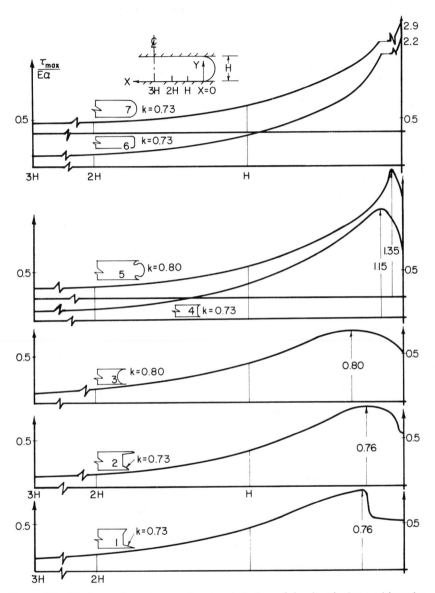

Figure 1.48. Maximum shear stresses along bonded edges of the shrunk plates subjected to biaxial restraint.

the cartesian shear stresses along the interface for both longitudinal and biaxial restraints have been obtained. An illustration is shown in Figure 1.48.

A summary of the stress concentrations factor (in terms of maximum shear, and also in terms of maximum cartesian shear) is shown in Figures 1.49 and 1.50.

These analyses of stresses are obviously very important in the study of crack propagation at interfaces. More information can be found in [61, 62].

CORNERS WITH PEAK τ_{max} ON INTERFACE			
TYPE OF CORNER $k = \dfrac{transv.\ restraint}{longit.\ restraint}$	k	PEAK VALUE OF $\tau_{max}/E\alpha$	PEAK VALUE OF INTERFACE SHEAR $\tau_{xy}/E\alpha$
⑦	0.73	2.9	2.7
⑥	0.73	2.2	2.1

CORNERS WITH PEAK τ_{max} ON FREE BOUNDARY		
TYPE OF CORNER	k	PEAK VALUE OF $\tau_{max}/E\alpha$
⑤	0.80	1.9
①	0.73	1.8
②	0.73	1.5
④	0.7	1.4
③	0.80	1.0

α = free shrinkage

Figure 1.49. Comparison of stresses in the neighborhood of corners (plates subjected to biaxial restraint).

CORNERS WITH PEAK τ_{max} ON INTERFACE		
TYPE OF CORNER	PEAK VALUE OF $\tau_{max}/E\alpha$	PEAK VALUE OF INTERFACE SHEAR $\tau_{xy}/E\alpha$
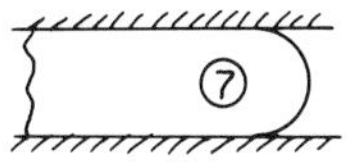 ⑦	1.6	1.0
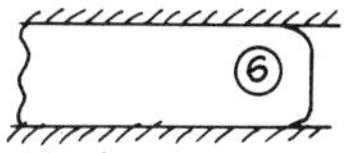 ⑥	1.7	1.4

CORNERS WITH PEAK τ_{max} ON FREE BOUNDARY	
TYPE OF CORNER	PEAK VALUE OF $\tau_{max}/E\alpha$
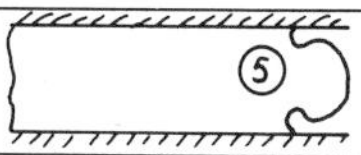 ⑤	1.0
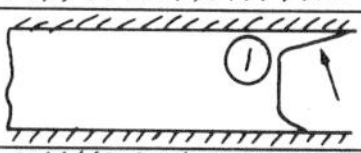 ①	0.9
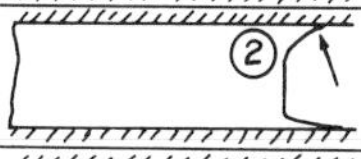 ②	0.6
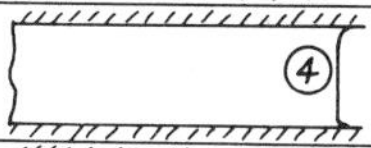 ④	1.0
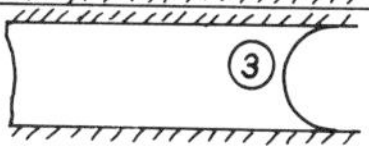 ③	0.6

α = free shrinkage

Figure 1.50. Comparison of stresses in the neighborhood of corners (plates subjected to longitudinal restraint).

Stress concentrations at corners of slabs with different edge geometries, when bonded on one face and shrunk. The situation is found when flat surfaces of two bodies are bonded, cemented or welded to each other and subjected to thermal or mechanical differential shrinkage [63]. The boundary conditions are illustrated in Figure 1.51. The stress concentration factors for five different geometries are shown in Figure 1.52. Typical distribution of normalized stresses are shown in Figures 1.53 to 1.55.

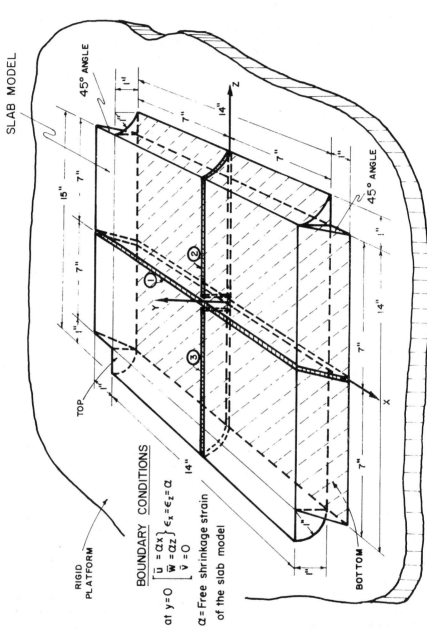

SLAB MODEL

45° ANGLE

45° ANGLE

TOP

BOTTOM

RIGID PLATFORM

BOUNDARY CONDITIONS

at $y=0$ $\begin{bmatrix} \bar{u} = \alpha x \\ \bar{w} = \alpha z \\ \bar{v} = 0 \end{bmatrix} \epsilon_x = \epsilon_z = \alpha$

α = Free shrinkage strain of the slab model

Figure 1.51. Square slab with different edge geometries, bonded on one face and shrunk.

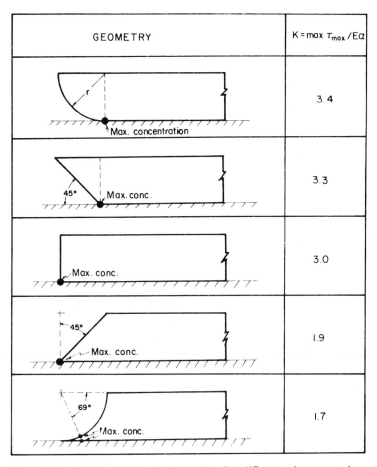

GEOMETRY	$K = \max \tau_{\max} / E\alpha$
	3.4
	3.3
	3.0
	1.9
	1.7

Figure 1.52. Stress concentration factors for five different edge geometries.

Considerations on analysis at corners. The corner at the intersection of the bonded and free edges is particularly difficult to analyze. The free edge requires that $\sigma_n = 0$ and $\tau_n = 0$. If it is assumed that on the bonded edge $\varepsilon_x = a$, for a square corner Hooke's law then requires a negative tangential stress and strain on the free boundary, $\sigma_y = -Ea/\nu$ and $\varepsilon_y = -a/\nu$. Analysis of the isochromatics along the free edge shows the fringes to represent a positive σ_y everywhere at the free boundary (increasing to a high positive value as the free boundary approached the bonded boundary). This discrepancy (between a negative and a positive

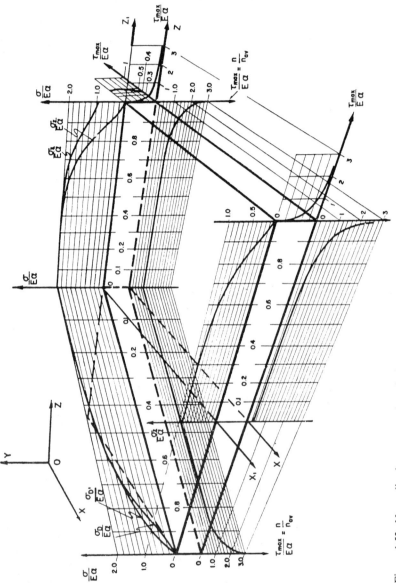

Figure 1.53. Normalized stresses at the principal planes of the top and bottom faces of the square slab bonded on one face and shrunk.

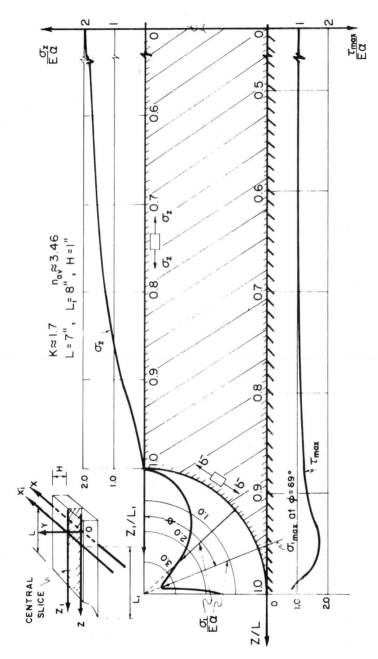

Figure 1.54. Normalized stresses at the boundaries of the central slice of a slab, bonded on one face and shrunk. (Geometry: concave corner).

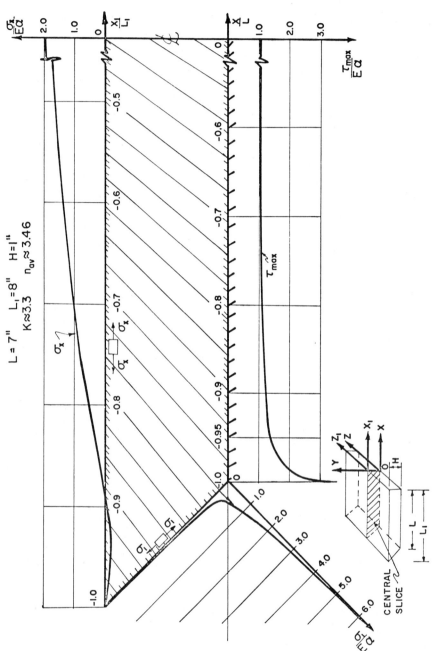

Figure 1.55. Normalized stresses at the boundaries of the central slice of a slab, bonded on one face and shrunk. (Geometry: outward corner).

value for σ_y) suggests that very near the corner a complicated situation develops. Thus, for example, if on the free boundary within less than a thousandth of an inch of the corner, σ_y changes to negative, there would be an isotropic point which may be very difficult to detect.

The following comments should also be considered: (1) the corner geometry cannot be ideally square and probably had a radius of the order of 0.001 in., (2) the optical and photographic techniques used do not permit a higher resolution than 0.001 in., (3) the thickness of the plate used was 0.125 in. In the neighbourhood of the corner (closer than 0.125 in.) the elastic phenomenon is probably three-dimensional and only an approximate analysis can be conducted with the two-dimensional techniques. Results obtained using three-dimensional methods (Fig. 1.54) are more representative in this respect.

It is also possible that the boundary condition $\varepsilon_x = \alpha$ is not valid all the way up to the corner. To satisfy the condition of a positive tangential stress, it is necessary that ε_x be compressive; however, the existence of a compressive ε_x at the corner could not be substantiated by a moiré analysis using 1000 lines per inch along the bonded edge. Again, as with photoelasticity, it is possible that very near the corner the sensitivity of the methods used may not be sufficiently high to record the phenomenon.

The theoretical solutions do not appear helpful in this respect. Solutions given by Duffy [64] and Theocaris and Dafermos [65] show the corner as a singular point at which all the stresses approach infinity, which is physically impossible. In particular, the tangential stress, σ_y is shown to have an unbounded positive value. One theoretical solution by Aleck [66] shows a finite positive value of σ_y, but does not satisfy the condition $\varepsilon_x = \alpha$ near the corner.

It is recognized that there is a considerable distortion at the corner. This suggests another possible explanation of the difficulty. The distortion at the corner may be sufficient to produce a geometry that *does* allow positive values of tangential stress with positive values of ε_x. To see what the geometric relation of the free boundary to the bonded boundary would have to be for resolving the difficulty, consider the combined Mohr circle of stress and strain (plane-stress) shown in Figure 1.56.

The tangential stress and normal stress on the free boundary are given as σ_t and σ_n ($\sigma_n = 0$). Circles of strain are drawn for $v = 1/2$ and $v = 1/3$. It can be seen that to obtain positive values of strain along the bonded edge, the angle β between the bonded edge and the free edge can be no greater than 55° for the $v = 1/2$ or 60° for $v = 1/3$.

A. J. Durelli

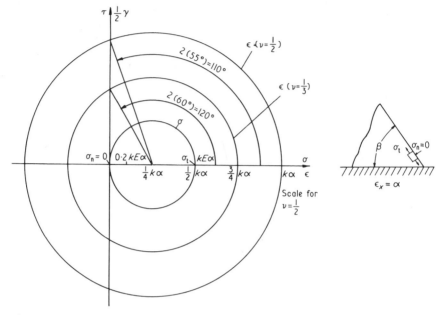

Figure 1.56. Mohr's circle to relate stresses and strains at a corner of a bonded and a free edge.

If the value of $\sigma_t = 10\,E\alpha$, for $v = 1/2$ the angle needed to produce $\varepsilon_x = \alpha$ on the bonded edged is about 50°. If the reasoning were correct, however, this angle should appear as soon as any amount of shrinkage is applied to the restrained boundary.

Stress concentrations associated with notches in a restrained shrinking disk. Stress concentrations associated with notches in infinite, semi-infinite and finite plates have been studied with great detail, as shown in some of the sources reviewed in Section 1.4. Few systematic studies have been conducted however when the plates have more complicated boundary conditions. Here results are given for the case of a circular disk with a diametral slot of variable length, and of variable radius at the notch end [67]. The outer perimeter is subjected to a restrained shrinkage α, as when bonded materials with different thermal coefficients of expansion are subjected to a change in temperature. The problem was solved using photoelasticity and Figure 1.57 shows the geometry of the several cases investigated. All the results obtained are shown

CUT	1	2	3	4	5	6	7	8	9	10	11	12	13	14	15	16	17
L/D	.086	.167	.333	.333	.333	.417	.417	.500	.500	.583	.583	.667	.667	.833	.833	.940	.940
W/D	.086	.048	.048	.048	.144	.048	.144	.048	.144	.048	.144	.048	.144	.048	.144	.048	.144

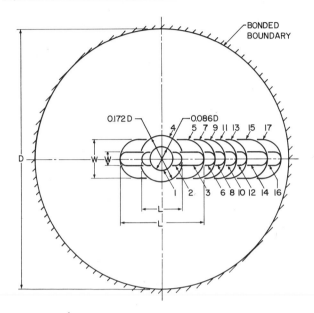

Figure 1.57. Cutting plan to obtain a series of isochromatics patterns from a single urethane rubber disk bonded to a rigid ring.

parametrically in Figure 1.58. It is interesting to note that, as is the case for the cracks usually considered in fracture mechanics, the isochromatics axis do not follow the axis of the slot. The maximum values are on either side of the tip.

Stress concentrations around elliptical holes in shrunk plates with bonded boundaries and discontinuities in bond. If the boundaries of a plate are bonded partially or totally to a rigid frame and the plate is then shrunk, a biaxial stress field is set. The plate may have straight or semicircular boundaries. If a hole is cut tangent to the boundary, a stress concentration will be induced along the free boundary of the hole, and the plate will fail in this region when the shrinkage becomes too large. In specimen I all plate boundaries were bonded, and one elliptical hole was located longitudinally at one apex of the plate and another was located

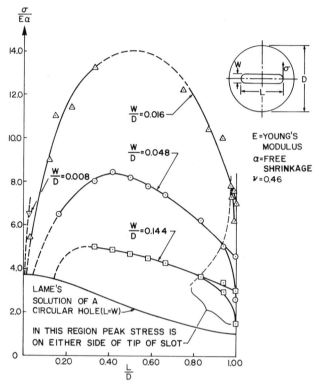

Figure 1.58. Stress concentration factors for slots with semi-circular ends in a disk restrained at its outer boundary and shrunk.

transversely at the other apex. In specimen II the plate was bonded only along the straight boundaries, and the circular boundaries were free. Two elliptical holes were located longitudinally at the point of discontinuity between bonded and unbonded boundaries at one end of the plate, and two more were located transversely at similar points on the other end. The sizes of the holes were varied systematically. All elliptical holes had a 2:1 ratio of major axis a to minor axis b (Figure 1.59).

Figure 1.60 shows typical isochromatic patterns near elliptical holes located longitudinally and transversely at the point of discontinuity between bonded and unbonded boundaries [68]. The stress concentration factor $K = \sigma_{max}/E\alpha$ obtained in this study is restricted to locations at the free boundaries of the holes. Here σ_{max} is the maximum tangential stress at the free boundary, E is Young's modulus, and α is the free shrinkage.

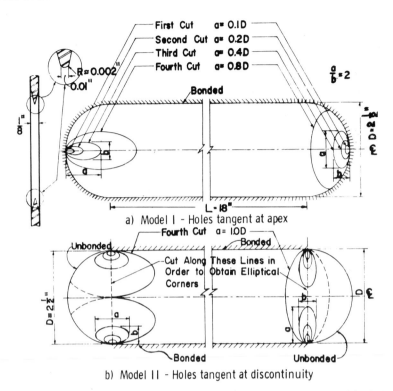

Figure 1.59. Geometry of plates with bonded boundaries and with discontinuities in bond.

Figure 1.61 shows the results for the two photoelastic models made with an incompressible material. For specimen I, Figure 1.61a the maximum stress associated with elliptical holes with major axes parallel to the transverse direction exceeds that for holes with major axes parallel to the longitudinal direction. In both cases, the peak K occurs at $a/D = 0.3$. For model II Figure 1.61b the relative values of K for the two orientations of the holes was opposite to the previously mentioned result.

If the semicircular ends in the second model were cut straight along its diameter, as indicated in Figure 1.59 the plate would become a long rectangular plate with free, short, straight edges. The maximum stresses at the elliptical corners obtained in this case have values very close to those of the original model with a maximum deviation of 6%. Therefore, the results given in Figure 1.61 can be also used for such cases.

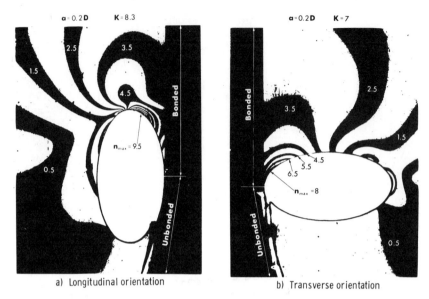

a) Longitudinal orientation   b) Transverse orientation

Figure 1.60. Enlargements of isochromatics in neighborhood of elliptical holes at discontinuity of bond.

Curves for circular holes $(a = b)$ also were obtained and are in-
corporated in Figures 1.61 and 1.62. Figure 1.62 shows K vs A/A_0, the
ratio of hole area $(A = \pi ab/4)$ to equivalent cross-sectional area $(A_0 =
\pi D^2/4)$, which may be significant in cases of cylinders. Finally, Figure
1.63 shows k vs a/b for various values of A/A_0. All curves seemed to
have minimum values near $a/b = 1$ (a circular hole).

*Thermal stress concentration at the interfaces in three-ply
laminates.* The middle layer of the laminate is made of a material
which differs from that of the top and the bottom layers, and may tend
to expand or shrink, when the temperature changes, to an amount
different from the amount of expansion or shrinkage of the exterior
layers. The geometry of the laminate used for the photoelastic tests and
the stress distribution are shown in Figure 1.64. Fracture starts at small
regions of the interface, and at a short distance from the corner. The
location of these regions correlates well with the position of the stress
concentration. More information can be found in [69].

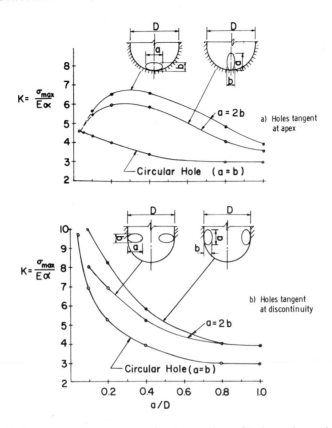

Figure 1.61. Parametric stress concentration factors (k vs a/D) for various elliptical holes in plates with bonded boundary, subjected to restrained shrinkage.

Stress concentration at the angular corners of long strips bonded on one end and shrunk. The configuration of the strip used for the tests is shown Figure 1.65. One end of the strip is a plane inclined at different angles ϕ with respect to the longitudinal axis and has a radius R at the corner. Three different corners with radius $R = \frac{1}{8}$ in., $\frac{1}{32}$ in. and a radius of less than $\frac{1}{10,000}$ in. were used. While the first two were formed by using $\frac{1}{4}$ and $\frac{1}{16}$ in. diam routers, respectively, the last one was obtained using a stainless-steel plastic-coated razor blade. The end angle $\phi = 0$ deg (equivalent to a crack) was obtained by forcing the blade into the rubber along the knife edge of the bonding bar.

The configuration of the tip of the cut made with the razor blade is

A. J. Durelli

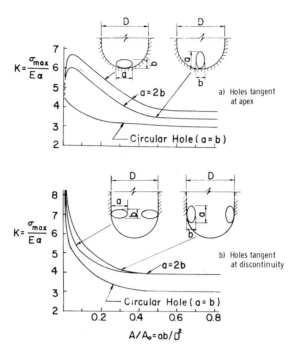

Figure 1.62. Parametric stress concentration factors (K vs A/A_0) for various elliptical holes in plates with bonded boundary, subjected to restrained shrinkage.

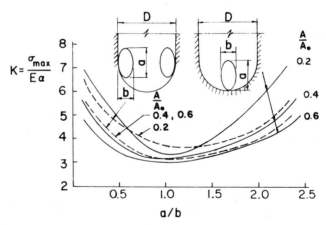

Figure 1.63. Parametric stress concentration factors (K vs a/b) for various ratios of elliptical holes in plates with bonded boundaries, subjected to restraint shrinkage.

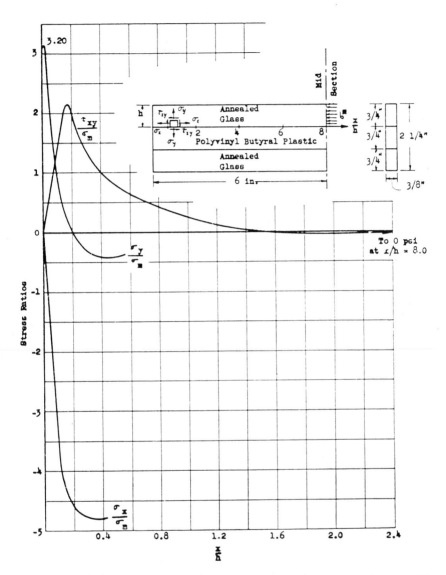

Figure 1.64. Normal and shear-stress distribution along glass-plastic interface at—20 F. (Positive stress ratios denote either tension or type of shear shown in sketch.)

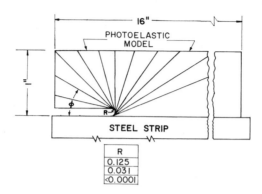

Figure 1.65. Dimensions of the photoelastic model subjected to restrained shrinkage
showing progression of cuts to obtain various angles.

probably not circular. It has been estimated, however, that the crack tip
has a curvature of less than 0.0001 in.

Stress concentration factors for end angles ϕ varying from 0 to
180 deg and corners radii $R = \frac{1}{8}, \frac{1}{32}$ and $<\frac{1}{10,000}$ in. are plotted in Figure
1.66. It is interesting to see that peaks of stress concentration occur at
approximately $\phi = 90$ deg as well as $\phi = 0$. The optimum acute angle
seems to be in the neighborhood of 60 deg.

In the results shown in Figure 1.66, the triangles and the solid circles
and squares are data points obtained using two different machining techniques. The scattering of the data can be seen in the figure.

The position of the point of maximum stress is not located at the
interface but somewhere above. This position changes as the angle ϕ
changes. Figure 1.67 shows the variation for $R = 0.125$. It is interesting
to know that for this case, the position and value of the s.c.f. practically
do not change when ϕ changes from 0° to 110°.

More information can be found in [70]. Enlarged isochromatics are
shown in Figure 1.68.

*Stress concentration associated with cracks near a rigid boundary and
subjected to restrained shrinkage.* If the circular end of the slots shown
in Figure 1.58 is made smaller and smaller, the matrix stress increases.
Figure 1.69 shows the geometry of a slot made with the smallest width
of all the slots that ended with a simple radius (and were tested in the
program). Figure 1.70 shows the isochromatic pattern obtained from this
model. The K factor is 9.2.

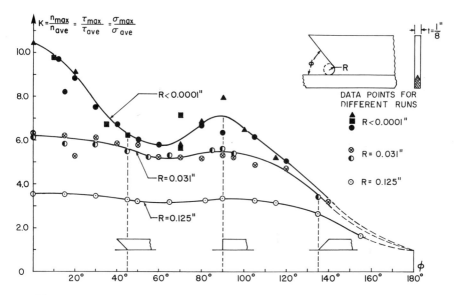

Figure 1.66. Parametric stress concentration factors for various angular corners.

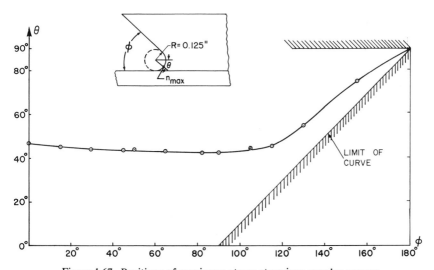

Figure 1.67. Positions of maximum stress at various angular corners.

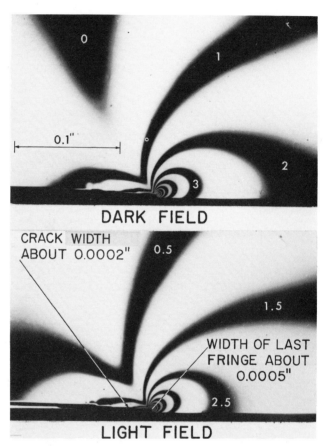

Figure 1.68. Enlargements of isochromatic field in the neighborhood of the corner of a crack.

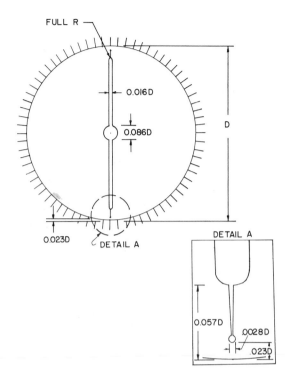

FULL R

0.016D

0.086D

D

0.023D

DETAIL A

DETAIL A

0.057D

.0028D

.023D

Figure 1.69. Circular disk with a slot ended with a small circular hole and subjected to restrained shrinkage.

The analysis of this type of model may be questioned. First the thickness of the model is about 12 times the radius at the tip of the slot. The surfaces of the model are in a condition of plane stress. The central portion of the model through the thickness can be thought to approach plane strain. The photoelastic analysis gives the result of an integral effect along the path of light. This is probably not a serious limitation. There is evidence to show that the difference between the two extreme situations of plane stress and plane strain ($\varepsilon_z = 0$) is not too large.

A more serious difficulty is the high level of the stress. As the radius becomes smaller, the same load produces higher stress and at some point the assumption of a linear relation between stress, strain and birefringence breaks down. This non-linear response can be avoided to some extent by reducing the load. It is believed that in the pattern shown in Figure 1.70 the fringe orders are not sufficiently great to exhibit

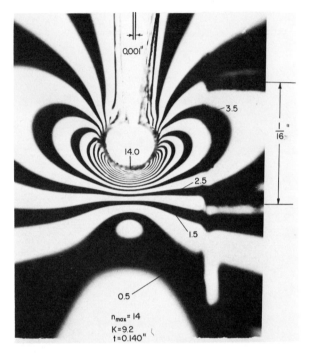

Figure 1.70. Isochromatics at the tip of a slot ended with a small hole in a circular disk subjected to restrained shrinkage.

marked non-linearity and it can be seen that the radius is not appreciably increased.

When these ideas are extended to the case of a crack, similar difficulties are encountered. In the isochromatic patterns around the slot it can be seen that most of the response takes place within a distance from the slot boundary, of the order of one slot width. It follows that a model of a crack must be viewed within approximately one crack width to obtain an estimate of the stresses at the tip.

Figure 1.71 shows the pattern around one crack with a shape similar to the one shown in Figure 1.69. The crack was started with a knife, and propagated spontaneously to the position shown in the figure. At the tip there may exist a much higher fringe order than the 10 or so fringes that can be seen in the photograph, and the three-dimensional and non-linear effects may be important.

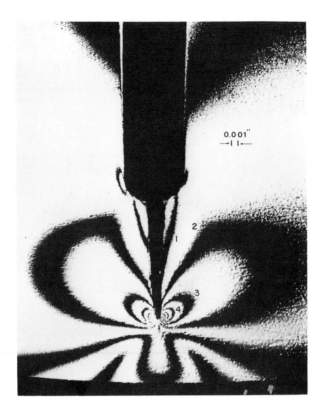

Figure 1.71. Isochromatics at the tip of a crack cut with a knife and propagated spontaneously in a circular disk subjected to restrained shrinkage.

It is to be noted that the maximum stress that is at the axis of the crack in Figure 1.70, shifts to the side of the crack in Figure 1.71. The distance between the bottom of the crack and the rigid boundary is 0.05 in. in the first case, and 0.015 in the second. The situation is also shown, in dimensionless manner, in Figure 1.58.

Stress concentrations in a rectangular plate with circular perforations along its two bonded edges and subjected to restrained shrinkage. In some cases stress concentrations associated with a hole or a fillet may be decreased by repeating the discontinuity. The geometry of the problem to be dealt with here is shown in Figure 1.72. The stresses are

94 A. J. Durelli

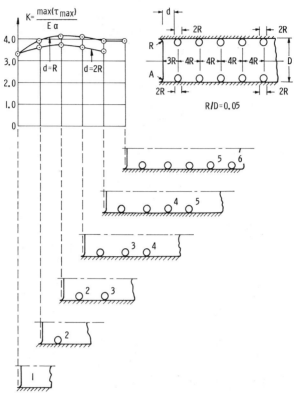

Figure 1.72. Stress concentration in rectangular plates with rows of holes along two bonded boundaries, subjected to restrained shrinkage.

determined for the case of no hole, one hole, and successively for 2, 3, 4 and 5 holes. The maximum stress occurs always at point A as shown in the figure. The presence of holes tangentially to the boundaries did not decrease the s.c.f. Lengthening the distance between the first hole and the fillet reduced the s.c.f. but not to a level lower than the one present when there were no holes [71].

1.6 Stress concentrations in some specific problems

Most of the problems studied and the solutions of which are reviewed in Section 1.3 as well as most of the stress concentrations reported in Sections 1.4 and 1.5 correspond to more or less idealized geometries. So will be the situation of the three-dimensional and the dynamic cases to be dealt in Sections 1.7 and 1.8. In this section some solutions will be presented for cases which still are of general interest, but corresponding to more specific applications.

Stress concentrations at the fillet of internal flanges. When a machine element contains an internal flange the resultant loading in the cylinder often is parallel to the axis, but eccentric (Figure 1.73). It is also frequently the case that these flanges have to be very narrow which requires that the fillet connecting the flange to the shell to be small, giving rise to severe stress concentrations. The stress distribution has been studied [72] for various positions of the load P along the radial line CO. The influence of the cylinder thickness t and the flange thickness T are considered. Figures 1.74 to 1.78 show the results obtained, parametrically as function of eccentricity and for several ratios of the diameter of the cylinder to the thickness of the cylinder and of the flange.

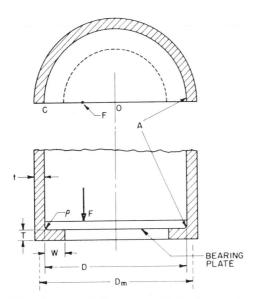

Figure 1.73. Internal flange at the end of a hollow cylinder. The axial and eccentric load is transmitted to the flange by means of a plate.

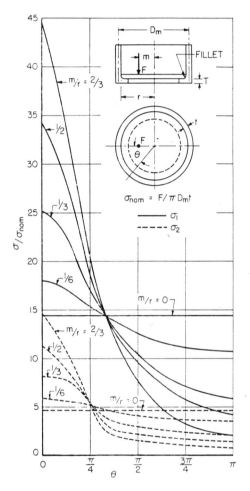

Figure 1.74. Principal stresses at the bottom of the fillet for different eccentricities of load. $D_m/t = 13$, and $D_m/T = 25$.

Stress concentrations in pressurized circular shells with discontinuities, with or without stiffeners. This is the summary [73] of a series of research programs dealing with the experimental determination of stresses, strains and displacements in pressurized circular vessels with a circular hole. The hole can be either plain or reinforced by a ring. The vessels were either of constant thickness or stiffened by transverse internal ribs. The solutions obtained correspond to internal pressure

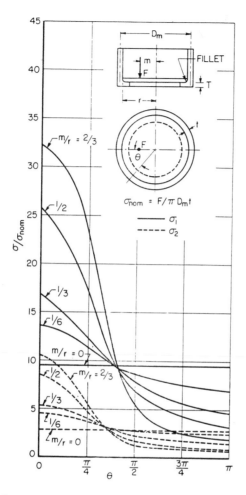

Figure 1.75. Principal stresses at the bottom of the fillet for different eccentricities of load. $D_m/t = 21$, and $D_m/T = 25$.

loading but can easily be adapted to external pressure loading, with the exception of local variations due to the manner of plugging the hole.

Six geometric variations were studied:

Case 1 plain hole-unstiffened shell
Case 2 plane hole-stiffened shell

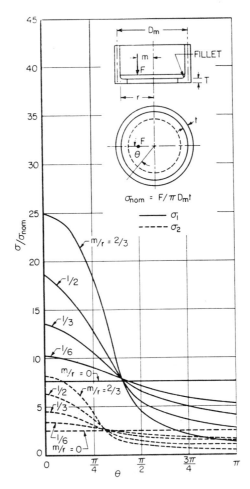

Figure 1.76. Principal stresses at the bottom of the fillet for different eccentricities of load. $D_m/t = 30$, and $D_m/T = 25$.

Case 3 small reinforced hole-unstiffened shell
Case 4 large reinforced hole—unstiffened shell
Case 5 reinforced hole—stiffened shell—hole between ribs
Case 6 reinforced hole—stiffened shell—hole through ribs.

The geometries of the vessels analyzed are represented in Figure 1.79. The hole size varied from $1-\frac{1}{4}$ in. to $2-\frac{3}{8}$ in. The edge of the hole

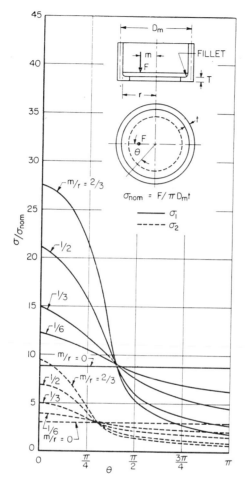

Figure 1.77. Principal stresses at the bottom of the fillet for different eccentricities of load. $D_m/t = 30$, and $D_m/T = 35$.

was reinforced by means of a ring, the outer diameter of which was $\frac{1}{2}$ in. larger than hole diameter for cases 3, 4 and 5 and $\frac{5}{8}$ in. larger for case 6. The height of the reinforcing ring was $\frac{1}{2}$ in. for cases 3 and 5, $\frac{9}{16}$ in. for case 4, and $1-\frac{3}{4}$ in. for case 6. The height of the reinforcing ring for case 6 was such that it reached the inner radius of the rib through which the hole was passed and thus served to replace the rib in the region at the hole.

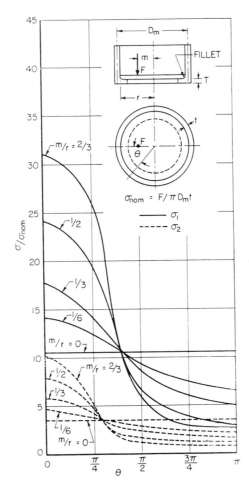

Figure 1.78. Principal stresses at the bottom of the fillet for different eccentricities of load. $D_m/t = 30$, and $D_m/T = 50$.

The radii of the fillets between rib and shell, between reinforcing ring and shell, and between rib and reinforcing ring were all $\frac{1}{16}$ in. with the single exception of the fillets in case 4, which were $\frac{1}{8}$ in.

Internal pressure was applied to the vessels. In order to contain the pressure, plugs made of silicone rubber were placed in the holes. The plugs had a shoulder that rested on the inner surface and were kept in place by the pressure.

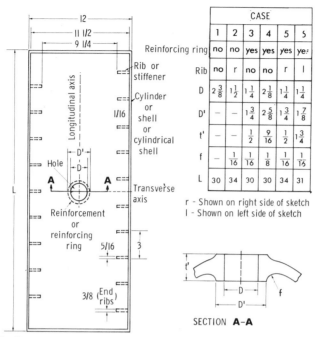

	CASE					
	1	2	3	4	5	5
Reinforcing ring	no	no	yes	yes	yes	yes
Rib or stiffener (Rib)	no	r	no	no	r	l
Cylinder or shell or cylindrical shell D	$2\frac{3}{8}$	$1\frac{1}{2}$	$1\frac{1}{4}$	$2\frac{1}{8}$	$1\frac{1}{4}$	$1\frac{1}{4}$
D'	–	–	$1\frac{3}{4}$	$2\frac{5}{8}$	$1\frac{3}{4}$	$1\frac{7}{8}$
t'	–	–	$\frac{1}{2}$	$\frac{9}{16}$	$\frac{1}{2}$	$1\frac{3}{4}$
f	–	$\frac{1}{16}$	$\frac{1}{16}$	$\frac{1}{8}$	$\frac{1}{16}$	$\frac{1}{16}$
L	30	34	30	30	34	31

r - Shown on right side of sketch
l - Shown on left side of sketch

SECTION **A-A**

Figure 1.79. Geometry of six cylindrical vessels with a hole and various reinforcements.

The stress analysis was conducted using brittle coatings, electrical resistance strain gages and three-dimensional photoelasticity. Figure 1.80 shows a typical detail of the isochromatic patterns obtained. Figure 1.81 includes a table showing the position and the value of the stress concentration factors for all the cases studied. The complete details of these analyses are given in [74–78].

The stresses in Figure 1.81 are normalized by dividing the actual stresses obtained from the tests by the factor pR/t, where p is test pressure, R the radius of the middle surface of the shell, and t the shell thickness. For the case of a thin-walled cylinder with no reinforcement under pressure, pR/t is well known to be the theoretical stress in the circumferential (hoop) direction, $\sigma_{\theta\theta}$. The longitudinal (axial) stress for the same case is also well known to be $\sigma_{zz} = pR/2t$. Thus the lowest possible maximum stress in a thin unstiffened cylinder would have the normalized stress value of 1.0 for $\sigma_{\theta\theta}/pR/t$. In the case of a cylinder stiffened with ribs, equilibrium would require that the axial stress be at least $\sigma_{zz}/(pR/t) = 0.5$. These minima should be kept in mind in considering the values reported in Figure 1.81.

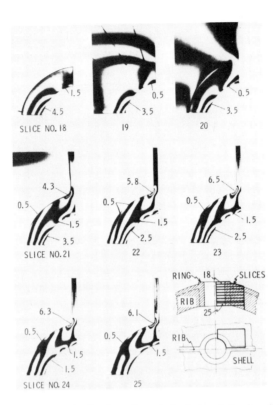

Figure 1.80. Isochromatics in slices through the hole reinforcing ring of case 6.

Stress concentrations at or near a square corner in the neighborhood of a hydrostatically loaded circular hole. Several papers have been published recording stress distributions in square plates and strips of finite width, with a pressurized hole [79–82]. An illustrative result will be shown here. Figure 1.82 shows the geometry of the plate and the variable position of the hole, and the stress distribution along the boundary of the hole. Figure 1.83 shows the value of the maximum stress anywhere in the plate. It is significant that the stress concentration changes location, from the boundary of the hole to the boundary of the plate, as the edge of the hole approaches the edge of the plate. The results are given in dimensionless form taken the pressure as reference (see Section 1 on definition of stress concentration factors).

Surface	Stress	Locations of Stress Concentrations	Case 1 Plain shell with a nonreinforced hole	Case 2 Ribbed shell with a nonreinforced hole	Cases 3 and 4 Plain shell with two reinforced holes		Case 5 Ribbed shell with a reinforced hole in bay	Case 6 Ribbed shell with a reinforced hole on rib
		D/2R	0.202	0.128	0.106	0.181	0.106	0.106
		ring-shell fillet radius / t			0.25	0.50	0.25	0.25
Outer Surface of Shell	$\frac{\sigma_{\theta\theta}}{pR/t}$	At edge of hole	3.90	2.30	1.70	2.10	1.45	1.70
	$\frac{\sigma_{\theta\theta}}{pR/t}$	At fillet of ring and shell			1.10	1.60	1.10	0.98
	$\frac{\sigma_{zz}}{pR/t}$	At edge of hole	2.20	2.10	0.52	0.60	0.85	0.40
	$\frac{\sigma_{zz}}{pR/t}$	At fillet of ring and shell			0.85	1.00	0.95	0.70
Inner Surface of Shell	$\frac{\sigma_{\theta\theta}}{pR/t}$	At edge of hole	3.65	2.70	1.45	1.55	1.15	0.35
	$\frac{\sigma_{\theta\theta}}{pR/t}$	At fillet of ring and shell			1.10	1.15	1.00	1.25
	$\frac{\sigma_{f}}{pR/t}$	At fillet of ring and rib						1.95
	$\frac{\sigma_{zz}}{pR/t}$	At edge of hole	-1.25	-0.55	0.50	0.50	0.49	-1.10
	$\frac{\sigma_{zz}}{pR/t}$	At fillet of ring and shell			0.50	0.65	0.45	
	$\frac{\sigma_{f}}{pR/t}$	At fillet of ring and shell					0.13*	1.80
	$\frac{\sigma_{f}}{pR/t}$	At fillet of rib nearest hole		1.65			1.25	1.05
	$\frac{\sigma_{f}}{pR/t}$	At fillet of rib through hole						1.60

* At this fillet no well-defined maximum was recorded. The stress increased continuously over the fillet.

Figure 1.81. Stress concentration factors associated with circular holes in cylindrical shells.

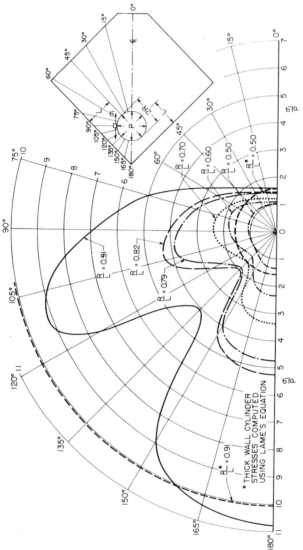

Figure 1.82. Hydrostatically loaded circular hole in the neighborhood of a corner. Stress distribution along the boundary of the hole.

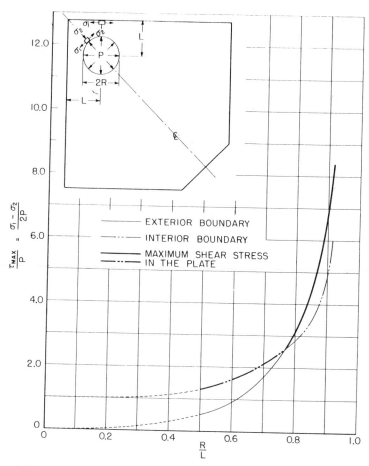

Figure 1.83. Hydrostatically loaded circular hole in the neighborhood of a corner. Maximum shear stress in the plate.

1.7 Stress concentrations in three-dimensional problems

The determination of stress concentrations in three-dimensional problems is considerably more difficult than the determination in the case of two-dimensional problems, both theoretically and experimentally. As could be 'expected, most of the contributions refer to stress concentrations on free surfaces of three-dimensional bodies because although the method used may be three-dimensional, the stress distribution problem is essentially two-dimensional. Some three-dimensional solutions that may be of interest, either in industrial applications or for the methodology, will be presented in this section. Solutions to other similar problems have been summarized in [86].

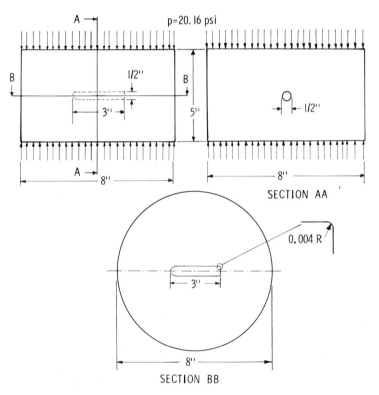

Figure 1.84. Circular bar embedded in a matrix and loaded perpendicular to axis of bar.

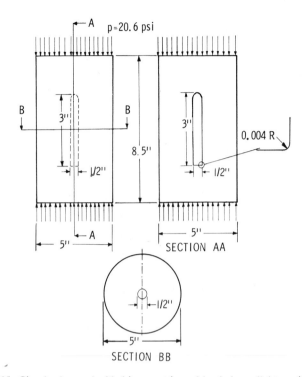

Figure 1.85. Circular bar embedded in a matrix and loaded parallel to axis of bar.

Stress concentration at flat and spherical ends of a circular bar embedded in a matrix in a triaxial stress field. The problem is defined in Figures 1.84 and 1.85. An illustration of the physical evidence is shown in Figure 1.86.

The most significant results related to stress concentrations are shown in Figures 1.87 and 1.88. The stresses are given in a dimensionless manner making the results applicable to all cases of rigid bars embedded in softer matrices. The peak tangential shear stresses on the interface at the hemispherical end take place at approximately an angle of 55° with the axis (measured from the tip of the hemispherical end). They are shown in Figure 1.89. Further information can be found in [83].

Stress concentration at the ends of pressurized cylindrical holes. The experiments were conducted using cylinders the diameter of which was

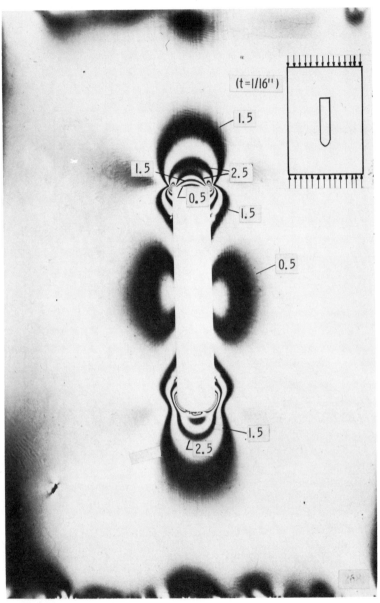

Figure 1.86. Light field isochromatic pattern for meridian slice.

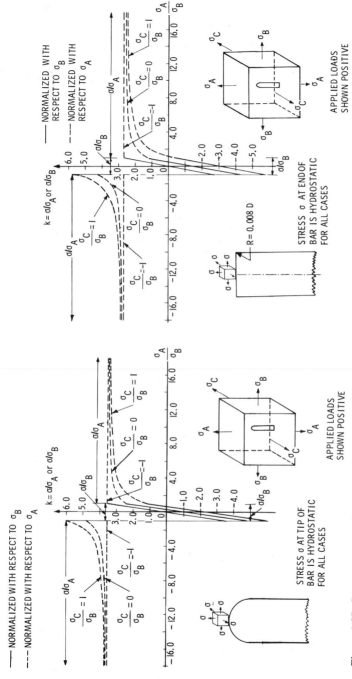

Figure 1.87. Stress-concentration factor at the hemispherical end of a circular bar embedded in a matrix and subjected to triaxial loading.

Figure 1.88. Stress-concentration factor at center of flat end of circular bar embedded in a matrix and subjected to triaxial loading.

109

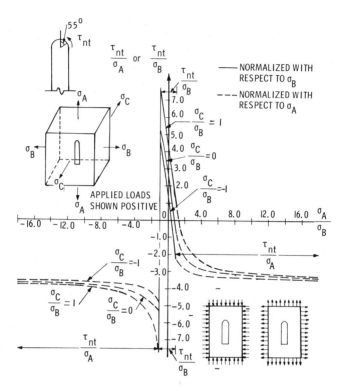

Figure 1.89. Normalized tangential shear stress τ_{nt} on the hemispherical end of an embedded circular bar at a point 55° from the axis of the bar and the σ_B stress plane.

about four times the diameter of the hole [84]. The results obtained can be used as first approximation for any large cylinders with small axial holes or for the infinite medium. Three hole end geometries were investigated. One had a hemispherical end, the other two had flat ends with fillets (radius 0.58 and 0.17 of the radius of the hole) connecting the ends and the sides. The stress distributions are shown in Figures 1.90 to 1.92 in dimensionless form, referred to the pressure. The circumferential component of the stress is little sensitive to the fillet radius changes, while the meridional component is much more dependent on the fillet radius. However, the increase in the fillet radius sharpness from 0.58 to 0.17 of the cylinder radius does not increase appreciably the stress concentration factor.

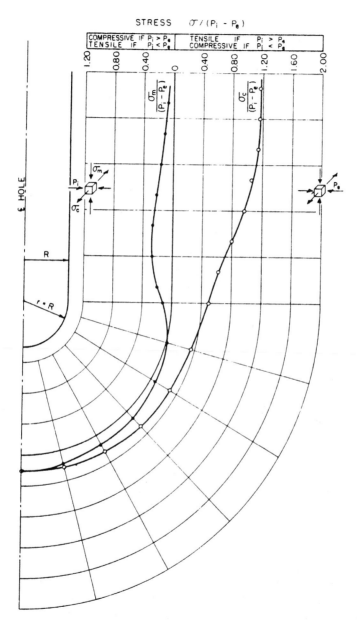

Figure 1.90. Distribution of the two principal stresses tangential to the surface of the hole with hemispherical end.

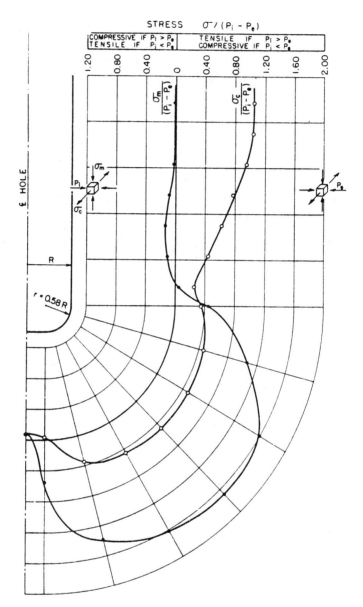

Figure 1.91. Distribution of the two principal stresses tangential to the surface of the hole with flat end and fillet radius equal to 0.58 the radius of the hole.

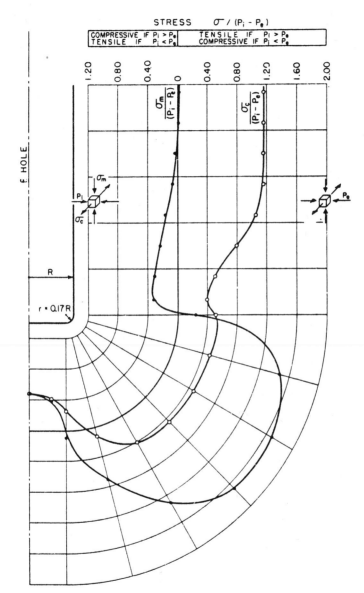

Figure 1.92. Distribution of the two principal stresses tangential to the surface of the hole with flat end and fillet radius equal to 0.17 the radius of the hole.

Stress concentrations associated with the restrained shrinkage of cylinders with toroidal cavities. Three hollow cylinders (one of them with a dome-shaped end) were bonded on the outside surface to shells. The geometry of each cylinder and its corresponding shell is defined in the sketches in Figures 1.93 to 1.94. The meaning of the dimensionless quantities $2L/2b$, b/a, $d/(b-a)$ and t/b are also indicated there. That part of the external surface which is bonded is also shown in the figures. Three-dimensional photoelasticity was used to "freeze" the photoelastic and moiré response. The shells are several times more rigid than the material of the cylinders at the time of 'freezing' the isochromatics. Displacement components were obtained using moiré effects, by photographing a cross-grating on slices of the cylinders, annealing them, and analyzing with a one-way grating. Illustrative examples of the records obtained are shown in Figures 1.93 to 1.98. More information can be found in [85]. The stress concentrations are tabulated in Table 1.1.

TABLE 1.1

Peak values of normalized maximum stresses in the three cylinders

Cylinder	Geometry			$\dfrac{F_\sigma}{E\alpha_z^*}$ For 1/8″ slice Units: fringe^{-1}	Interface restraint at $z=0$	At $\begin{cases} r=a \\ z=0 \end{cases}$	Maximum stresses	
	$\dfrac{2L}{2b}$	$\dfrac{b}{a}$	$\dfrac{d}{b-a}$		$\bar{\varepsilon}_z = \alpha_z^* \quad \dfrac{\bar{\varepsilon}_\theta}{\bar{\varepsilon}_z}$	$\dfrac{\sigma_\theta}{E\alpha_z^*}$	$\dfrac{\max \tau_{\max}}{E\alpha_z^*}$	$\dfrac{\max \sigma_1}{E\alpha_z^*}$
I	2.60	2.70	0.25	0.91	0.0032 1.84	16.1	6.6	13.2
II	1.25	2.40	0.33	0.81	0.0033 1.97	9.7	5.7	12.9
III	1.29	2.33	0.29	0.63	0.0055 0.88	6.1	4.0	13.8

Stress concentration at the interface of an elbow embedded in a shrinking matrix. The geometry of the square bar is shown in Figure 1.99. The bar is much more rigid than the matrix. The maximum shear strains, and the shear strains γ_{nt} in the plane of symmetry, are shown in Figures 1.100 and 1.101. Three-dimensional photoelasticity was used for the analysis. More information can be found in [86].

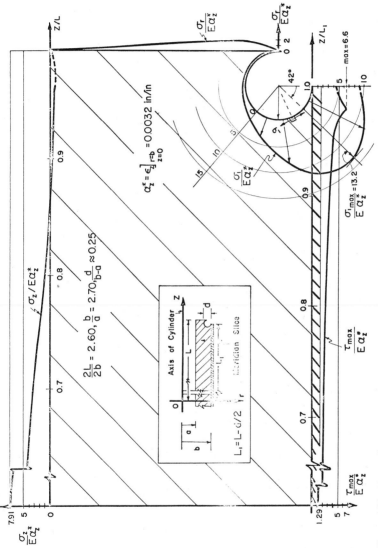

Figure 1.93. Normalized stresses at the boundaries of a meridian plane in hollow cylinder I, bonded on part of the outside surface to a shell and subjected to restrained shrinkage.

116

A. J. Durelli

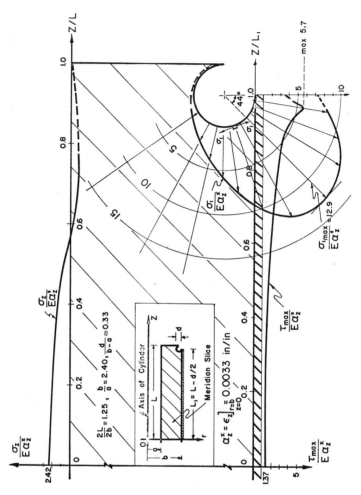

Figure 1.94. Normalized stresses at the boundaries of a meridian plane in hollow cylinder II, part of the outside surface to a shell and subjected to restrained shrinkage.

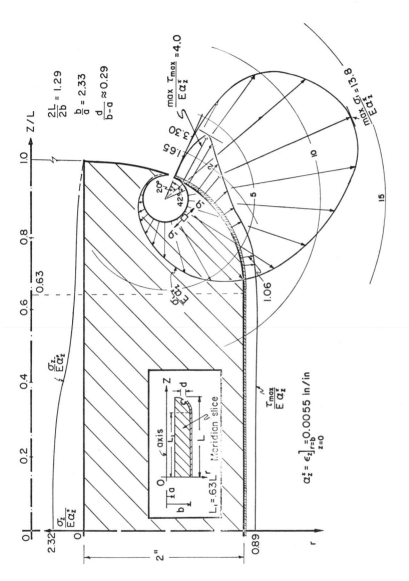

Figure 1.95. Normalized stresses at the boundaries of a meridian plane in a hollow cylinder with dome shape end (cylinder III), bonded on part of the outside surface to a shell and subjected to restrained shrinkage.

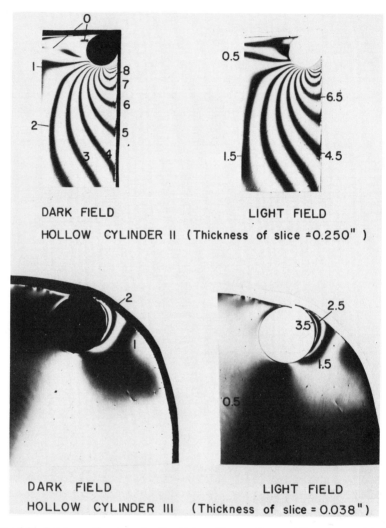

Figure 1.96. Isochromatics near the toroidal cavities on the meridian planes of hollow cylinders II and III.

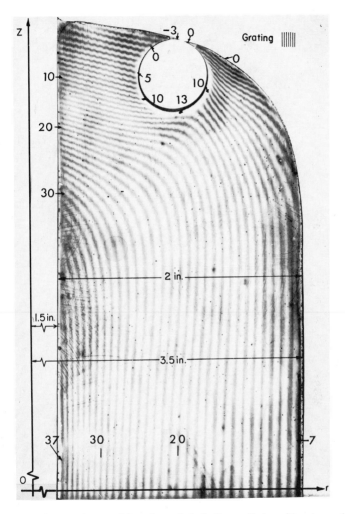

Figure 1.97. Isothetics u_r in a meridian plane of the hollow cylinder with a dome shape end (cylinder III). Note: The radial displacements are given by the expression $u_r = 0.001$ $(n + 2r)$ where n is the fringe order and u and r are expressed in inches.

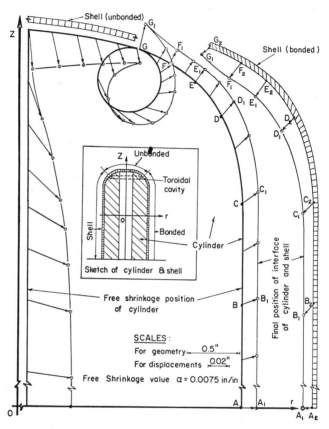

Figure 1.98. Boundary displacements for shell and hollow cylinder with a dome shape end (cylinder III).

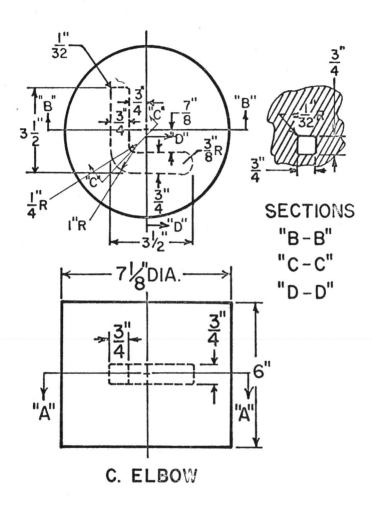

C. ELBOW

MATERIAL : INSERTS — HYSOL 4290 EPOXY
MATRIX — SAMPSON'S EPOXY

Figure 1.99. Geometry of the rigid elbow embedded in the shrinking matrix.

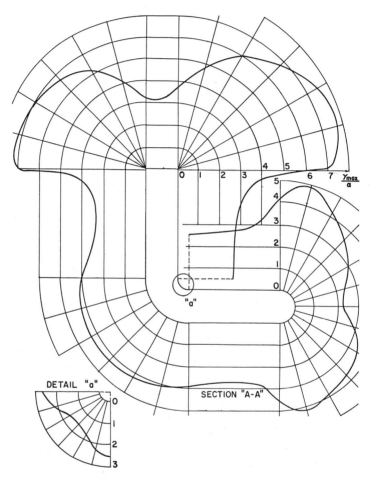

Figure 1.100. Maximum shear strain in the plane of symmetry at the interface between a matrix and an embedded elbow with a flat and a circular end. (restrained shrinkage).

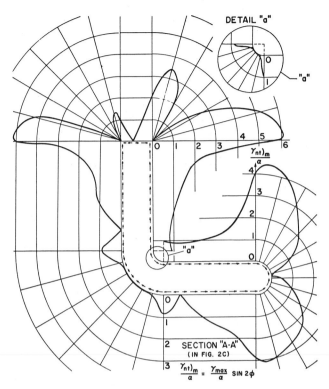

Figure 1.101. Shear strains in the plane of symmetry and tangential to the interface between a matrix and an embedded elbow with a flat and a circular end (restrained shrinkage).

Stress concentration at the interface of two circular bars embedded in a shrinking matrix. In one of the cases studied the two bars were aligned [86], in the other the axis of the bars made an angle of 14°. The geometry is shown in Figure 1.102. Three-dimensional photoelasticity was used for the analysis. The maximum shear strains at the interface are shown in Figures 1.103 and 1.104.

Stress concentrations in cylinders with small perforations and outside surface partially bonded to a rigid case, when subjected to restrained shrinkage. The boundary conditions are defined in Figures 1.105 and 1.106. The stress concentration factors and the position of the points at which they take place are shown in Figures 1.107 and 1.108.

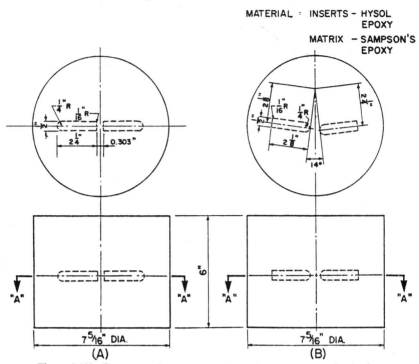

Figure 1.102. Geometry of two models of matrices cast about circular bars.

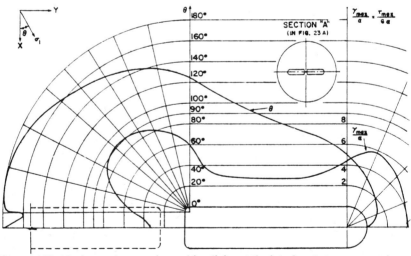

Figure 1.103. Maximum shear strains and isoclinics at the interface between a matrix and two aligned circular bars separated a short distance by the matrix. (restrained shrinkage).

124

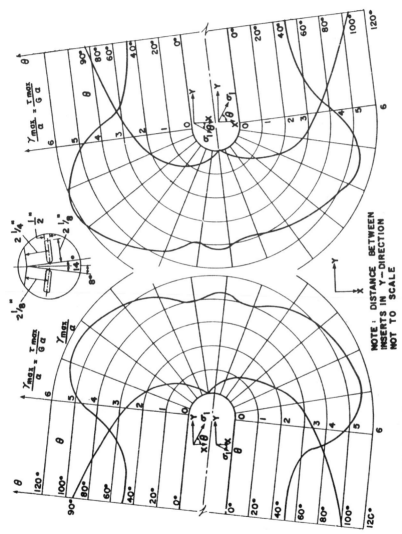

Figure 1.104. Maximum shear strains and isoclinics at interfaces between a matrix and two non-aligned circular bars separated a short distance by the matrix.

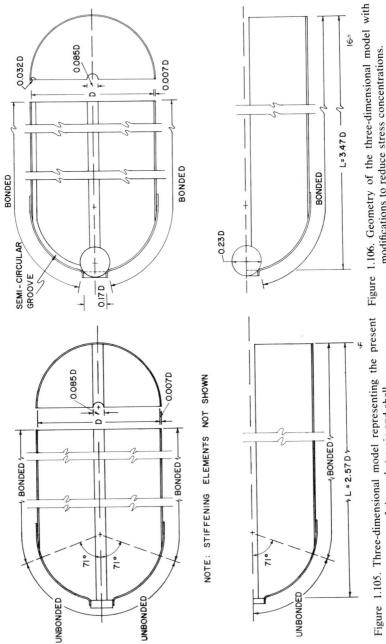

Figure 1.105. Three-dimensional model representing the present geometry of the rocket grain and shell.

Figure 1.106. Geometry of the three-dimensional model with modifications to reduce stress concentrations.

SLICE NO.	LOCATION	$K = \dfrac{\text{slice}(\tau_{max})}{E\,\alpha^*}$
1	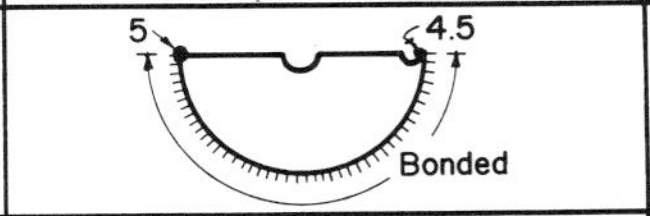	5, 4.5
2	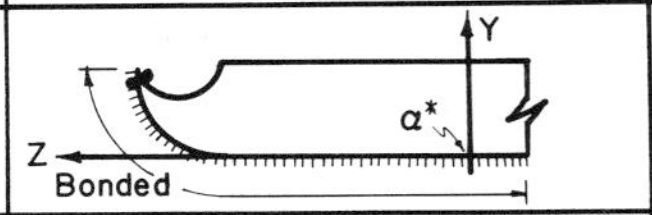	3
3	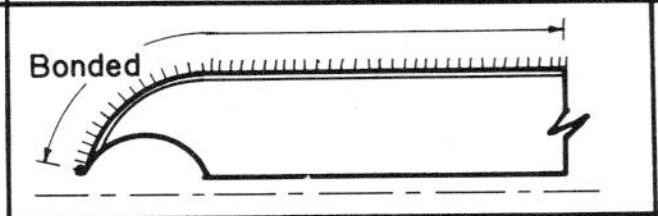	2.7
3A	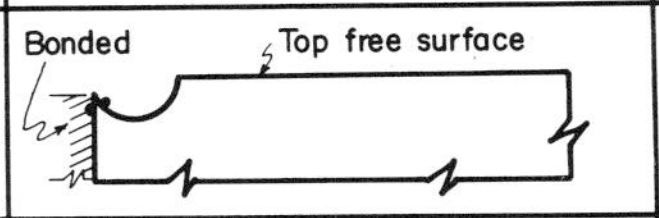	2.5
3B	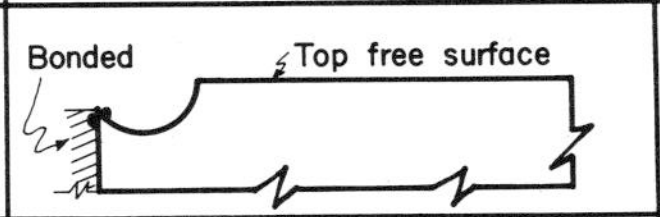	2.4
3C	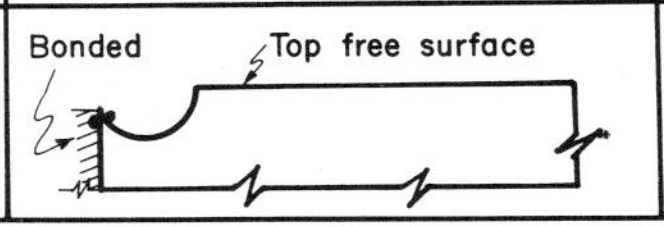	2.5

$$\epsilon_z\big|_{x=y=z=0} = \bar{\epsilon}_z = \alpha^*$$

The stress concentration takes place at the points marked by a dot.

Figure 1.108. Stress concentration factors for the modified geometry of the rocket grain.

SLICE NO.	LOCATION	$K = \dfrac{\text{slice}(\tau_{max})}{E\,\alpha^*}$
1	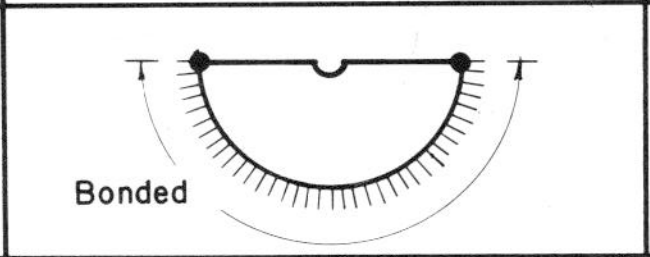	5.8
2, 2A		6.3, 5
3	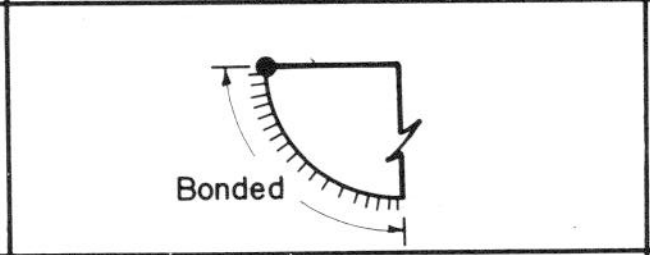	3
4, 4A		7.5, 4.1

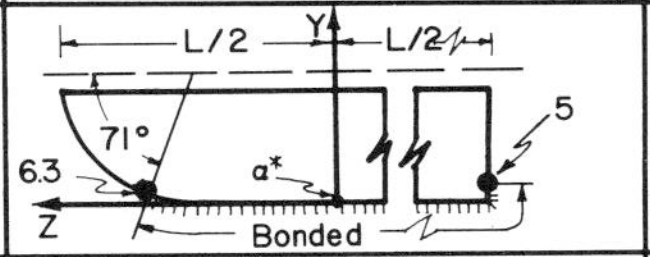

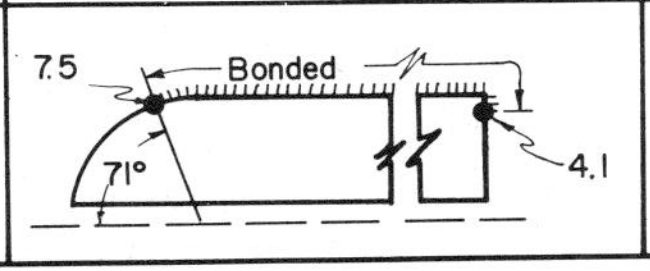

$$\epsilon_z\big|_{x=y=z=0} = \bar{\epsilon}_z = \alpha^*$$

The stress concentration takes place at the points marked by a dot.

Figure 1.107. Stress concentration factors for the present geometry of the rocket grain.

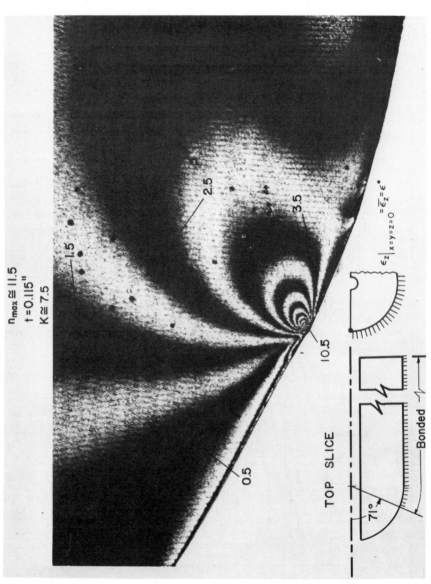

Figure 1.109. Isochromatics in the top slice (slice 4) of the epoxy model corresponding to the present geometry of the grain.

The isochromatic pattern in Figure 1.109. is particularly significant in fracture mechanics. The separation between the shrinking body and the bonded case developed spontaneously. The width of the gap in the neighborhood of the point of concentration is of the order of 0.001 in. The thickness of the observed slice was 0.115 in. The pattern indicates a finite s.c.f.

Stress concentrations in a cubic box subjected to concentrated loads. The geometry and boundary conditions are defined in Figure 1.110. The distribution of the two principal stresses along the diagonal on the inside and outside surfaces is shown in Figures 1.111 and 1.112. Distributions of other stresses in the box can be found in [87]. The results

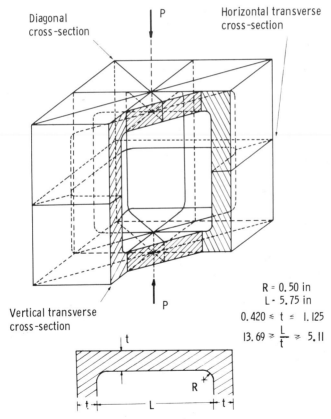

Diagonal cross-section

P

Horizontal transverse cross-section

Vertical transverse cross-section

P

R = 0.50 in
L = 5.75 in
$0.420 \leqslant t \leqslant 1.125$
$13.69 \geqslant \dfrac{L}{t} \geqslant 5.11$

t

R

t ─────── L ─────── t

Figure 1.110. Geometry and loading of a cubic box subjected to a concentrated load at the center of two opposite faces.

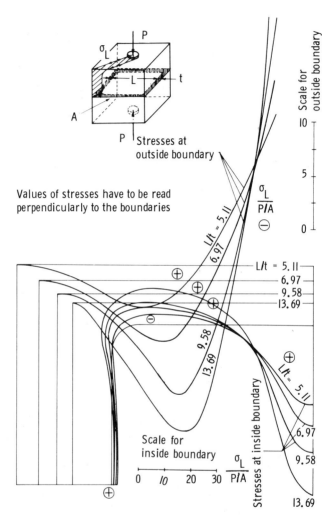

Figure 1.111. Stresses in the longitudinal direction at the intersection of the diagonal cross-section with the inside and outside surfaces, on a cubic box subjected to a concentrated load at the center of two opposite faces.

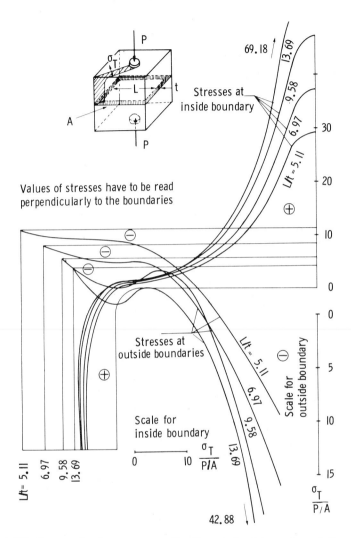

Figure 1.112. Stresses in the transverse direction at the intersection of the diagonal cross-section with the inside and outside surfaces, on a cubic box subjected to a concentrated load at the center of two opposite faces.

are equally applicable to the case of loads applied to the inside boundary by changing the sign of the stresses shown in the figures.

Stress concentrations in a cubic box subjected to pressure.   The same box showed in Figure 1.110 has been subjected to internal pressure. The distribution of stresses at some typical cross sections is shown in Figures 1.113 to 1.114. The distribution of other stresses in the box can be found in [88]. Results are equally applicable to the case of external pressure by changing the sign of the stresses shown in the figures.

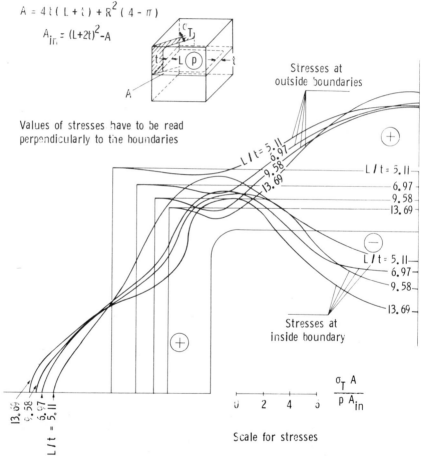

Figure 1.113. Stresses in the transverse direction at the intersection of the diagonal cross-section with the inside and outside surfaces, on a cubic box subjected to internal pressure.

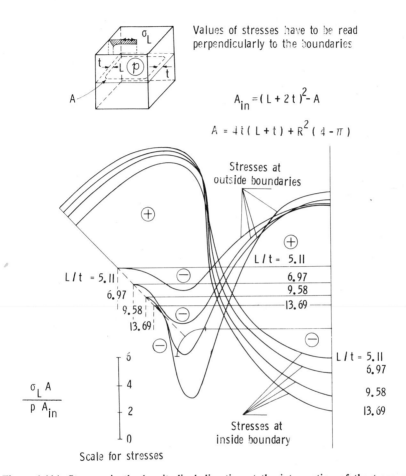

Figure 1.114. Stresses in the longitudinal direction at the intersection of the transverse cross-section with the inside and outside surfaces on a cubic box subjected to internal pressure.

1.8   Dynamic stress concentration factors*

The deformations and stresses generated, when dynamic loads are applied to a body, propagate through the body in the form of waves. At discontinuities within the body, the waves are reflected and refracted, that is, scattered, which may result in a local intensification of the stresses known as a dynamic stress concentration.

The theory associated with the propagation of stress waves in solids was developed during the last half of the 19th century as an extension of the theory of elasticity. Since no experimental means existed at that time for checking the theory, the subject was for the most part neglected during the first half of this century. With the development of electronic equipment and new plastic materials during and after World War II, experimental studies of wave propagation became possible and interest in the subject revived.

Most of the first experimental work in wave propagation was conducted using electrical resistance strain gages. This method gives a continuous time record of a dynamic event but limits observations to discrete points on the surface of the body. This lack of full-field information provided the motivation for studies in dynamic photoelasticity.

Theoretical dynamic stress concentration studies.   One of the earliest theoretical investigations of dynamic stress concentrations (d.s.c.f.) was made by Nishimura and Jimbo [89]. They determined the stresses in the vicinity of a spherical inclusion in an elastic solid which is subjected to a dynamic tensile-compressive force with harmonic time variation. Stress concentrations around a spherical cavity, a rigid sphere, and an elastic spherical inclusion were discussed in detail. They showed that at certain wave frequencies, the d.s.c.f. are larger than the static ones.

In an important theoretical paper, which is widely used for experimental comparisons, Baron and Matthews [90] presented the solution for an infinitely long cylindrical cavity in an infinite elastic medium which is acted on by a plane shock wave whose front is parallel to the axis of the cavity. An integral transform technique was used to determine the stress field produced in the medium as the incoming shock wave is diffracted by the cavity. Although the problem is considered for pressure waves with

---

* The author is indebted to Prof. W. Riley, of Iowa State University, for preparing this section.

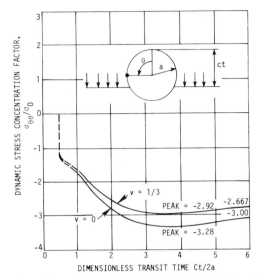

Figure 1.115. Dynamic s.c.f. as a function of dimensionless transit time for different values of Poisson's ratio, for an infinitely long cylindrical cavity in an infinite medium (Baron and Matthews).

a step distribution in time, the results obtained for this case may be used as influence coefficients to determine, by Duhamel integrals, the stress field produced by waves with time-varying pressures. Numerical results are presented in Figure 1.115 for the normalized hoop stress $\sigma_{\theta\theta}/\sigma_0$ at a point on the cavity boundary 90 degrees from the point of initial contact of the wave front with the boundary as a function of dimensionless transit time $Ct/2a$ for two different values of Poisson's ratio ν. These results indicated that the hoop stress on the cavity boundary is amplified by the dynamic loading in the ratio of 3.28 to 3 or approximately 9 percent for the case of Poisson's ratio $\nu = 0$. For the more realistic case of Poisson's ratio $\nu = \frac{1}{3}$, the results are 2.92 to 2.667 which corresponds to an increase in s.c.f. for the dynamic loading of approximately $9 - \frac{1}{2}$ percent.

In a more recent study, Pao [91] determined stress concentrations around a circular cavity in an infinitely extended, thin, elastic plate during passage of plane compressional waves. The problem is the dynamic counterpart of Kirsch's static problem discussed earlier. Briefly, the d.s.c.f. $\sigma_{\theta\theta}/\sigma_0$ were found to be dependent on the incident wave length λ and on Poisson's ratio ν for the plate material. At certain

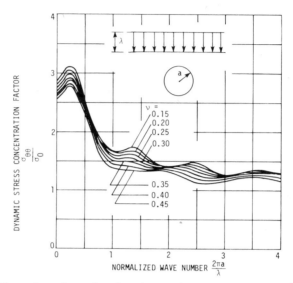

Figure 1.116. Dynamic s.c.f. as a function of normalized wave number for different values of Poisson's ratio, for a circular cavity in an infinite thin plate (Pao).

wave lengths, the d.s.c.f. were larger than those encountered under static loading as shown in Figure 1.116. For the seven values of Poisson's ratio (between 0.15 and 0.45) shown in the figure, the ratios of the d.s.c.f. to the static s.c.f. ranged between 1.09 and 1.10. Thus, the maximum amplification in stress due to dynamic loading appears to be about 10 percent. The results of this study clearly indicate that at very short wave lengths, a significant reduction in s.c.f. occurs under dynamic loading conditions.

Finally, the transient response for a rigid spherical inclusion in an elastic medium can be found in a paper by Mow [92]. An extensive theoretical treatment of diffraction of elastic waves and dynamic stress concentrations can be found in a monograph by Mow and Pao [93].

Experimental dynamic stress concentration studies. One of the first experimental investigations of a dynamic stress problem was made by Wells and Post [94] who determined the dynamic stress distribution surrounding a running crack started from an edge of a plate in tension. A multiple spark technique of the type originated by Schardin was employed to obtain full field photographs of the photoelastic (relative retardation) fringe patterns at various times during the dynamic event. In addition, absolute retardation measurements were made in similar models with running cracks by means of an optical interferometer so than individual principal stresses could be determined. Even though the s.c.f. at the tip of the crack could not be determined, the results of this study indicated that the dynamic stress distributions in the vicinity of the crack were essentially the same as static distributions.

The first experimental determination of a dynamic stress concentration factor was reported by Durelli and Dally [95] who studied the stress distribution around a central circular hole in a strut subjected to an axial impact type of loading. The model was machined from a low modulus photoelastic material having a modulus of elasticity of approximately 600 psi. The photoelastic fringe patterns during the dynamic event were recorded with a Fastax camera operating at approximately 12,000 frames per second. A selection of frames from the Fastax record are shown in Figure 1.117. The results of the study indicated that the d.s.c.f. was not significantly different from the static s.c.f. if the nominal stress for the dynamic determination is taken as the stress which would have been present in the strut at the center of the hole at the same instant if the hole had not been present. For the

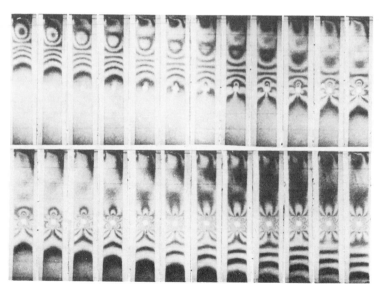

Figure 1.117. Propagation of a stress wave past a central circular hole in an axially loaded bar. (Durelli and Dally).

dynamic loading used, the ratio of pulse length to hole diameter was approximately 16.

In a related study, Dally and Halbleib [96] determined d.s.c.f. in axially loaded struts with central circular holes having hole diameter to strut width ratios d/w of 0.15, 0.33, 0.50 and 0.67. The results indicated that the d.s.c.f. were essentially the same as the static factors for d/w ratios less than 0.35. For d/w ratios larger than 0.35, the ratio of the d.s.c.f. to the static s.c.f. became progressively smaller than 1 as the d/w ratio was increased.

During the same interval of time that the strut studies were conducted (1959–1963), Durelli and Riley [97–101] conducted a series of dynamic photo-elasticity investigations to determine d.s.c.f. associated with circular, elliptical, and square holes in large plates subjected to impact and explosive loadings. In the plate models, the discontinuities were located along a radial line 30° from the centerline of the plate as shown in Figure 1.118. The loading of the model was accomplished by dropping a weight to impact on the top edge of the plate or by detonating a small charge of lead azide explosive. The ratio of pulse length to hole diameter for the falling weight loading [97–100] was approximately 40 if the pulse length is taken as twice the rise time from zero to peak multiplied by the wave

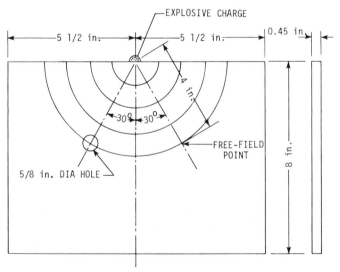

Figure 1.118. Sketch of a plate showing the location of the hole and the symmetric free-field point, for study of wave propagation around a hole. (Durelli and Riley).

front velocity. The results of the studies using falling weight loadings indicated that the dynamic compressive stresses on the hole boundary (circular, elliptical, or square) were larger than they would be in an equivalent static field. The ratio of maximum dynamic compressive stress to maximum static compressive stress was between 1.05 and 1.20 for the different models. The dynamic tensile stresses on the hole boundary were always smaller than they would be in an equivalent static field. The portion of the stress pulse studied in these models was approximately 75 percent of the front of the pulse. Thus, the peak stresses on the hole boundary were not observed. A study of a greater portion of the pulse would be possible only with larger plates or pulses of shorter duration. This was accomplished by using the explosive loadings.

When an explosive is detonated on the edge of a plate [101], both dilatational and distortional stress waves are produced. These waves propagate radially from the point of detonation of the explosive with different velocities. Thus, as the stress wave propagates, the two waves which were simultaneously generated tend to separate. Sufficient separation was achieved in the plate studies with explosive loadings to permit determination of dynamic stress concentrations due to stress fields associated with both types of waves. The rise from zero to peak of the dilatational wave indicates that the ratio of pulse length to hole diameter was approximately 10 for the explosive loading studies. A representative set of photoelastic fringe patterns for a plate with a circular hole resulting from detonation of an explosive on the edge of the plate are shown in Figure 1.119. Stress distributions on the hole boundary and at the symmetric free-field point during passage of the dilatational wave are shown in Figure 1.120. A comparison of the static and dynamic distributions shown in the figure indicates that the dynamic stresses are approximately 15 percent greater than the static stresses computed using the free-field stresses and the Kirsch [30] solution. At a later time, the hole is located in a stress field associated primarily with the distortional wave. As shown in Figure 1.121, the static and dynamic stress distributions are in good agreement at these times.

In a related series of studies, Durelli and Daniel [102] placed circular holes at different locations with respect to the point of impact of the falling weight. The results are in agreement with the preceeding discussion.

A further study of dynamic stress distributions around a circular hole in a large plate was conducted by Daniel and Riley [103]. In this study a

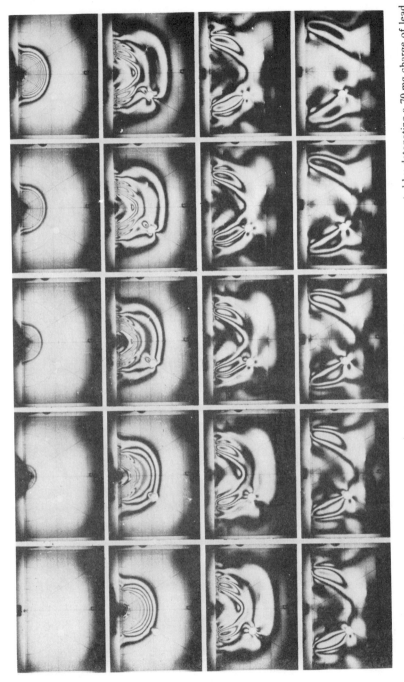

Figure 1.119. Stress wave propagation past a circular hole in a plate. The stress wave was generated by detonating a 70 mg charge of lead azide on the top edge of the plate. Photographs were taken with a Fastax camera at 6780 frames per second. (Durelli and Riley).

140

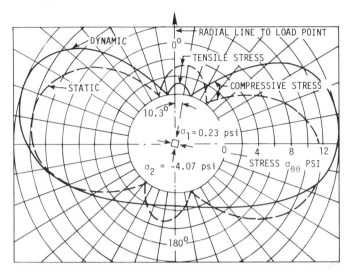

Figure 1.120. Static (free field, dotted line) and dynamic stress distributions on the hole boundary resulting from the dilational wave. (Durelli and Riley).

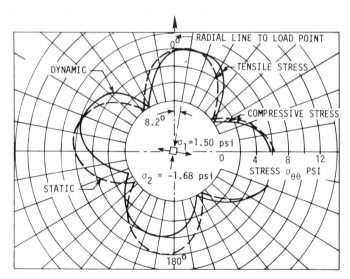

Figure 1.121. Static (free field, dotted line) and dynamic stress distributions on the hole boundary resulting from the distortional wave. (Durelli and Riley).

plane pressure (dilatational wave) was generated in the plate passing an air shock wave of fast rise and slow decay along an edge of the plate. A series of photographs of the photoelastic fringe patterns produced by the plane pressure wave propagating past the hole are shown in Figure 1.122. Detailed photographs of the photoelastic fringe patterns in the vicinity of the hole during passage of the front of the pressure pulse are shown in Figure 1.123. The results indicate that the d.s.c.f. is much lower than the static one initially as the front of the pressure pulse moves past the hole Figure 1.124. At later times, the dynamic factor exceeds the static factor by approximately 13 percent as shown in Figure 1.124. The velocity of propagation of the zero order fringe (wave front) was measured at 4400 inches per second. The results of the study are in agreement with the theoretical predictions of Baron and Matthews [90]. A study of dynamic stress distributions around a rigid circular inclusion in a large plate due to a plane pressure wave was conducted by Riley [104]. The dynamic distributions were not significantly different from static distributions computed using free-field stresses.

Photoelastic studies of dynamic stress distributions around a circular hole in a large plate due to a plane pressure wave generated by the detonation of an explosive charge of lead azide on the edge of the plate were performed by Fang, Hemann, and Achenbach [105]. Measurements were carried out for three ratios of hole diameter to pulse length. Dimensionless hoop stress as a function of time for one of the models is shown in Figure 1.125. Theoretical and experimental values of dimensionless maximum hoop stress $\sigma_{\theta\theta}/\sigma_0$ as a function of hole diameter to pulse length ratio $2a/\lambda$ are shown in Figure 1.125. The experimental determinations are in excellent agreement with the theoretical results of Mow and Pao [93].

Electrical resistance strain gage methods have been used by Shea [106] to evaluate dynamic stress concentration factors as a function of hole diameter to pulse length ratio. Thin plexiglas plates with central circular holes were used for the study. A plane pressure pulse was generated in the plates by detonating a length of 'Primacord' explosive along an edge of the plate. Results of the study are shown in Figure 1.126. These results indicate that there is a pronounced variation of the dynamic stress-concentration factor with pulse frequency. In regions of pulse length shorter than the diameter of the hole, the stress concentration factor drops off significantly. The theoretical predictions of Pao [91] are shown in the figure for comparison.

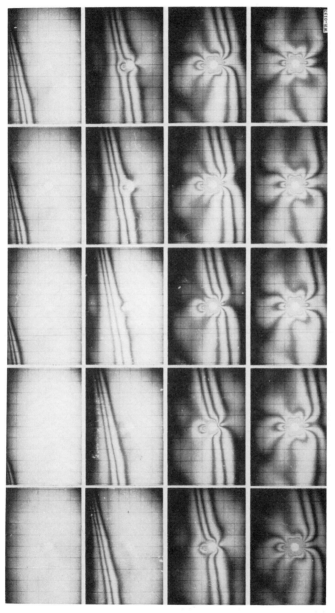

Figure 1.122. Stress wave propagation past a circular hole in a plate. The plane stress wave was generated by an air shock wave moving across the top edge of the plate. (Daniel and Riley).

Figure 1.123. Photoelastic fringe patterns in the vicinity of the hole as the plane stress wave moves past the hole. (Daniel and Riley).

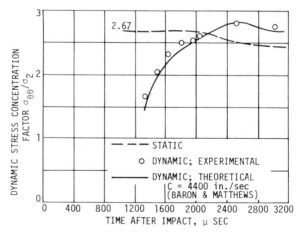

Figure 1.124. Comparison of experimental and theoretical values of maximum dynamic tangential stress on the hole boundary. (Daniel and Riley).

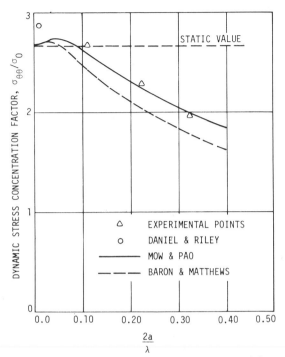

Figure 1.125. Comparison of theoretical and experimental values of the maximum dimensionless tangential stress for different hole diameter to pulse length ratios. (Fang, Hemann, and Achenbach).

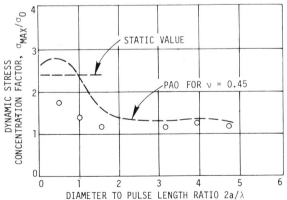

Figure 1.126. Experimental and theoretical s.c.f. as a function of the hole diameter to pulse length ratio in a plexiglass plate with central circular hole. (Shea).

1.9   Unconventional approaches to the study of stress concentrations

It is well known that the strength of brittle materials and the fatigue strength of ductile materials, in general, decreases as the size of the body increases. It is also well known that when strain gradients are present the strength of brittle materials, and the fatigue strength of ductile materials also decrease as the gradient decreases. It is also true that the onset of static yielding in the presence of stress concentrations occurs at higher loads than might be expected on the basis of the stress concentration factors calculated using theory of elasticity. On the other hand, it is also true that the presence of flaws frequently weakens the strength of a body. All these facts raise the question of the significance of the stress concentration factors as usually determined. Different schools of thought have tried to look at the situation and have taken different points of view.

Even more important than the significance of the theoretical concept of stress in mechanics, is the previous problem that develops from the fact that it is impossible to measure a stress, since by definition stress is a limiting process. Stress has to be determined indirectly, from measurements of force, or from measurements of force and displacements.

Attempts have been made at averaging the stress obtained theoretically, over the structural elements of physical bodies. Two of the contributions on the subject are one due to Neuber and the other to Gurney. Considerations on the effect of possible flaws have given origin to recent development in fracture mechanics. Theoretical developments in elasticity have included the possible influence of other elasticity constants besides E and v. Finally the idea has come of determining the stress, not from displacement measurements and strain determinations but from values of the fracture of brittle materials. Several of these 'unconventional' approaches to the determination of stress concentration factors will be reviewed in this section.

Neuber's particle theory.   After showing the failure of the traditional theory in the zone of stress concentrations, Neuber [38, 51] states that it is necessary to accept a particle of finite dimensions as a basic concept, rather than the infinitesimal parallelepiped, but that is only necessary to do so in the zone of disturbance. He suggests that the size of the particle (width ε) became a new constant of the material. The stresses acting on this particle are the stresses computed using the classic theory of elasticity, but averaged over the surface of the particle. Neuber had developed simplified equations using the classic theory, and applying his 'theory of pointed notches' to the 'particle' at the bottom of the notch he arrives to the conclusion that the expressions giving the stress concentration factors have the same form in both cases, but that in the pointed notch theory the value of the radius of curvature ρ at the bottom of the notch has to be replaced by ρ' which is equal to one half the dimension of the particle.

This new stress concentration factor α_s is a 'technical' stress concentration factor, and has to be conceived as a limit value as the radius becomes smaller and smaller. The limit value of the stress concentration factor for a sharp shallow circumferential notch under shear or torsion is then

$$\alpha_s = \sqrt{\frac{2t}{\varepsilon}} = \sqrt{\frac{t}{\rho'}} \tag{1.60}$$

rather than the value of the infinitesimal theory $\alpha_f = \sqrt{t/\rho}$ which tends to infinity.

Similarly for deep sharp circumferential notches under shear or torsion the limiting technical factor is:

$$\alpha_s = \frac{2}{\pi}\sqrt{\frac{2a}{\varepsilon}} = \frac{2}{\pi}\sqrt{\frac{a}{\rho'}} \tag{1.61}$$

rather than the value of the infinitesimal theory

$$\alpha_k = \frac{2}{\pi}\sqrt{\frac{a}{\rho}} \tag{1.62}$$

and for the case of the deep sharp notch on both sides of a plate subjected to axial load:

$$\alpha_s = \frac{4}{\pi}\sqrt{\frac{2a}{\varepsilon}} = \frac{4}{\pi}\sqrt{\frac{a}{\rho'}} \tag{1.63}$$

rather than

$$\alpha_k = \frac{4}{\pi}\sqrt{\frac{a}{\rho}} \tag{1.64}$$

Neuber adds finally two considerations: (1) that the flank angle of the notch, which he neglected in all his developments, becomes very important when the radius of curvature at the bottom of the notch is very small. The stress concentration factors drops as the flank angle increases; (2) the deformation at the bottom of the notch is appreciable for sharp notches and its effect is to decrease the stress concentration. All these influences are taken into account using ρ the half of the length of the particle, and ω the flank angle, in the final equation:

$$\alpha_k = 1 + \frac{\alpha'_{k-1}}{1 + \frac{\pi}{\pi - \omega}\sqrt{\frac{\rho'}{\rho}}} \tag{1.65}$$

where α_k is the 'technical' stress concentration factor, and α'_k the infinitesimal one.

It is regrettable that an attempt was made at substantiating the particle theory with experimental strain analysis measurements. The author's conclusion that the particle size in steel is about 0.96 mm, (in the second edition a somewhat different reasoning gives 0.30 mm) and that the value also applies to photoelastic materials does not seem to take into consideration the accuracy and precision of mechanical and optical measurements. The selected Fischer's 'Strain Measurements' published in 1932, were conducted using mechanical strain gages on steel specimens with radii as small as 2.5 mm. A relatively high level of error could have been expected. On the other hand, the way Armbruster's tests, which contradict the theory, are disposed of, shows a regrettable tendency to fit facts to theories. Reference to the classic book of Coker and Filon would have avoided unfortunate comparisons.

The particle theory is still a good assumption to interpret rupture. It does not seem that it can be verified with usual birefringence measurements, nor with measurements obtained from mechanical strain gages. An attempt at verifying the couple-stress theory with photoelastic measurements has been recently made by Shiryaer [115] but it does not seem to prove the point either. The author claims that the stress concentration factor decreases from 2.7 to 2 when a hole in a plate is reduced in size from a radius of 4 mm to a radius of 0.5 mm. Techniques to determine fringe values, however, at the edge of holes 0.5 mm. in diameter are very delicate and subjected to numerous errors when the thickness of the specimen is 10 times the radius of the discontinuity. The gradient of the stresses becomes very high requiring high resolution of the fringe, the 'space' effect at the edge is relatively more pronounced, and so is the three dimensional effect. It is quite possible that the apparent decrease in s.c.f. is due to the fact that fringes cannot be detected, rather than to the effect of 'particles' or 'couple stresses'.

Gurney's stress averaging over an atom. The particle considered by Neuber is not physically defined, and its size is computed indirectly. From the context, and the comparisons with experimental deter-minations, it seems to correspond approximately to the grains in some metals, for instance the chromium–chromium carbide shown in Figure 1.6, or the inclusions in the rocket solid propellant shown in Figure 1.5. No attempt was made to analyze what may happen at the level of the atomic structure.

Gurney [107] tries to average the stresses predicted by elasticity theory over the volume occupied by an atom. The computation is conducted for a small length ε at the end of an elliptical crack of length $2a$, having radius ρ. (Neuber's notation is used for easier continuity in the development of this way of thinking.) Because of simplicity in handling the equations, the stress concentration factors are computed for the case of equal biaxiality. Gurney concludes that whereas the real value is $2t/s$, t and s being the semiaxes of the ellipse, when the radius equals the length over which the stress is averaged, the effective stress concentration (or average effect) or 'technical' value is only $1.1 \, t/s$. If the radius of curvature is made zero, the effective stress concentration factor is

$$1.8\sqrt{\frac{t}{\varepsilon}} \tag{1.66}$$

Therefore decreasing ρ at the end of the crack from ε to zero, only increases the average stress over ε by about 70%.

Cox's studies on microscopic flaws. If a plate containing a hole with its axis perpendicular to the plane of the plate is stressed in the plane of the plate, the maximum stress is tangential to the free edge of the hole, and is the only one with non-zero value of three principal stresses at the point. A material with a system of such holes would fail under a uniaxial state of stress, irrespective of the law of failure of the material and irrespective of the loading conditions in the plane of the plate. It is also true that the ratio of the ultimate stress under shear and the ultimate stress under uniaxial tension applied to the plates as a whole would depend only on the geometry of the hole and the ratio between the strength of the material under uniaxial tension and compression. Cox [108] adds that if by experiments materials do not conform to this reasoning, it would mean that the rate of variation of stress in the neighborhood of the point of maximum stress must have also an influence on failure. (This conclusion could of course have been obtained otherwise.)

The author considers also the case of cylindrical holes oriented at random and representative of flaws that could be microscopic or even submicroscopic in size, provided that their size is still large by comparison with the atomic structure of the material.

He concludes that internal flaws are unlikely to account for the mechanical properties of engineering materials. He adds that experimental results do not support the hypothesis that the low strength of materials in relation to estimates of their strength based on thermodynamic data, may be explained by assuming that all materials contain numerous sub-microscopic flaws. The conclusion seems to go against the present trends in the analysis of fracture.

Notch sensitivity. Discussions on 'particles' on which the stress is averaged, on the effect of stress on 'atoms,' or on microscopic or submicroscopic 'flaws' present in a material, are sometimes motivated by theoretical thinking. Theoreticians try to better simulate the physical reality. Frequently, however, these studies belong to the attempts of engineers to make the results of the theory of elasticity useful to solve problems in design.

In few words the fact is that materials seldom if ever fail with a reduction in strength of the magnitude of the stress concentration factor. Theoretically inclined people may look at the situation talking about the influence of stress on atoms. Engineers try to explain it using words like 'notch sensitivity,' or 'volume of material' subjected to the maximum stress, or to some percentage of it. Contributions of Peterson (2) and Lipson [40] can be included in this last group of studies.

The notch sensitivity index can be expressed as:

$$q = \frac{K_f - 1}{K_t - 1} \tag{1.67}$$

with values varying from 0 for materials ignoring the presence of the concentration, to 1 for materials behaving ideally as could be predicted from theory of continuum. In equation (1.67), K_f is the so-called 'fatigue stress concentration factor' given by:

$$K_f = \frac{\text{endurance limit of specimens without a notch}}{\text{endurance limit of specimens with a notch}} \qquad (1.68)$$

and K_t is what has been called here s.c.f.

The concept of 'notch sensitivity' does not hide completely the fact that s.c.f. are determined using loads at failure. From equation (1.67), it follows that:

$$K_t = \frac{K_f - 1}{q} + 1 \qquad (1.69)$$

This approach determines therefore the s.c.f. by measuring two loads at failure, one acting on the notched body and the other acting on the plain body.

Size and stress concentration. Concepts like 'notch sensitivity' try to explain the difference in behavior of different materials in which the same notch or groove is present. But another fact complicating behavior has also to be considered. It is the fact that the same material, subjected to the same level of stress may behave differently depending on its size. Reference is made here to the classic work of Weibull, recently reviewed by Freudenthal [109]. A contribution has also been made in [110] to explain at the same time the size and gradient of stress influence, and the sensitivity of the material to the presence of the discontinuity. It is sufficient to test the material to failure using two specimens with different stress gradients. The relationship between stress and volume v of material subjected to 95 percent of the maximum stress, or more, is given by:

$$\sigma = \sigma_s \left(\frac{v_s}{v}\right)^{\ln(\sigma_r/\sigma_s)/\ln(v_r/v_s)} \qquad (1.70)$$

where σ is the maximum stress at the point of failure, v is the volume of material subjected to 95 per cent or more of the maximum stress, and r and s correspond to tests conducted on the different specimens. When the material has been calibrated by conducting tests on two specimens,

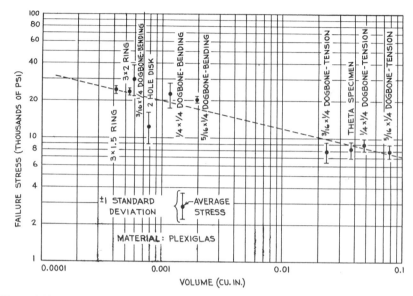

Figure 1.127. Value of σ_1 at failure vs. volume subjected to 95% of σ_1, or more, for plexiglas specimens.

and σ_r, σ_s, v_r and v_s are known, the tensile strength can be obtained from

$$\sigma = c(v)^{-k} \tag{1.71}$$

where c and k are constants. The relationship comes directly from the plot of the straight line on log-log paper. An illustration is shown in Figure 1.127. The value 95% has been selected for convenience, but could be related to statistical concepts.

Here again s.c.f. is computed by measuring two loads at failure, P that produces failure in a body of volume v and P_s that produces failure in a body of volume v_s. It is sufficient to replace in equation (1.70) $\sigma/\sigma_s = P/P_s$ thus:

$$\text{s.c.f.} = \frac{\sigma}{\sigma_s} = \frac{P}{P_s} = \left(\frac{v_s}{v}\right)^{(\ln(\sigma_r/\sigma_s)/\ln(v_r/v_s))} \tag{1.72}$$

This approach emphasizes the concept that the s.c.f. is not a constant quantity, ideally determined, but depends on the specimen taken as reference. It also suggests that the reference could be made absolute if a maximum limit like $E/10$, or a fundamental particle with a size of order 0.0001 in. could be determined and its concept clarified.

Couple stresses and micropolar elasticity. Recent developments in the classical theory of elasticity also attempt, in a different way, to bring stress concentrations factors down to the values obtained from experiments. Among the most important, the theories of 'couple stresses' and 'micropolar elasticity' should be mentioned. Both require the introduction of new material constants, with dimensions of a length, and give results which depend on the relative size of the discontinuity and the characteristic new constants of the material. These new constants in a more elaborated and refined way do a work similar to the one attempted by Neuber with the use of 'particles' (Section 1.9). An illustration of the variation in the stress concentration factors associated with an elliptical hole in an infinite plate subjected to uniaxial loading is shown in Figure 1.128 as obtained by Kim and Eringen [111] using micropolar elasticity. A similar illustration of the variation in the stress concentration factor associated with a circular hole in an infinite plate subjected to uniaxial loading is shown in Figure 1.129, as obtained by Mindlin [112] using the theory that includes couple stresses in the equations of equilibrium. Sternberg and Muki studied the problem of the crack [113] and Hartranft

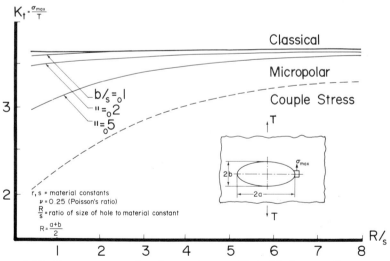

Figure 1.128. Stress concentration factors for an elliptical hole, according to several theories. (Kim and Eringen).

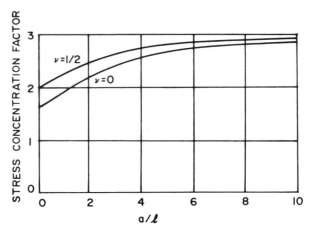

Figure 1.129. Stress concentration factor at a circular hole in a field of simple tension: variation with ratio of radius of hole to material constant 1. (couple-stress theory, by Mindlin).

and Sih [114] the problem of the circular inclusion, also using couple-stress theory. Shiryaer [115] claims to have proved experimentally the decrease in the value of the s.c.f. as the size of a hole decreases, using photoelasticity. As pointed out earlier the evidence is not convincing.

Stress intensity factors. The so-called stress intensity factors have been rationalized and determined, independently, by several authors. They all have the same basic form: a constant times the square root of a length (usually half the length of a crack)

$$K = p\sqrt{\pi t} \tag{1.73}$$

The stress at the end of the major axis of an elliptical hole, subjected to a single load perpendicular to that major axis, in terms of the radius of curvature at that point is given by equation (1.48). When ρ is very small in relation to t, equation (1.48) can be approximated by:

$$\sigma \simeq 2p\sqrt{\frac{t}{\rho}} \tag{1.74}$$

(Recall that t is half the axis of the ellipse and ρ the radius of curvature at the base of the notch.)

This equation is of the same form as the one giving the typical stress intensity factor, equation (1.73). The relationship is obvious:

$$\sigma = p \frac{2}{\sqrt{\rho}} \sqrt{t}; \quad K = p\sqrt{\pi} \sqrt{t} \qquad (1.75)$$

Catalogs of values of stress intensity factors for numerous problems have been published by Cartwright and Rooke [116]. Sih [117, 118], Tada, Paris and Irwin [119], and Rooke and Cartwright [120]. Most of the solutions reported have been obtained analytically, or using numerical methods.

Conclusions. It follows from the previous considerations that the mathematical concept of stress concentration factor cannot be used in a direct simple manner to explain or to predict fracture. In a micro scale, matter is not continuous and cannot be divided indefinitely without changing its properties. It is true that in general a body with a stress concentration of k does not fail when the point at which the concentration is located is subjected to a stress equal to $k\sigma_{av} = \sigma_u$, σ_u being the ultimate stress in calibrating specimens tested at the same temperature, at the same strain rate and to the same number of loading cycles. It usually takes higher stress than $k\sigma_{av}$, which leads to the studies summarized above trying to explain why the load at fracture may be higher than the theoretically predicted.

References

[1] Roark, R. J. and Young, W. C., *Formulas for Stress and Strain*, Fifth edition, McGraw-Hill (1975). The first edition was published in 1938.
[2] Peterson, R. E., *Stress Concentration Factors*, John Wiley and Sons, New York, pp. 316 (1974). A first version: Peterson, R. E.: 'Stress Concentration Design Factors' was published by Wiley in 1953.
[3] Griffel, W., *Handbook of Formulas for Stress and Strain*, Frederick Ungar Publishing Co., pp. 418 (1966).
[4] Tetelman, A. S. and McEvily, A. J., *Fracture of Structural Materials*, Wiley (1967).
[5] Savin, G. N., *Stress Distribution Around Holes*, Kiev (1968), Transl. NASA, TTF-607, p. 997 (Nov. 1970). The first edition: Savin, G. N., Concentration of Stress Around Holes was published in 1951 by the State Publishing House of Technical Theoretical Literature. Trans. Pergamon Press (1961).
[6] Durelli, A. J., Phillips, E. and Tsao, C., *Introduction to the Theoretical and Experimental Analysis of Stress and Strain*, McGraw-Hill (1958).
[7] Durelli, A. J., *Applied Stress Analysis*, Prentice-Hall (1967).

[8] Durelli, A. J., The Difficult Choice, Feature Article in *Applied Mechanics Reviews*, pp. 1167–1178 (Sept. 1977).

[9] Collatz, L., *The Numerical Treatment of Differential Equations*, 3rd. Ed. Trans. by P. G. Williams, Springer-Verlag, N. Y. (1966).

[10] Hulbert, L. E., The Numerical Solution of Two-Dimensional Problems of the Theory of Elasticity, *Ph.D. Dissertation*, Ohio State University (1963).

[11] Muskhelishvili, N. T., *Some Basic Problems of the Mathematical Theory of Elasticity*, Translated by J. R. M. Radok, P. Noordhoff Ltd., Groningen, The Netherlands (1963).

[12] Wilson, H. B. Jr., Mathematical Determination of Stresses and Displacements in Star-perforated Grains, *AIAA Journal* 2, pp. 1247–1253 (July 1964).

[13] Wilson, H. B., Jr. and Becker, E. B., *An Effective Numerical Method for Conformal Mapping of Simply Connected Regions*, Developments in Mechanics, Vol. 3, Part I: Solid Mech. and Mat., John Wiley and Sons, Inc., pp. 403–414 (1967).

[14] Richardson, M. K. and Wilson, H. B., Jr., *A Numerical Method for the Conformal Mapping of Finite Doubly Connected Regions*, Developments on Theoretical and Applied Mechanics, Vol. 3, pp. 305–321, edited by W. A. Shaw, Pergamon Press (1967).

[15] Laura, P. A., Conformal Mapping of a Class of Doubly Connected Regions, *Ph.D. Dissertation*, The Catholic University of America (1965).

[16] Rao, A. K., Review of Continuum, Finite Element and Hybrid Techniques in the Analysis of Stress Concentration in Structures, *Nuclear Engineering and Design*, Vol. 31, pp. 427–433 (1974).

[17] Dunham, R. S. and Becker, E. B., TEXGAP the Texas Grain Analysis Program. *TICOM Report 73–1*, The University of Texas at Austin (Aug. 1973).

[18] Bushnell, D., *Difference Models Versus Finite-elements Models*, Numerical and Computer Methods in Structural Mechanics, Academic Press, New York, pp. 291–336 (1973).

[19] Key, S. and Krieg, R., *Comparison of Finite-element and Finite-difference Methods*, Numerical and Computer Methods in Structural Mechanics, Academic Press, New York, pp. 337–352 (1973).

[20] Perrone, N. and Kao, R., A General Finite Difference Method for Arbitrary Meshes, *International Journal of Computers and Structures*, 5, pp. 45–57 (April 1975).

[21] Zienkiewicz, O. C., in volume The Mathematics of Finite-elements and Applications, edited by J. R. Whiteman. *Proceedings of Brunel University Conference, April 1972*, Academic Press (1973).

[22] Oden, J. T., Finite-element Applications in Nonlinear Structural Analysis, in *Proceedings of ASCE Symposium on Application of Finite Element Methods in Civil Engineering*, Nashville (1969) ed. by W. H. Rowan, Jr. and R. M. Hackett, Vanderbilt Univ. (1969).

[23] Durelli, A. J., Parks, V. J. and Feng, H., Evaluation of Grid Data to Determine Displacements and Strains, *Journal of Strain Analysis*, Vol. 2, No. 3, pp. 181–187 (July 1967).

[24] Fischer, G., Versuche ueber die Wirkung von Kerben an elastische beanspruchten Biegestaeben, *Dissertation Aachen* (1932), V. D. I. Verlag, Duesseldorf (1932).

[25] Durelli, A. J., Phillips, E. and Tsao, C., *Introduction to the Theoretical and Experimental Analysis of Stress and Strain*, McGraw-Hill (1958).

[26] Parks, V. J. and Durelli, A. J., Moiré Patterns of Partial Derivatives of Displacement Components, *Journal of Applied Mechanics*, Vol. 33, pp. 901–906 (1966).

[27] Erf, R. K. (ed.) *Holographic Non-destructive Testing*, Academic Press, (1974).

[28] Erf, R. K. (ed.) *Speckle Metrology*, Academic Press (1978).

[29] Bickle, L. W., The response of Strain Gages to Longitudinally Sweeping Strain Pulses, *Experimental Mechanics*, Vol. 10, No. 8, pp. 333–337 (August 1970).

[30] Kirsch, G., Ueber einige Eigenschaften des ebenen Problems der Elastizitaetstheorie, *Zeit. des Ver. Deutscher Ing.* Vol. 42, pp. 787–807 (1890).

[31] Inglis, C. E., Stresses in a Plate Due to the Presence of Cracks and Sharp Corners, *Trans. Inst. Naval Arch.* London, England, 1913; *Engineering*, Vol. 95, p. 415 (1913).

[32] Durelli, A. J., Phillips, E. A. and Tsao, C. H., *Introduction to the Theoretical and Experimental Analysis of Stress and Strain*, McGraw-Hill (1958).

[33] Kosolov, G. V., *On One Application of the Theory of Complex Variable Functions to Plane Problems of the Mathematical Theory of Elasticity*, Yurger (1909).

[34] Isida, M., On the Tension of a Semi-infinite Plate with an Elliptic Hole, *Science Papers of Faculty of Engineering*, Tokushima University, Vol. 5, No. 1 (Feb. 1955).

[35] Durelli, A. J., Parks, V. J. and Feng, H. C., Stresses Around an Elliptical Hole in a Finite-Plate Subjected to Axial Loading, *Journal of Applied Mechanics*, pp. 192–195 (March 1966).

[36] Coker, E. G. and Filon, L. N. G., *A Treatise on Photo-elasticity*, Cambridge University Press (1931).

[37] Lehr, E., *Spannungsverteilung in Konstruktionselementen*. Auswertung der bisherigen Forschungsergebnisse für die praktische Anwendung. VDI-Verlag, Berlin (1934).

[38] Neuber, H., *Theory of Notch Stresses: Principles for Exact Stress Calculation*. Transl. of Kerbspannungslehre: Grundlagen fuer genaue Spannungsrechnung, Springer, Berlin (1937), by David Taylor Model Basin, Washington, D. C. Nov. 1945.

[39] Roark, R. J. and Young, W. C., *Formulas for Stress and Strain*: Fifth ed., McGraw-Hill (1975). The first edition was published in 1938.

[40] Lipson, C., Noll, G. and Clock, L., *Stress and Strength of Manufactured Parts*, McGraw-Hill, (1950).

[41] Savin, G. N., *Stress Distribution Around Holes*, Kiev (1968), Transl. NASA, TTF-607, pp. 997 (Nov. 1970). A first version: Savin, G. N., Concentration of Stress Around Holes, was published in 1951 by the State Publishing House of Technical Theoretical Literature. Trans. Pergamon Press (1961).

[42] Foeppl, L. and Sonntag, G.: *Tafeln und Tabellen zur Festigkeitslehre*. R. Oldenburg, Muenchen (1951).

[43] Peterson, R. E., *Stress Concentration Factors*: John Wiley and Sons, New York, pp. 316 (1974). A first version: Peterson, R. E.: Stress Concentration Design Factors was also published by Wiley in 1953.

[44] Griffel, W., *Handbook of Formulas for Stress and Strain*, Frederick Ungar Publishing Co., p. 418 (1966).

[45] Nisitani, H.: *Solutions of Notch Problems by Body Force Method*. Mechanics of Fracture, ed. by G. C. Sih, Noordhoff Intern. Publish. Alphen aan den Rijn., Vol. 5, pp. 1–68.

[46] Heywood, R. B., *Photoelasticity for Designers*, Pergamon Press (1969).

[47] Hogan, M. B., A Survey of the Literature Pertaining to Stress Distribution in the Vicinity of a Hole and the Design of Pressure Vessels. Bull. No. 48, *Utah Engineering Experimental Station*, Vol. 41, No. 2 (Aug. 1950).

[48] Neuber, H. and Hahn, H. G., Stress Concentration in Scientific Research and Engineering, *Applied Mechanics Reviews*, Vol. 19, No. 3 (March 1966).

158   A. J. Durelli

[49] Sternberg, E., Three-dimensional Stress Concentrations in the Theory of Elasticity, *Applied Mechanics Reviews*, Vol. 11, pp. 1–4 (Jan. 1958).
[50] Durelli, A. J., The Difficult Choice, *Applied Mechanics Reviews*, Vol. 30, No. 9, pp. 1167–1178 (Sept. 1977).
[51] Neuber, H., *Theory of Notch Stresses*: Principles of Exact Calculation of Strength with Reference to Structural Form and Material. Transl. of Kerbspannungslehre: Grundlage für genaue Festigkeit Berechnung mit Beruecksichtigung von Konstruktion's form und Werkstoff. Springer, Berlin (1958) by U.S. Atomic Energy Comm. AEC-tr-4547, Chap. X, (June 1961).
[52] Durelli, A. J., Parks, V. J. and Lopardo, V. J., Stresses and Finite Strains Around an Elliptic Hole in Finite Plates Subjected to Uniform Load, *International Journal of Non-Linear Mechanics*, Vol. 5, pp. 397–411 (1970).
[53] Durelli, A. J., Parks, V. J., Lopardo, V. J. and Chen, T. L., Normalized Stresses Around an Elliptic Hole in a Finite Plate of Linear Material Subjected to Large Uniform In-plane Loading, *Experimental Mechanics*, Vol. 13, No. 10 pp. 441–444 (Oct. 1973).
[54] Chen, T. L. and Durelli, A. J., Stress Field in a Sphere Subjected to Large Deformations, *International Journal of Solids and Structures*, Vol. 9, pp. 1035–1052 (1973).
[55] Chen, T. L. and Durelli, A. J., Displacements and Finite-Strain Fields in a Hollow Sphere Subjected to Large Elastic Deformations, *International Journal of Mechanical Science*, Vol. 16, pp. 777–788 (1974).
[56] Durelli, A. J. and Mulzet, A. P., Large Strain Analysis and Stresses in Linear Materials, *Journal of Engineering Mechanical Division*, Proc. ASCE., June 1965, pp. 65–91.
[57] Durelli, A. J., Parks, V. J. and Feng, H., Experimental Methods of Large Strain Analysis, *International Journal of Non-Linear Mechanics*, Vol. 2, pp. 387–404 (1967).
[58] Durelli, A. J., Parks, V. J. and Chen, T. L., Stress and Finite-Strain Analysis of a Circular Ring Under Diametral Compression, *Experimental Mechanics*, pp. 210–214 (May 1969).
[59] Durelli, A. J., Parks, V. J. and Bühler Vidal, J. O., Linear and Non-Linear Elastic and Plastic Strains in a Plate with a Large Hole Loaded Axially in its Plane. *International Journal of Non-Linear Mechanics*, Vol. 11, pp. 207–211 (1976).
[60] Durelli, A. J. and Parks, V. J., Photoelastic Stress Analysis in the Bonded Interface of a Strip with Different End Configurations, *American Ceramic Society Bulletin*, Vol. 46, No. 6, pp. 586 (June 1967).
[61] Durelli, A. J., Parks, V. J. and Bhadra, P., Experimental Determination of Stresses and Strains in a Rectangular Plate Subjected to Biaxial Restrained Shrinkage, *British Journal of Applied Physics*, Vol. 17, pp. 917–926 (1966).
[62] Parks, V. J. and Durelli, A. J., Stress Distribution at Variously Shaped Corners in Plates, Bonded on Two Long Edges, and Subjected to Restrained Shrinkage, Mech. Behavior Working Group, Interagency Chemical Rocket Propulsion Group, *CPIA Bulletin*, No. 119, Vol. 1, pp. 365–392 (Nov. 1966).
[63] Durelli, A. J., Parks, V. J., and del Rio, C. J., Stresses in Square Slabs with Different Edge Geometries, when Bonded on One Face to a Rigid Plate and Shrunk, *Experimental Mechanics*, Vol. 7, No. 11, pp. 481–484 (Nov. 1967).
[64] Duffy, J., The Elastic Wedge under Mixed Boundary Conditions, *Technical Report*, No. 1, Div. Engng., Brown Univ., DA-G4-1-April (1961).

[65] Theocaris, P. S. and Dafermos, K., The Elastic Strip Under Mixed Boundary Conditions, *Applied Mechanics*, 31, (Trans. Amer. Soc. Mech. Engrs., 86, pp. 714–716) (Dec. 1964).

[66] Aleck, B. J., Thermal Stresses in a Rectangular Plate Clamped Along an Edge, *Journal of Applied Mechanics*, 16, pp. 118–22 (1949).

[67] Parks, V. J., Durelli, A. J. and Chen, T. L., Stresses in a Restrained Shrinking Disk with a Central Slot, *AIAA Journal*, Vol. 6, No. 6, pp. 1181–1182 (June 1968).

[68] Durelli, A. J., Parks, V. J. and Lee, Han-Chow, Stress Concentrations Around Elliptical Perforations in Shrunk Plates with Bonded Boundaries, *Journal of Spacecraft and Rockets*, Vol. 5, No. 2, pp. 1499–1501 (Dec. 1968).

[69] Durelli, A. J. and Tsao, C. J., Determination of Thermal Stresses in Three-Ply Laminates., *Journal of Applied Mechanics*, Vol. 22, No. 2, pp. 190–193 (June 1955).

[70] Parks, V. J., Chiang Fu-pen and Durelli, A. J., Maximum Stress at the Angular Corners of Long Strips Bonded on One Side and Shrunk. *Experimental Mechanics*, pp. 278–281 (June 1968).

[71] Durelli, A. J., Parks, V. J. and Chen, T. L., Stress Concentrations in a Rectangular Plate with Circular Perforations Along its Two Bonded Edges and Subjected to Restrained Shrinkage, *Strain*, Vol. 5, No. 1, pp. 1–4 (Jan. 1969).

[72] Hofer, K. E. and Durelli, A. J., Stress Distribution at the Fillet of an Internal Flange, *Journal of Applied Mechanics*, Vol. 28, E, No. 4. pp. 618–623 (Dec. 1961).

[73] Durelli, A. J. and Parks, V. J., Stresses in Pressurized Circular Shells with Discontinuities, with or without Stiffeners, *Journal of Ship Research*, Vol. 16, No. 2, pp. 140–147 (1972).

[74] Durelli, A. J., del Rio, C. J., Parks, V. J. and Feng, H., Stresses in a Pressurized Cylinder with a Hole, *Journal of Structure Division Proceedings ASCE*, Vol. 93, paper No. 5524, pp. 383–399 (Oct. 1967).

[75] Durelli, A. J., Parks, V. J. and Lee, H. C., Stresses in a Perforated Ribbed Cylindrical Shell Subjected to Internal Pressure, *International Journal of Solids and Structures*, Vol. 5, No. 6, pp. 573–586 (June 1969).

[76] Durelli, A. J., Parks, V. J. and Lee, H. C., Stresses in a Pressurized Cylindrical Shell with Two Unequal Diametrically Opposite Reinforced Circular Holes, *Acta Mechanica*, Vol. X, No. 3–4, pp. 161–179 (1970).

[77] Durelli, A. J., Parks, V. J. and Lee, H. C., Stresses in a Pressurized Ribbed Cylindrical Shell with a Reinforced Hole, *Report to Naval Ship Research and Development Center*, (Feb. 1969).

[78] Durelli, A. J. et al. Stresses in a Pressurized Ribbed Cylindrical Shell with a Reinforced Circular Hole Interrupting a Rib, Trans. ASME, *Journal of Engineering for Industry*, Series B, Vol. 93, No. 4, pp. 897–904 (Nov. 1971).

[79] Durelli, A. J. and Barriage, J. B., Stress Distribution in Square Plates with Hydrostatically Loaded Central Circular Holes, *Journal of Applied Mechanics*, Trans. ASME, Vol. 77, pp. 539–544 (1955).

[80] Durelli, A. J. and Riley, W. R., Stress Distribution in Strips with Hydrostatically Loaded Central Circular Holes, Proc. Mid-Western Conf. on Solid Mechanics, pp. 81–93 (Oct. 1955).

[81] Durelli, A. J. and Kobayashi, A. S., Stress Distributions Around Hydrostatically Loaded Circular Holes in the Neighborhood of Corners, *Journal of Applied Mechanics*, Vol. 26, No. 1, pp. 150–152 (March 1959).

[82] Riley, W. F., Durelli, A. J. and Theocaris, P. S., Further Stress Studies on a Square Plate with a Pressurized Central Circular Hole., Proc. 4th Annual Conf. on Solid Mech., Univ. of Texas, Austin, Texas, Sept. 1959, pp. 91–107.

[83] Parks, V. J., Durelli, A. J., Chandrashekhara, K. and Chen, T. L., Stress Distribution Around a Circular Bar, with Flat and Spherical Ends, Embedded in a Matrix in a Triaxial Stress Field, *Journal of Applied Mechanics*, pp. 578–586 (Sept. 1970).

[84] Riley, W. F. and Durelli, A. J., Boundary Stresses at the Ends of Pressurized Cylindrical Holes Axially Located in a Large Cylinder, *Proceedings SESA*, Vol. XVII, No. 1, pp. 115–126 (Dec. 1959).

[85] Durelli, A. J., Parks, V. J. and Del Rìo, C. J., *Stresses, Strains and Displacements Associated with the Restrained Shrinkage of Cylinders with Torodial Cavities.* Recent Advancements in Engineering Science Ed. A. C. Eringen, Vol. 3, pp. 521–540. Gordon and Breach., Sc. Publishers, NY (1968).

[86] Durelli, A. J., Parks, V. J. Feng, H. C. and Chiang, F., Strains and Stresses in Matrices with Inserts. Mechanics of Composite Materials. Proc. 5th Symposium on Naval Structural Mechanics, Ed. by Wendt, Liebowitz and Perrone, Pergamon, 1970, pp. 188–335.

[87] Durelli, A. J., Pavlin, V., Bühler Vidal, J. O. and Ome, G., Elastostatics of a Cubic Box Subjected To Concentrated Loads. *Strain*, Vol. 13, No. 1, pp. 7-12 (Jan. 1977).

[88] Durelli, A. J., Pavlin, V. and Bühler Vidal, J. O., Elastostatics of Cubic Boxes Subjected to Pressure. *Journal of Pressure Vessel Techniques*, Vol. 99, pp. 568–574 (Nov. 1977).

[89] Nishimura G. and Jimbo, Y. A Dynamical Problem of Stress Concentration, *Journal of the Faculty of Engineering*, University of Tokyo, Japan, Vol. 24, p. 101 (1955).

[90] Baron, M. L. and Matthews, A. T. Diffraction of a Pressure Wave by a Cylindrical Cavity in an Elastic Medium, *Journal of Applied Mechanics*, Series E, Vol. 28, No. 3, pp. 347–354 (1961).

[91] Pao, Y. H. Dynamical Stress Concentration in an Elastic Plate, *Journal of Applied Mechanics*, Series E, Vol. 29, No. 2, pp. 299–305 (1962).

[92] Mow, C. C. Transient Response of a Rigid Spherical Inclusion in an Elastic Medium, *Journal of Applied Mechanics*, Vol. 32, No. 3, pp. 637–000 (September 1965).

[93] Mow, C. C. and Pao, Y. H. The Diffraction of Elastic Waves and Dynamic Stress Concentrations, RAND Corporation, Report No. R-482-PR, (April 1971).

[94] Wells, A. A. and Post, D. The Dynamic Stress Distribution Surrounding a Running Crack—A Photoelastic Analysis, *Proceedings of SESA*, Vol. 16, No. 1, pp. 69–92 (1957).

[95] Durelli, A. J. and Dally, J. W. Stress Concentration Factors Under Dynamic Loading Conditions, *Journal of Mechanical Engineering Science*, Vol. 1, No. 1, pp. 1–5 (June 1959).

[96] Dally, J. W. and Halbleib, W. F. Dynamic Stress Concentrations at Circular Holes in Struts, *Journal of Mechanical Engineering Science*, Vol. 7, No. 1, pp. 23–27 (March 1965).

[97] Durelli, A. J. and Riley, W. F. Stress Distribution on the Boundary of a Circular Hole in a Large Plate During Passage of a Stress Pulse of Long Duration, *Journal of Applied Mechanics*, Series E, Vol. 28, No. 2, pp. 245–251 (1961).

[98] Riley, W. F. and Durelli, A. J. Stress Distribution on the Boundary of an Elliptical Hole in a Large Plate During Passage of a Stress Pulse of Long Duration (Major Axis Tangent to the Wave Front) *International Journal of Mechanical Sciences*, Vol. 2, No. 4, pp. 213–223 (1961).

[99] Durelli, A. J. and Riley, W. F. Stress Distribution on the Boundary of an Elliptical Hole in a Large Plate During Passage of a Stress Pulse of Long Duration (Major Axis Normal to the Wave Front) *Journal of Applied Physics*, Vol. 32, No. 7, pp. 1255–1260 (July 1961).

[100] Durelli, A. J. Riley, W. F. and Carey, J. J. Stress Distribution on the Boundary of a Square Hole in a Large Plate During Passage of a Stress Pulse of Long Duration, Proceedings of the International Symposium on Photoelasticity, Pergamon Press, N.Y., 1963, pp. 251–263.

[101] Riley, W. F. and Durelli, A. J. Stress Distribution on the Boundary of a Circular Hole in a Large Plate During Passage of a Stress Pulse of Short Duration, *Journal of Mechanical Engineering Sciences*, Vol. 3, pp. 62–68 (1961).

[112] Durelli, A. J. and Daniel, I. M. Stress Distribution Around a Circular Hole in a Semi-Infinite Plate under Impact at Different Points on the Edge, *Developments in Mechanics*, Vol. 1, Plenum Press, N.Y., pp. 268–285 (1961).

[103] Daniel, I. M. and Riley, W. F. Stress Distribution on the Boundary of a Circular Hole in a Large Plate Due to an Air Shock Wave Traveling Along an Edge of the Plate, *Journal of Applied Mechanics*, Series E, Vol. 31, No. 3, pp. 402–408 (1964).

[104] Riley, W. F. Photoelastic Study of the Interaction Between a Plane Stress Wave and a Rigid Circular Inclusion, *Journal of Mechanical Engineering Science*, Vol. 6, No. 4, pp. 311–317 (1964).

[105] Fang, S. J. Hemann, J. H. and Achenbach, J. D. Experimental and Analytical Investigation of Dynamic Stress Concentrations at a Circular Hole, *Journal of Applied Mechanics*, Series E, Vol. 41, No. 2, pp. 417–422 (1974).

[106] Shea, R. Dynamic Stress-concentration Factors, *Experimental Mechanics*, Vol. 4, No. 1, pp. 20–24 (1964).

[107] Gurney, C., The Effective Stress Concentration at the End of a Crack Having Regard to the Atomic Constitution of Materials. Aero. Research Council. Report No. 2285, Dec. 1945. His Majesty Stationery, London (1948).

[108] Cox, H. L., Four Studies in the Theory of Stress Concentration. Aero. Research Council. His Majesty Stationery. R and M. No. 2704, London (1953).

[109] Freudenthal, A. M., *Fracture*, ed. by Liebowitz, H., Vol. II, Chap. 6, Academic Press (1968).

[110] Durelli, A. J. and Parks, V. J., Influence of Size and Shape on the Tensile Strength of Brittle Materials. *British Journal of Applied Physics*, Vol. 18, pp. 387–388 (1967).

[111] Kim, B. S. and Eringen, A. C., Stress Distribution Around an Elliptic Hole in an Infinite Micropolar Elastic Plate, *Letters in Applied and Engineering Science*, Vol. 1, pp. 381–390, Pergamon (1973).

[112] Mindlin, R. D., Influence of Couple Stresses on Stress Concentrations, *Experimental Mechanics*, Vol. 3, No. 1, pp. 1–7 (January 1963).

[113] Sternberg, E. and Muki, R., The Effect of Couple-Stresses on the Stress Concentration Around a Crack, ONR report, April 1966 Contract Nonr-220(58)NR-064-431.

[114] Hartranft, R. J. and Sih, G. C., The Effect of Couple Stresses on the Stress Concentration of a Circular Inclusion, *Journal of Applied Mechanics*, (June 1965).

[115] Shiryaer, Ya M., Stress Concentration Near an Inhomogeneity and Experimental Clarification of the Couple-Stress Effect. *Zhurnal Prikladnoi Mekhaniki i Tekhnicheskoi Fiziki*, No. 4, pp. 142–144, (July–Aug. 1976). Trans. in *Journal of Applied Mechanics and Technical Physics*, Vol. 17, No. 4, pp. 569–571 (May 1977).

[116] Cartwright, D. J. and Rooke, E. P., Approximate Stress Intensity Factors Com-

pounded From Known Solutions, *Engineering Fracture Mechanics*, Vol. 6, p. 563 (1974).

[117] Sih, G. C. (Ed.), *Methods of Analysis and Solutions of Crack Problems*, Noordhoff Int. Publish. (1973).

[118] Sih, G. C., *Handbook of Stress Intensity Factors for Researchers and Engineers*, Inst. of Fract. and Sol. Mech., Lehigh Univ. Bethlehem, PA. 1973.

[119] Tada, H., Paris, P. and Irwin, G., *The Stress Analysis of Cracks Handbook.*, Del. Research Corp., Hillertown, PA (1973).

[120] Rooke, D. F. and Cartwright, D. J., *Compendium of Stress Intensity Factors*, Ministry of Defence, Her Majesty's Stationery Office, London (1974).

C. W. Smith

2 | Use of photoelasticity in fracture mechanics

2.1   Introduction

A number of transparent amorphous materials which are optically isotropic become optically anisotropic when stressed and exhibit characteristics similar to crystals such as the property of double refraction.
Upon unloading the effect disappears. This effect was first observed by
Sir David Brewster in 1816 and is known as temporary double refraction. When such materials are loaded and observed in a polarized light
field, temporary double refraction produces interference bands known as
isochromatics, or stress fringes, each of which represents the locus of
points of the same maximum shearing stress in the plane of the model
which is normal to the incident light beam. The fringe order, or number, is
proportional to the maximum shearing stress. Thus, if we observe a
point in a transparent model in a polarized light field as it is loaded, we
will observe (with monochromatic light) the change in color from dark to
bright to dark, representing one complete optical cycle. The initial dark
band would be fringe order zero in a polariscope set for extinction of
light, the second fringe order one, then two, three, etc. These bands
form contour-like patterns across the model. Closely spaced narrow
bands indicate regions of high stress gradients. Broad, widely separated
bands indicate regions of low stress gradients. Such effects are shown in
Figure 2.1 for a plate with two holes under vertical uniaxial tension. The
fringe order n' has been experimentally related to the maximum in plane
shearing stress for two dimensional problems through the Stress-Optic
Law as follows:

$$\tau_{max} = \frac{n'f}{2t'} \tag{2.1}$$

where

$\tau_{max} = (\sigma_{max} - \sigma_{min})/2$, i.e., half the difference in the in-plane principal stresses

$n' = $ fringe order

$f = $ material fringe value (i.e., maximum shear stress per order of fringe per unit thickness of model)

$t' = $ model thickness.

When stress fringes are broad, or when only a few fringes are present, it is necessary to utilize special methods for measuring fractional fringe orders. Several such methods are available. Compensation and fringe multiplication methods require special optical equipment in addition to a standard photoelastic polariscope. One technique, known as the Tardy Method, does not require special equipment. However, this method does require the use of a second type of interference fringe known as an isoclinic (locus of points of constant principal stress direction). A typical transmission polariscope is pictured in Figure 2.2. The polarizer and analyzer pass light only along one plane, i.e., the light is plane polarized. By inserting two quarter wave plates with optic axes crossed and at 45°

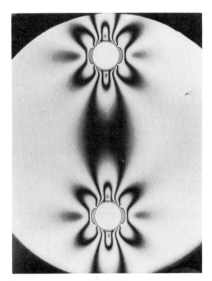

Figure 2.1. Fringe pattern for plate with holes. (courtesy D. Post)

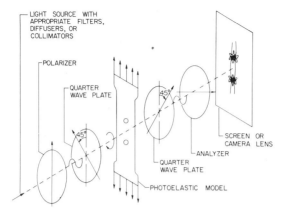

Figure 2.2. Typical transmission polariscope.

to the plane of the polarizer, a field of circularly polarized light is produced. When polarizer and analyzer are crossed also, then no light passes the analyzer. The same is true when the crossed quarter wave plates are removed, leaving a crossed plane polariscope. The crossed plane polariscope is used to produce isoclinics in the stressed model and the crossed circular polariscope produces isochromatics, or stress fringes.

In order to use the Tardy Method, the polarizer and analyzer are aligned with the principal stresses at a point and the analyzer is rotated in the crossed circular polariscope. This will change the fringe order at the point an amount proportional to the angle of rotation in radians divided by π. It is to be emphasized that the present discussion presumes working knowledge of a standard transmission polariscope. For those desiring further details on the photoelastic method, Chapter 13 of Reference [1] is recommended. Photoelasticity has been used extensively in solving and verifying solutions to two dimensional elasticity problems since publication of a treatise on the subject in 1931 [2]. This method was, in fact, the one used by Wells and Post [3] to study a propagating crack in 1958 and, it was the observation of the stress fringes around the crack tip such as shown in Figure 2.3 that led Irwin to postulate the concept of the stress intensity factor (SIF) for describing stress fields ahead of the crack.

The preceding discussion was confined to the principle of temporary double refraction. In 1937, Oppel [4] introduced the concept of "frozen stress". It capitalized upon the observation that certain materials exhibit

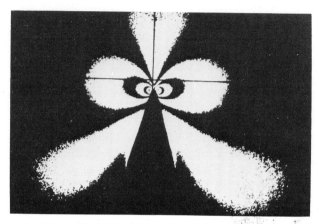

Figure 2.3. Mode I fringes for artificial crack.

diphase optical and mechanical behavior. Such materials exhibit a time dependent response to load at room temperature, but when loaded above a certain temperature called the "critical" temperature, the time dependency is suppressed to the extent that the material behaves as a linearly elastic, incompressible ($\nu \simeq 0.5$)* material. Above critical temperature, the material's elastic modulus is approximately 0.2% of its room temperature value and its material fringe value is about 5% of its room temperature value indicating twenty times more sensitivity to fringes than at room temperature. Frozen stress models may be produced by the following procedure:

(1) Heat photoelastic model to above the critical temperature.
(2) Apply live loads.
(3) Cool slowly to room temperature.
(4) Remove loads.

Since the material is so much more sensitive to stresses and deformations above critical temperature than at room temperature, negligible recovery of fringes and deformations result upon unloading at room temperature and so the stress fringes and deformations produced above critical temperature remain in the model after unloading at room temperature. Moreover, the model may be sliced into sub-sections without disturbing these fields. In fact, one may consider the term "frozen stress" a misnomer in the sense that it is the deformation field and the

* ν = Poisson's Ratio.

fringe field, rather than stresses which are frozen into the body. Finally, slices may be viewed under normal incidence in a crossed circular polariscope and fringes so observed may be interpreted exactly as in the two dimensional case except that f in equation (2.1) is the value at critical temperature instead of at room temperature. This can be done because the stress induced double refraction responds only to in-plane stresses.

2.2   Analytical foundations for cracked bodies

As noted in the previous section, the primary quantity obtained from photoelastic analysis is the maximum shearing stress in a plane normal to the direction of travel of the polarized light field. In this section, we shall develop the analytical background for converting this information into SIF values. It has been shown [5] that the stresses near the border of an elliptical crack, when expressed in terms of a set of local moving rectangular cartesian coordinates in a plane perpendicular to the flaw border, have the same form as the stresses in a plane perpendicular to the border of a straight front crack. Consider a half·space containing a surface flaw at an angle β to the boundary with remote uniform tension parallel to the z' direction (Figure 2.4). The local moving orthogonal coordinate system *tnz* is always oriented such that t is tangent to the flaw border and n is normal to the flaw border but both n and t are in

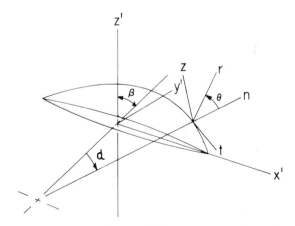

Figure 2.4. General mixed mode flaw geometry and notations.

the flaw plane. The z axis is normal to the flaw plane. In such a problem, all three local modes of near field deformation (i.e., Modes I, II and III) will be present as we move around the flaw border. We note that at $\alpha = 0$, however, Mode III will be absent.

The stress distribution near the part-through crack in the data zone, corresponding to the opening mode of deformation can be taken as

$$\sigma_{nn} = \frac{K_1}{(2\pi r)^{1/2}} \cos\frac{\theta}{2}\left[1 - \sin\frac{\theta}{2}\sin\frac{3\theta}{2}\right] - \sigma_{nn}^1 \tag{2.2a}$$

$$\sigma_{zz} = \frac{K_1}{(2\pi r)^{1/2}} \cos\frac{\theta}{2}\left[1 + \sin\frac{\theta}{2}\sin\frac{3\theta}{2}\right] - \sigma_{zz}^1 \tag{2.2b}$$

$$\tau_{nz} = \frac{K_1}{(2\pi r)^{1/2}} \sin\frac{\theta}{2}\left[\cos\frac{\theta}{2}\cos\frac{3\theta}{2}\right] - \sigma_{nz}^1 \tag{2.2c}$$

where K_1 is the Mode I SIF and the coordinates r and θ are shown in Figure 2.4. σ_{ij}^1 represent the contribution of the Mode I regular stress field to the measurement zone which is taken far enough from the crack tip to avoid a non-linear zone very near the tip. While they may generally be regarded as expressible in Taylor Series Expansions, it turns out that only the leading terms of said series are normally necessary so that σ_{ij}^1 are constants for a given point in the flaw border but vary from point to point. Stresses in the data zone corresponding to the Mode II can be taken as

$$\sigma_{nn} = -\frac{K_2}{(2\pi r)^{1/2}} \sin\frac{\theta}{2}\left[2 + \cos\frac{\theta}{2}\cos\frac{3\theta}{2}\right] - \sigma_{nn}^2 \tag{2.3a}$$

$$\sigma_{zz} = \frac{K_2}{(2\pi r)^{1/2}} \sin\frac{\theta}{2}\cos\frac{\theta}{2}\cos\frac{3\theta}{2} - \sigma_{zz}^2 \tag{2.3b}$$

$$\tau_{nz} = \frac{K_2}{(2\pi r)^{1/2}} \cos\frac{\theta}{2}\left[1 - \sin\frac{\theta}{2}\sin\frac{3\theta}{2}\right] - \tau_{nz}^2 \tag{2.3c}$$

where K_2 and σ_{ij}^2 are analogous to K_1 and σ_{ij}^1.

And finally the stresses in the data zone corresponding to the Mode III loading can be taken as

$$\tau_{nt} = -\frac{K_3}{(2\pi r)^{1/2}} \sin\frac{\theta}{2} - \tau_{nt}^3 \tag{2.4a}$$

$$\tau_{zt} = \frac{K_3}{(2\pi r)^{1/2}} \cos \frac{\theta}{2} - \tau_{zt}^3 \qquad (2.4b)$$

with K_3 and σ_{ij}^3 analogous to K_1 and σ_{ij}^1.

If the above modes of loading are superimposed, one gets the following stress distribution in a plane perpendicular to the crack front (the n-z plane)

$$\sigma_{nn} = \frac{K_1}{(2\pi r)^{1/2}} \cos \frac{\theta}{2} \left[1 - \sin \frac{\theta}{2} \sin \frac{3\theta}{2} \right]$$
$$- \frac{K_2}{(2\pi r)^{1/2}} \sin \frac{\theta}{2} \left[2 + \cos \frac{\theta}{2} \cos \frac{3\theta}{2} \right] - \sigma_{nn}^0 \qquad (2.5a)$$

$$\sigma_{zz} = \frac{K_1}{(2\pi r)^{1/2}} \cos \frac{\theta}{2} \left[1 + \sin \frac{\theta}{2} \sin \frac{3\theta}{2} \right]$$
$$+ \frac{K_2}{(2\pi r)^{1/2}} \sin \frac{\theta}{2} \cos \frac{\theta}{2} \cos \frac{3\theta}{2} - \sigma_{zz}^0 \qquad (2.5b)$$

$$\tau_{nz} = \frac{K_1}{(2\pi r)^{1/2}} \sin \frac{\theta}{2} \cos \frac{\theta}{2} \cos \frac{3\theta}{2}$$
$$+ \frac{K_2}{(2\pi r)^{1/2}} \cos \frac{\theta}{2} \left[1 - \sin \frac{\theta}{2} \sin \frac{3\theta}{2} \right] - \tau_{nz}^0 \qquad (2.5c)$$

which are independent of equations (2.4); (i.e., the local field equations for Modes I and II are completely separate from those for Mode III).

While σ_{ij}^0 has no influence upon the singular stress field itself, it does alter the isochromatic fringe pattern, which is proportional to the maximum in-plane shearing stress.

From the stress field given in equations (2.5), the maximum shearing stress in the plane perpendicular to the crack front, nz, can be obtained using

$$(\tau)_{max}^{nz} = \left[\left(\frac{\sigma_{zz} - \sigma_{nn}}{2} \right)^2 + \tau_{nz}^2 \right]^{1/2} \qquad (2.6)$$

and, truncating to the same order as equations (2.5), one gets

$$(\tau)_{max}^{nz} = \frac{A}{r^{1/2}} + B \qquad (2.7)$$

for fringe loops approaching the shape of Fig. 2.5* where

$$A = \left\{ \frac{1}{8\pi}[(K_1 \sin \theta + 2K_2 \cos \theta)^2 + (K_2 \sin \theta)^2] \right\}^{1/2} \tag{2.8}$$

and

$$B = B(\sigma^0_{ij})$$

The maximum shearing stress in the nz plane (the left side of equations (2.6) and (2.7) is determined photoelastically.

Now, in general, the effect of σ^0_{ij} involves both a folding and a change in eccentricity of the fringe loops (Figure 2.5). If folding occurs, θ_m, the angle along which the distance to a fringe from the crack tip is greatest, will vary with the fringe order n' and one must plot θ_m vs r/a and extrapolate to the origin in order to obtain θ^0_m, the value of θ_m associated with K_1 and K_2. In the present problem θ_m was constant over the data range in the fashion indicated qualitatively by Figure 2.5. Upon com-

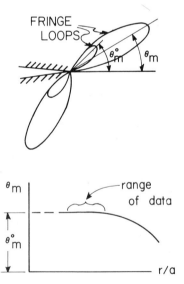

Figure 2.5. Determination of θ^0_m for mixed mode loading.

* For cases where $K_2 > K_1$, Eq. 2.7 and Fig. 2.5 may require modification.

puting

$$\text{Lim}_{\substack{r_m \to 0 \\ \theta_m \to \theta_m^0}} \left\{ (8\pi r_m)^{1/2} \frac{\partial (\tau)_{\max}^{nz}}{\partial \theta} (K_1, K_2, r_m, \theta_m, \sigma_{ij}^0) \right\} = 0 \qquad (2.9)$$

one obtains

$$\left(\frac{K_2}{K_1}\right)^2 - \frac{4}{3}\left(\frac{K_2}{K_1}\right) \cot 2\theta_m^0 - \frac{1}{3} = 0 \qquad (2.10)$$

Since θ_m^0 can be measured experimentally, (K_2/K_1) can be calculated from equation (2.10). Then by combining the Stress-Optic Law with a modified form of Equation 2.7

$$(\tau)_{\max}^{nz} = \frac{fn'}{2t'} = \frac{K_{AP}^*}{(8\pi r)^{1/2}} \qquad (2.11)$$

where $K_{AP}^* = (\tau)_{\max}^{nz}(8\pi r)^{1/2}$ is defined as the "apparent" SIF. Hence

$$K_{AP}^* = [(K_{1AP} \sin \theta_m + 2K_{2AP} \cos \theta_m)^2 + (K_{2AP} \sin \theta_m)^2]^{1/2} \qquad (2.12)$$

and one can solve for the individual values of K_1 and K_2.
In order to do this, one must obtain

$$K^* = [(K_1 \sin \theta_m^0 + 2K_2 \cos \theta_m^0)^2 + (K_2 \sin \theta_m^0)^2]^{1/2} \qquad (2.13)$$

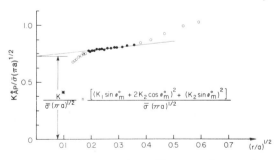

Figure 2.6. Estimating $K^*/\bar{\sigma}(\pi a)^{1/2}$ from typical test data.

from K_{AP}^* by plotting $K_{AP}^* = (\tau)_{max}^{nz}(8\pi r)^{1/2}$ vs $(r/a)^{1/2}$ identifying a linear zone, and extrapolating to the origin. This will yield K^*. A typical set of fringe data illustrating such a determination is given in Figure 2.6. Once K^*, K_2/K_1, and θ_m^0 are known, K_1 and K_2 can be calculated which then can be normalized using proper quantities.

Note that the above approach utilizes a two parameter (A, B) model since the linear zone can be located experimentally (Figure 2.6). However, if one cannot locate such a zone experimentally then additional terms leading to an equation of the form

$$\tau_{max} = \frac{A}{r^{1/2}} + \sum_{n=0}^{\infty} B_n r^{n/2} \tag{2.14}$$

with suitable truncation criteria must be considered. Since such criteria are not yet established, this latter approach is avoided where possible and was not necessary in the studies described in the sequel.

The stress distribution σ_{zz}, acting in a plane perpendicular to the crack surface and tangent to the crack front (zt plane) or in a plane parallel to the zt plane can be found from equations (2.5).

$$\sigma_{zz} = \frac{K_1}{(2\pi r)^{1/2}} \cos\frac{\theta}{2}\left[1 + \sin\frac{\theta}{2}\sin\frac{3\theta}{2}\right]$$
$$+ \frac{K_2}{(2\pi r)^{1/2}} \sin\frac{\theta}{2}\left[\cos\frac{\theta}{2}\cos\frac{3\theta}{2}\right] - \sigma_{zz}^0 \tag{2.15}$$

In order to arrive at a value for σ_{zz}, prior experiments by the authors indicate that the usual assumption of plane strain (for the plane problem) may not be valid here. However, if one assumes a state of nearly *generalized* plane strain such that the value of ε_{tt} can be considered constant over a portion of the length of the flaw border, then the observed state of varying transverse constraint along the flaw border can apparently be approximated rather well. Thus, we assume

$$\varepsilon_{tt} = \frac{\sigma_{tt} - \nu(\sigma_{nn} + \sigma_{zz})}{E} \simeq \bar{\varepsilon} \tag{2.16}$$

whence $\sigma_{tt} \simeq E\bar{\varepsilon} + \nu(\sigma_{nn} + \sigma_{zz})$ where $\bar{\varepsilon}$ may be adjusted at intervals along the flaw border and where, again from equation (2.5),

$$\sigma_{nn} = \frac{K_1}{(2\pi r)^{1/2}} \cos \frac{\theta}{2} \left[1 - \sin \frac{\theta}{2} \sin \frac{3\theta}{2} \right]$$
$$- \frac{K_2}{(2\pi r)^{1/2}} \sin \frac{\theta}{2} \left[2 + \cos \frac{\theta}{2} \cos \frac{3\theta}{2} \right] - \sigma_{nn}^0 \qquad (2.17)$$

For $\nu = (1/2)$ (as in the present experiments) we then have

$$\sigma_{tt} = \frac{K_1}{(2\pi r)^{1/2}} \cos \frac{\theta}{2} - \frac{K_2}{(2\pi r)^{1/2}} \sin \frac{\theta}{2} - \sigma_{tt}^0 + E\bar{\varepsilon} \qquad (2.18)$$

Moreover, from Mode III, we have

$$\tau_{zt} = \frac{K_3}{(2\pi r)^{1/2}} \cos \frac{\theta}{2} - \tau_{zt}^0 \qquad (2.19a)$$

$$(\tau)_{max}^{zt} = \left[\left(\frac{\sigma_{zz} - \sigma_{tt}}{2} \right)^2 + \tau_{zt}^2 \right]^{1/2} \qquad (2.19b)$$

Consider the line normal to the crack surface which passes through the crack tip in the zt plane. For this case $\theta = (\pi/2)$, and when substituted in equations (2.15), (2.18) and (2.19), there results

$$\sigma_{zz} = \frac{1}{4(\pi r)^{1/2}} [3K_1 - K_2] - \sigma_{zz}^0 \qquad (2.20a)$$

$$\sigma_{tt} = \frac{1}{2(\pi r)^{1/2}} [K_1 - K_2] - \sigma_{tt}^0 + E\bar{\varepsilon} \qquad (2.20b)$$

$$\tau_{zt} = \frac{K_3}{2(\pi r)^{1/2}} - \tau_{zt}^0 \qquad (2.20c)$$

then for a two parameter model as before,

$$(\tau)_{max}^{zt} = \frac{C}{r^{1/2}} + D \qquad (2.21)$$

where

$$C = \frac{1}{(4\pi)^{1/2}} \left[\frac{1}{16}(K_1 + K_2)^2 + K_3^2 \right]^{1/2} \qquad (2.22)$$

and

$$D = D(E\bar{\varepsilon} \text{ and } \sigma_{ij}^0) \tag{2.23}$$

Now by combining the Stress-Optic Law with a modified form of equation (2.21)

$$(\tau)_{\max}^{zt} = \frac{fn'}{2t'} = \frac{K_{AP}^{**}}{(8\pi r)^{1/2}} \tag{2.24}$$

where

$$K_{AP}^{**} = \sqrt{2}\left[\frac{1}{16}(K_{1AP} + K_{2AP})^2 + K_{3AP}^2\right]^{1/2} \tag{2.25}$$

the value of K_3 can be obtained.

In order to do this, one must obtain

$$K^{**} = \sqrt{2}\left\{\frac{1}{16}(K_1 + K_2)^2 + K_3^2\right\}^{1/2} \tag{2.26}$$

from K_{AP}^{**} by plotting a $K_{AP}^{**} = (\tau)_{\max}^{zt}(8\pi r)^{1/2}$ vs $(r/a)^{1/2}$ curve, identifying a linear zone, and extrapolating to the origin. This will yield K^{**}.

At the points where the flaw border intersects the boundary of the plate, SIF values are uncertain and require a boundary layer analysis for an accurate evaluation.

2.3   Experimental considerations

The photoelastic method may be applied directly to two dimensional problems but the frozen stress method is required in order to obtain SIF distributions in three dimensional problems. We shall discuss the two cases separately. Another consideration which concerns both the direct and frozen stress methods is the type of crack. When the shape of the crack front is prescribed, the crack can be produced (in limited configurations) by a machined slot. From studies by the writer [6, 7] and others [8], it has been concluded that a narrow slot terminating in a 30° Vee notch of root radius ≈ 0.025 mm will yield photoelastically determined SIF values to within 2% of those in natural cracks of the same border configuration. We call such slots "artificial" cracks. Natural cracks are made by fixing a sharp blade normal to a surface and striking the blade with a hammer. A starter crack is initiated dynamically and propagates into the material and arrests. In the frozen stress process, these cracks may be extended by applying monotonically increasing loads above critical temperature until crack growth occurs. When the desired crack size is reached, loads are reduced, terminating flaw growth, and the stress freezing cycle is completed by cooling under reduced load. The geometry of cracks produced in this manner cannot be controlled. The shapes may be quite complex [9, 10] but appear to be virtually identical to those produced by stable fatigue crack growth in metals [11] when only small scale yielding occurs surface crack closure effects are negligible and stress ratios are within a limited range above or below unity. This is an important observation since it implies that the frozen stress method can open the way for estimating SIF distributions in three dimensional cracked body problems where neither the SIF distribution nor the flaw shape are known a priori. This point will be amplified in a later section.

Two dimensional problems. The first application of photoelasticity to the stress analysis of crack problems was made by Post [12] in 1954 when he observed both isochromatic and isopachic (loci of points of constant principal stress sum) patterns for edge cracked models. Since that time, a number of studies have been conducted, several of which have been directed towards dynamic crack propagation and arrest [3, 13, 17]. For direct application of the photoelastic method to two dimensional static problems, the most important considerations involve minimizing creep effects under load and insuring that crack fronts are straight. However, these considerations are no different from those associated with ordinary two dimensional photoelastic analysis.

Frozen stress analysis. There are several factors which must receive special attention when applying the frozen stress method to cracked body problems. First of all, the loads must be kept small in order to minimize finite deformations near the crack tip. Secondly, the photoelastic material is incompressible above critical temperature, and the influence of the elevated value of Poisson's Ratio upon near field data needs to be accounted for. This effect in cracked body problems may be described as follows: When a through the thickness crack occurs in a body of finite thickness, a region of high constraint develops around the crack tip characterized by the root radius of the crack tip and the plate thickness, the latter always being orders of magnitude larger than the former. This constraint dissipates rapidly with increasing distance from the crack tip and away from the crack tip generalized plane stress then prevails in a "two-dimensional" problem. The presence of the constraint variation, however, is a three-dimensional effect which is present in all real bodies, and when measurements are made near the crack tip, they will be associated with a state of plane strain, while measurements away from the near tip zone, such as compliance measurements, will be for a zone in which generalized plane stress exists. G. R. Irwin and others have suggested [18, 19] that near tip measurements in such problems can be "converted" to two-dimensional results through the multiplying factor $(1 - v^2)^{1/2}$. This effect, while small for $v \simeq 0.3$, is substantial when $v \simeq 0.5$. The author and his associates have extended these ideas and suggested the factor $\{(1 - 0.3)^2/(1 - 0.5^2)\}^{1/2}$ for converting plate type frozen stress data for comparison with "two-dimensional" results where $v \simeq 0.3$. This method has been found to be remarkably accurate for plate shaped bodies. Details of this study are found in [20]. When cracked bodies are not two-dimensional (as with surface flaws) this effect cannot be estimated in the same way, but estimates of the author and his associates indicate a maximum elevation in the SIF from near field measurements of the order of the experimental error (i.e., about 5%).

Finally, quantitative data from the frozen stress method is obviously restricted to linear elastic behavior. The practical significance of this limitation is that when one introduces a sharp crack into the body (theoretically a singularity) one expects a local region of non-linearity very near the crack tip. Such a region is, in fact, observed photoelastically. This region may be due to non-linear material behavior, crack tip blunting, non-linear strain-optic relations or some combination of these effects. The important point is that this region must be identified and excluded from the SIF determination.

Most two-dimensional problems have yielded to analytical methods of solution (except for dynamic fracture and arrest studies) and significant progress is being made in adapting numerical methods of analysis to the three-dimensional cracked body problems, especially in the area of improved solution convergence in SIF values. However, when such methods are used, computer code verification is still desirable, if not mandatory, in order to insure that the mathematical modelling conforms to the physical problem. Recognizing this need, the writer and his associates have, over a period of a decade, evolved [21–23] an experimental method consisting of a marriage between the local stress field equations of linear elastic fracture mechanics and the frozen stress method. The method appears to have evolved into a useful technique for estimating SIF distributions in three-dimensional, finite, cracked body problems and appears to offer the potential for providing flaw shapes as well as SIF distributions when neither are known a priori. In the next section, the use of the method will be illustrated by applying it to several problems of substantial technological significance.

2.4   Application of the frozen stress method

In order to provide a window for viewing the breadth of applicability of the frozen stress method to cracked body problems, two basically different problems will be used for illustrative purposes. The first problem will utilize an artificial part circular surface flaw but will be subjected to mixed mode loading. The second problem will utilize a natural crack at the reentrant corner of a reactor vessel nozzle. The test procedure is basically the same for both cases and may be summarized as follows:

(1) Construct the model from stress free diphase transparent photoelastic material by inserting either a machined artificial flaw to exact size or tapping in a natural starter crack much smaller than desired size.

(2) For complex shapes, parts are then glued together with no crack tips near glue lines.

(3) The model is then placed in a stress freezing oven in an appropriate loading rig, heated above critical temperature, and loaded in the same fashion as the prototype. It is important that the test rig not restrain the model during the heating cycle since the thermal coefficient of expansion of the model material is an order of magnitude higher than for structural metals. For natural cracks, sufficient load is applied to extend the starter crack to several times its initial size and then reduced in order to terminate flaw growth.

(4) The model is cooled under load to room temperature and sliced in planes mutually orthogonal to the crack plane and the crack border. Slices should be stored in moisture free air if the material is sensitive to time-edge effect [1].

(5) Slices are coated with a liquid of the same refractive index as the model material and analyzed in a crossed circular polariscope using white light which will produce colored isochromatics instead of dark and bright fringes, reading tint of passage between red and blue, and utilizing the Tardy Method for measuring fractional fringe orders.

(6) Fringe order data are then fitted to a simple least squares com-

puter program to establish a linear $K_{AP}^*/q(\pi a)^{1/2}$ vs $(r/a)^{1/2}$ zone and this line is used to obtain the normalized SIF for a slice taken at a given location along the flaw border. q represents a load parameter such as remote stress $(\bar{\sigma})$ or internal pressure (p).

Problem No. 1—A part circular surface flaw under mixed mode loading. The overall problem geometry is given in Figure 2.7 and the local geometry and notation is the same as in Figure 2.4 with loading a uniform tension parallel to the z' direction. It is clear from Figure 2.4 that, as one moves around the flaw border, all three local modes of deformation are present. However, as noted in Section 2.2, the local field equations for Modes I and II are completely separate from Mode III, and for convenience, only results for Modes I and II will be discussed here. A separate sub-slice and analysis would be required for Mode III which is present at all points in this problem except for $\alpha = 0$. A typical mixed mode fringe pattern is pictured in Figure 2.8. A typical set of raw data taken from a slice is given in Figure 2.6 and a typical set of results is presented in Figure 2.9 where $\bar{K}_1 = \bar{\sigma} \sin^2 \beta (\pi a)^{1/2}$; the Mode I SIF for a through crack inclined at the angle β to the remote tensile stress and $\bar{K}_2 = \bar{\sigma} \sin \beta \cos \beta (\pi a)^{1/2}$, the Mode II SIF for the same case. Figure 2.9

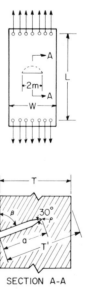

SECTION A-A

Figure 2.7. Test setup and problem geometry—problem no. 1.

Figure 2.8. Typical mixed mode fringe pattern—problem no. 1.

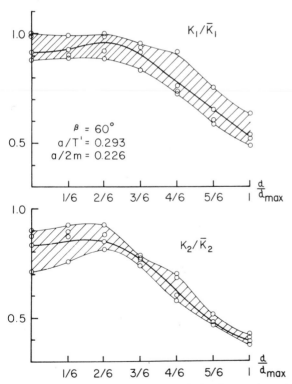

$K_1/\bar{K}_1$

$\beta = 60°$
$a/T' = 0.293$
$a/2m = 0.226$

$\frac{d}{d_{max}}$

$K_2/\bar{K}_2$

$\frac{d}{d_{max}}$

Figure 2.9. Typical test data for problem no. 1 showing maximum scatter.

179

reports the results from four identically oriented slices in order to reflect the maximum level of experimental scatter which might be encountered. At $\alpha/\alpha_{max} = 0$ the scatter in $K_1/\bar{K}_1 \simeq \pm 6\%$ while for $K_2/\bar{K}_2$, the scatter is about twice this value. This lower accuracy in $K_2/\bar{K}_2$ measurement is characteristic of Mode II measurements.

Another point of interest is that the data of Figure 2.9 suggest that the maximum value of $K_1/\bar{K}_1$ does not occur at the point of maximum flaw penetration. On the other hand, when a natural, semi-elliptic shaped crack is analyzed for the same shallow aspect ratio, the value of $K/\bar{K}_1$ is a maximum at the point of maximum flaw penetration. This observation indicates the importance to numerical analysts in utilizing the proper crack shapes in three dimensional cracked finite body analyses. In 3D problems, crack shapes may be quite complex, and the use of simple curves for assumed crack shapes may lead to SIF distributions quite different from the results being sought. Details of this study are found in [24].

Problem No. 2—Flaw shapes and stress intensity distributions for reactor vessel nozzle corner cracks. Nuclear reactors which employ water as a coolant require feed water inlet and exit nozzles at a number of locations around the primary pressure boundary of the vessel. These junctures may be described as the intersection between a thin walled cylinder (the vessel) and a reinforced thick walled cylinder (the nozzle), the latter consisting of a variable thickness wall. The reentrant corner of this juncture is a source of cracks which seem to occur in radial planes around the inner surface of the juncture. These cracks are believed to be initiated by thermal shock, but their growth is believed to be primarily due to fluctuations in the internal pressure.

To date, a number of numerical solutions [25–30] have been proposed for estimating the stress intensity factors (SIFs) for nozzle corner cracks. However, most of these solutions involve assumed flaw shapes and self-similar flaw growth, and some provide only a single SIF value for a given flaw size. One effort [31], currently underway, is utilizing real flaw shapes in order to obtain SIF distributions along such flaw borders. This general approach has been used before [30] but on a simpler geometry.

With the above ideas in mind, a series of experiments were designed for the purpose of characterizing the nozzle corner crack shapes and SIF distributions for typical boiling water reactor geometries. After inserting a single natural crack in each of eight nozzles in four vessels (each vessel, Figure 2.10, contained two diametrically opposite nozzles) the vessels were assembled and heated above critical temperature at which point they were pressurized to grow the cracks and, after reducing the pressure, were cooled to room temperature. Local problem geometry and notation is given in Figure 2.11. Slices parallel to the nz plane, when observed photoelastically, indicated the presence of only Mode I (as in Figure 2.3). By taking data along $\theta = \pi/2$, the expression for A in equation (2.8) simplifies to:

$$A = \frac{K_1}{(8\pi)^{1/2}} \tag{2.27}$$

and equations (2.7) and (2.11) become, upon normalization:

$$\frac{\tau_{\max}(8\pi r)^{1/2}}{p(\pi a)^{1/2}} = \frac{K_{AP}}{p(\pi a)^{1/2}} = \frac{K_I}{p(\pi a)^{1/2}} + \frac{B(\sigma_{ij}^1)(8\pi r)^{1/2}}{p(\pi a)^{1/2}} \tag{2.28}$$

or

$$\frac{K_{AP}}{p(\pi a)^{1/2}} = \frac{K_I}{p(\pi a)^{1/2}} + \frac{(8)^{1/2}B(\sigma_{ij}^1)}{p}\left(\frac{r}{a}\right)^{1/2} \tag{2.29}$$

Figure 2.10. Test vessel problem no. 2.

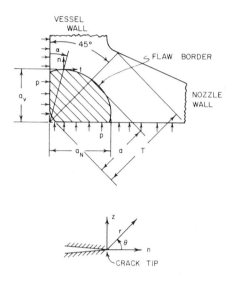

Figure 2.11. Local problem geometry and notation—problem no. 2.

Equation (2.29) suggests that there is, again, a linear relation between $K_{AP}/p(\pi a)^{1/2}$ and $(r/a)^{1/2}$ in the zone dominated by linear elastic fracture mechanics. A typical set of data from one slice is pictured in Figure 2.12. Flaw shapes and SIF distributions obtained from this study are given in Figure 2.13 and they reveal interesting features.

Among the numerical solutions cited earlier which have been proposed for the nozzle-corner crack problem, some [26, 30] had predicted dish shaped SIF distributions along the flaw border, and others [29] had predicted the reverse distribution. The experimental results show both distributions depending upon crack size. Two solutions [27, 28] widely used in the reactor pressure vessel industry predict only one average K_I value for a given crack, and assume quarter circular crack shapes with self-similar flaw growth. Figure 2.13 clearly shows that flaw shapes are not quarter circular, nor is flaw growth self-similar. Figure 2.14 presents a comparison between the average experimental SIF values and the results of [27] and [28] suggesting good general correlation except for shallow flaws despite the above noted discrepancies. In fact, the added conservatism of the analyses for shallow flaws may not be undesirable from a designer's viewpoint. Details of this study will be found in Reference [32].

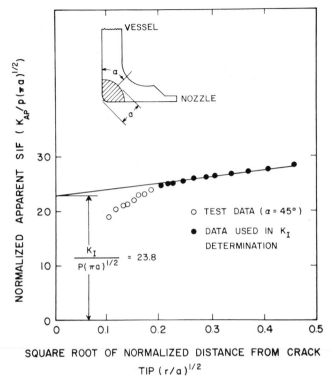

Figure 2.12. Typical test data and SIF estimation—problem no. 2.

Flaw shapes. In the preceding section it was implied that the flaw shapes obtained in the frozen stress method by extending the cracks under monotonically increasing pressure produced the "real" flaw shape which would be found in the metal structure. In a separate study [9, 11, 30] the writer found that flaw shapes produced in nozzle corners in flat photoelastic plates under remote uniaxial extension were virtually identical to those produced in reactor vessel steel models under tension-tension fatigue loads. This has led to the conjecture that, subject to certain constraints, (i.e., small scale yielding, stress ratio near unity, sufficient loading cycles to eliminate crack closure effects, etc.) the natural cracks grown by the frozen stress method will closely resemble those resulting from stable fatigue crack growth. Moreover, if these shapes are significantly different from those assumed in a numerical analysis, then the SIF distribution will also be markedly affected. In the

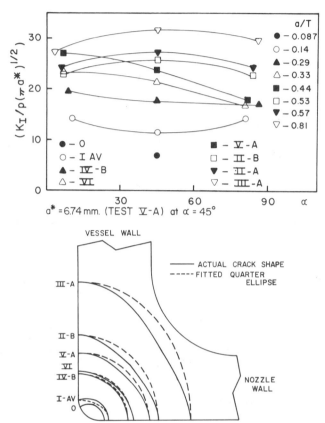

Figure 2.13. Flaw shapes and SIF distributions—problem no. 2.

studies just cited, Broekhoven [30] measured the crack dimensions a_V and a_N (Figure 2.11) and fitted a quarter elliptic curve to these measured dimensions in constructing a finite element solution for the SIF distribution. This approach yielded accurate results until the flaw had penetrated far enough to begin to "flatten" in its middle part. Here the experimental SIF distribution diverged from the numerical result in the fashion indicated by Figure 2.15 (a is the flaw depth at $\alpha = 45°$ as in Figure 2.11). Such differences may be regarded as typical and lead to the general observation that an assumed flaw shape, if different from the natural flaw shape, will lead to higher SIF gradients along the flaw border than will be observed for the natural flaw shapes. However, even the natural flaw shape will not lead to a constant value of the SIF along the flaw border in problems of complex geometry.

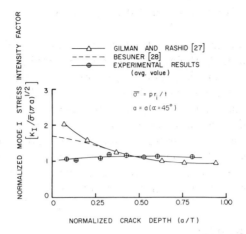

Figure 2.14. Comparison of average experimental results with theory.

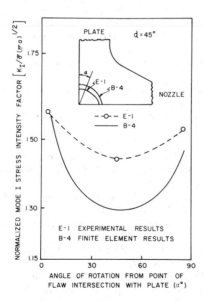

Figure 2.15. Theoretical vs. experimental SIF distributions for Nozzle corner crack in flat plate.

2.5 Summary and conclusions

The use of photoelasticity in fracture mechanics emphasizing a technique for estimating both flaw shapes and SIF distributions for complex, three dimensional cracked body problems has been presented. The method has its own limitations; it is restricted to elastic behavior of an incompressible material and certain loading constraints when comparing flaw shapes with those in metal prototypes. Nevertheless, it appears to be a powerful tool for assisting the numerical analyst in treating problems where neither flaw shape nor SIF distribution are known a priori and possesses the potential for opening the way to the development of sound design programs for a previously intractable set of problems.

References

[1] Dally, J. W. and Riley, W. F. *Experimental Stress Analysis* 2nd Ed. McGraw-Hill Co., New York (1978).
[2] Coker, E. G. and Filon, L. N. G. *A Treatise on Photoelasticity*, Cambridge University Press, London (1931).
[3] Wells, A. A. and Post, D., *Proceedings of the Society for Experimental Stress Analysis*, Vol. 16, No. 1, pp. 69–96 (1958). See also Discussion by G. R. Irwin.
[4] Oppel. G., NACA TM (Translation by J. Vanier) (1937).
[5] Kassir, M. and Sih, G. C., *Journal of Applied Mechanics*, Vol. 33, No. 3, pp. 601–611 (Sept. 1966).
[6] Schroedl, M. A. and Smith, C. W. *Journal of Engineering Fracture Mechanics*, Vol. 7, pp. 341–355 (1975).
[7] Smith, C. W., McGowan, J. J. and Jolles, M. *Journal of Experimental Mechanics*, Vol. 16, No. 5, pp. 188–193 (May 1976).
[8] Gross, B. E. and Mendelson, A. *International Journal of Fracture Mechanics*, Vol. 8, No. 3, pp. 267–376 (Sept. 1972).
[9] Smith, C. W. and Peters, W. H. *Proceedings of the 6th International Conference on Experimental Stress Analysis*, VDI Verlag GmbH Düsseldorf, pp. 861–864 (1978).
[10] Smith, C. W., Jolles, M. and Peters, W. H. *Journal of Experimental Mechanics*, Vol. 17, No. 12, pp. 449–454 (Dec. 1977).
[11] Smith, C. W. and Peters, W. H. III *Developments in Theoretical and Applied Mechanics*, V. 9, pp. 225–234 (May 1978).
[12] Post, D. *Proceedings of the Society for Experimental Stress Analysis*, Vol. 12, No. 1, pp. 99–116 (1954).
[13] Kobayashi, A. S., Wade, B. G. and Maiden, D. E. *Journal of Experimental Mechanics*, Vol. 12, No. 1, pp. 32–37 (Jan. 1972).
[14] Kobayashi, A. S., Wade, B. G., Bradley, W. B. and Chiu, S. T. *Journal of Engineering Fracture Mechanics*, Vol. 6, No. 1, pp. 81–92 (March, 1974).
[15] Kobayashi, A. S. and Wade, B. G. *Dynamic Crack Propagation*, Ed. G. C. Sih, Noordhoff Int. Leyden, pp. 663–678 (1973).

[16] Kobayashi, A. S. and Mall, S. *Journal of Experimental Mechanics*, Vol. 18, No. 1, pp. 11–18 (1978).
[17] Irwin, G. R., Dally, J. W., Kobayashi, T., Fourney, W. L., Etheridge, M. J. and Rossmanith, H. P. *Journal of Experimental Mechanics*, Vol. 19, No. 4, pp. 121–128 (Apr. 1979).
[18] Srawley, J. R., Jones, M. H. and Gross, B. NASA-TN-D 2396 (May 1964).
[19] Irwin, G. R., William Murray Lecture, SESA Fall Meeting (Oct. 1973).
[20] Smith, C. W., McGowan, J. J. and Jolles, M. *Journal of Experimental Mechanics*, Vol. 16, No. 5, pp. 188–193 (May 1976).
[21] Smith, C. W. *Experimental Techniques in Fracture Mechanics*, Society for Experimental Stress Analysis Monograph #2 (A. S. Kobayashi, Ed.), pp. 3–41 (1975).
[22] Jolles, M., McGowan, J. J. and Smith, C. W. *Computational Fracture Mechanics*, American Society of Mechanical Engineers—Applied Mechanics Division Special Publication Proceedings of Second National Congress on Pressure Vessels and Piping., (S. E. Benzley and E. F. Rybicki, Eds.), pp. 63–82 (1975).
[23] Smith, C. W., *Fracture Mechanics and Technology*, Vol. 1, (G. C. Sih and C. L. Chow, Eds.), Sijthoff and Noordhoff, Alphen aan den Rijn, The Netherlands, pp. 591–605 (1977).
[24] Smith, C. W., Peters, W. H. and Andonian, A. T., *Journal of Engineering Fracture Mechanics*, Vol. 13, pp. 615–629 (1979).
[25] Hellen, T. K. and Dowling, A. R., *International Journal of Pressure Vessels and Piping*, Vol. 3, pp. 57–74 (1975).
[26] Reynen, J., ASME Paper No. 75-PVP-20 (June 1975).
[27] Rashid, Y. R. and Gilman, J. D., *Proceedings of the First International Conference on Structural Mechanics in Reactor Technology*, Vol. 4, Reactor Pressure Vessels, Part G, Steel Pressure Vessels, pp. 193–213 (Sept. 1971).
[28] Besuner, P. M., Cohen, L. M. and McLean, J. L., *Transactions of the Fourth International Conference on Structural Mechanics in Reactor Technology*, Vol. G, Structural Analysis of Steel Reactor Pressure Vessels, Paper No. G 4/5 (August 1977).
[29] Schmitt, W., Bartholome, G., Gröstad, A. and Miksch, M., International Journal of Fracture, Vol. 12, No. 3, pp. 381–390 (June 1976).
[30] Broekhoven, M. J. G., *Proceedings of the Third International Conference on Pressure Vessel Technology*, Part II, Materials and Fabrication, pp. 839–852 (April 1977).
[31] Kathiresan, K. and Atluri, S. N., *Pressure Vessel Technology, Vol. 6, Materials, Fracture and Fatigue*, Preprint No. C 28/80, pp. 163–168. Institution of Mechanical Engineers, London (1980).
[32] Smith, C. W., Peters, W. H., Hardrath, W. T. and Fleischman, T. S., NUREG/CR-0640 ORNL/SUB/7015-2 Dist. Category NRC-5-VPI-E-79-2, 33 pp., Jan. 1979. See also *Transactions of the 5th International Conference on Structural Mechanics in Reactor Technology, Division G, Structural Analysis of Steel Reactor Pressure Vessels* (August 1979).

P. S. Theocaris

3 | *Elastic stress intensity factors evaluated by caustics*

3.1 Introduction

The practical usefulness of the concept of the stress intensity factor at a crack tip in an elastic medium in the theory of fracture mechanics has led to a variety of theoretical, as well as experimental techniques for the evaluation of these factors. Most of them concern the case of a crack inside an isotropic elastic medium under generalized plane stress or plane strain conditions, but several other cases have also extensively considered in the literature. Among the first theoretical treatments of crack problems in plane isotropic elasticity we can mention those included in the well-known monograph by Muskhelishvili [1], as well as in the well-known paper by Westergaard [2]. Today thousands of papers reporting elasticity solutions of crack problems and providing formulas and diagrams for the evaluation of stress intensity factors at crack tips can be found in the literature and are referenced in the recently appeared handbooks for the evaluation of stress intensity factors by Tada, Paris and Irwin [3] and Rooke and Cartwright [4]. Furthermore, the most important theoretical techniques for the solution of plane elasticity crack problems and the evaluation of stress intensity factors were reviewed by Paris and Sih [5] and Sih [6] in the first volume of this series.

On the other hand, a brief review of almost all theoretical and experimental methods for the solution of crack problems and the evaluation of stress intensity factors at crack tips, mainly inside plane isotropic elastic media, was made by Cartwright and Rooke [7]. Two more powerful methods for the evaluation of stress intensity factors at crack tips are missing from this review paper and have not gained much attention up to now. The first is the method of complex Cauchy type singular integral equations [8], which is capable of treating the general crack problems of arbitrary shape inside simple or composite elastic media of arbitrary geometry and boundary conditions too. Some of the applications of this method are referenced in [9], whereas some of the numerical techniques for the solution of Cauchy type singular integral equations are referenced in [10].

The second relatively unknown method for the evaluation of stress intensity factors at crack tips is the experimental method of caustics developed by the present author [11]. The application of this method to the experimental determination of stress intensity factors at crack tips in plane isotropic and anisotropic elastic media, as well as in elastic plates and shells, will be reviewed in this chapter and, as hoped, the powerfulness, accuracy and simplicity of this method will become clear. It is believed that the method of caustics is the ideal method for the simple and most accurate experimental determination of stress intensity factors at crack tips.

As regards the publications relevant to the method of caustics, we can mention references [11–20] (stress intensity factors in plane isotropic optically isotropic elastic media), [21–22] (stress intensity factors in plane isotropic optically birefringent elastic media), [23] (stress intensity factors in anisotropic elastic media), [24–30] (arrays of cracks), [31–33] (branched cracks), [34–35] (cracks along interfaces), [36–38] (stress intensity factors at *V*-notch tips), [39–41] (stress intensity factors in plates and shells), [42–43] (problems of holes), [44–46] (inclusion problems), [47–52] (loads along boundaries and pseudocaustics), [53–54] (stress intensity factors in viscoelastic media), [55–59] (stress intensity factors in plastic and similar media), [60–65] (stress intensity factors in dynamic crack problems), [66–68] (contact problems), [69–71] (plate problems), [72–76] (problems associated with reflectors), [77] (three-dimensional problems) and [78–79] (determination of properties of materials). Finally, references [80–85] constitute review papers on various applications of the method of caustics.

As is clear from the contents of the foregoing references, the applicability of the method of caustics is not restricted at all to the experimental determination of stress intensity factors. Yet, in this chapter the treatment will be restricted to this case. After an introductory presentation of the physical meaning of the method of caustics and the derivation of the fundamental equations, this method will be applied to the experimental determination of stress intensity factors at crack tips in isotropic and anistropic plane elastic media, as well as elastic plates and shells. The cases of arrays of cracks, branched cracks and cracks along interfaces will be also considered together with the problem of experimental determination of generalized stress intensity factors at *V*-notch tips. Finally, the method of caustics will be compared with other methods of determining stress intensity factors and several more applications of the method of caustics will be presented in brief.

3.2 The basic formulas

As is well-known, when speaking about a caustic, we mean a curve or a surface along which a high light intensity is observed. In other words, a light beam consisting of an infinity of light rays is said to form a caustic if there is a surface (the caustic) which constitutes an envelope of the light rays of the beam in the sense that many of these rays are tangent to this surface or pass very close to it.

In this way, because of the high light intensity on the caustic and in its neighborhood (on the one side only!), this surface can clearly be experimentally observed. Here we will not be concerned with caustic surfaces in space, but only with caustic curves formed as the intersection of a plane screen with a caustic surface. Moreover, no attention will be paid on the concentration of light intensity on and at the vicinity of a caustic, although it is assumed that the caustic will be sufficiently clear to be observed during an experiment. The use of a light beam from a laser, and not from an ordinary light source, is thus preferable, yet not necessary.

To get a first idea on caustics, let us consider the simple experimental arrangement of Figure 3.1. A parallel light beam with a direction towards the negative part of the *Oz*-axis impinges on a cylindrical surface *Sr* with an equation of the form

$$z = f(x), \tag{3.1}$$

independent of *y* in a Cartesian coordinate system *Oxyz*. Assuming the surface *Sr* to be smooth and polished enough to form a mirror, we obtain the reflected light rays in accordance with the well-known elementary reflection laws. Several impinging and the corresponding reflected light rays are also shown in Figure 3.1. By putting a screen *Sc* with an equation $z = z_0$ (along the positive *Oz*-axis) opposite the surface *Sr*, we observe that all rays impinge on the screen at points having the *x*-coordinate less than x_0, this last value corresponding to a limiting straight line *M* (parallel to the *Oy*-axis) on *Sc*. This line is shown as a point *M* in Figure 3.1. The existence of this line *M* becomes evident

P. S. Theocaris

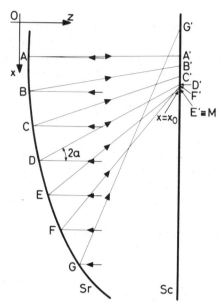

Figure 3.1. Geometrical interpretation of the formation of a caustic.

from the geometry of the surface *Sr*, the slope of which with respect to
the *Ox*-axis increases as *z*-increases. The fact that the density of light
rays impinging on the screen is high near *M* (of course in the region with
$x \leq x_0$) is also clear from Figure 3.1. The line *M* is a limiting curve on
the screen *Sc* in the sense that, if a light ray *EE'* impinges on *M*, then the
light rays *DD'* and *FF'* (with $x_D < x_E < x_F$ on the surface *Sr*) will
impinge at points *D'* and *E'* of the screen *Sc* lying in the same side of *M*.

Analytically, by taking into account equation (3.1) for the surface *Sr*
and the equation $z = z_0$ for the screen *Sc*, as well as the laws of
reflection, it can easily be seen that the correspondence between the
points of the surface *Sr* and the screen *Sc*, where the same light ray
impinges, is expressed by

$$X = x - [z_0 - f(x)] \tan 2\alpha, \tag{3.2}$$

where *x* corresponds to the points of the surface *Sr* and *X* to the points
of the screen *Sc*. In this equation α is the angle of incidence of the light
ray on *Sr*, evidently determined by

$$\tan \alpha = f'(x). \tag{3.3}$$

Thus we obtain

$$X = x - [z_0 - f(x)]2f'(x)/[1 + f'^2(x)]. \tag{3.4}$$

Of course, for the geometry of the surface *Sr* in Figure 3.1 the value of $f(x)$, as well as the value of $f'(x)$, are positive. Finally, by assuming that the coordinate $z = z_0$ of the screen *Sc* is much greater than the values $z = f(x)$ of the same coordinate for the surface *Sr*, we can approximate the difference $[z_0 - f(x)]$ in equation (3.4) by z_0. Moreover, if the slope of *Sr* with respect to the *Oxy*-plane is small, we can approximate the sum $[1 + f'^2(x)]$ in the same equation by unity. This is the case of initially plane specimens to be studied. In such a specimen the *Oxy*-plane is considered to coincide with the plane lying in the middle between the specimen surfaces. Thus, in the sequel, equation (3.4) will be replaced by

$$X = x - 2z_0 f'(x) \tag{3.5}$$

except where stated otherwise. Moreover, if the surface *Sr* is not a cylindrical surface but a surface of general shape, described by the equation

$$z = f(x, y) \tag{3.6}$$

and not by equation (3.1), then an equation similar to (3.5) will hold along the *Oy*-direction. Thus

$$X = x - 2z_0 \frac{\partial f(x, y)}{\partial x}, \; Y = y - 2z_0 \frac{\partial f(x, y)}{\partial y}, \tag{3.7}$$

where (X, Y) are the coordinates of the point of the screen *Sc* $(z = z_0)$ corresponding to a point (x, y) on the surface *Sr* $(z = f(x, y))$. In vectorial form equations (3.7) can be written as

$$(X, Y) = (x, y) - 2z_0 \operatorname{grad} f(x, y). \tag{3.8}$$

A second derivation of equations (3.7) can be made as follows: A wavefront of the parallel light beam impinging on the surface *Sr* in Figure 3.1 can be considered to be plane and given by

$$z = c, \tag{3.9}$$

where c a suitable constant. By taking into account the assumption that the shape of the surface Sr, described by equation (3.6), does not differ much from a plane, we deduce that a wavefront of the light beam, after its reflection on Sr, is determined by

$$z = c - [-2f(x, y)] = c + 2f(x, y) \quad \text{or} \quad z - 2f(x, y) = c, \tag{3.10}$$

the term $-2f(x, y)$ expressing the retardation Δs of the rays of the beam due to the shape of Sr with respect to the Oxy-plane and c being a suitable (although in principle arbitrary) constant. Furthermore, as is well-known, the gradient of a wavefront determines the direction of the light rays. This fact together with equation (3.10) and the assumption $f(x, y) \ll z_0$ leads to

$$(W_x, W_y) = z_0 \text{ grad } \Delta s \tag{3.11}$$

for the deviation ($W_x = X - x$, $W_Y = Y - y$) of a light ray along the Oxy plane between the points of its incidence on Sr, (x, y), and Sc, (X, Y). Since

$$\Delta s = -2f(x, y), \tag{3.12}$$

equation (3.11) can also be written as

$$(W_x, W_y) = -2z_0 \text{ grad } f(x, y), \tag{3.13}$$

which is equivalent to equations (3.7) since

$$W_x = X - x, \quad W_y = Y - y. \tag{3.14}$$

These considerations, making use of the concept of a wavefront, would be of no usefulness if we had to consider only reflection problems. However, this is not the case, as will become clear in the next section, whenever, besides the reflection, refraction of the light rays will be also considered.

Finally, in Figure 3.2 the usually used experimental setup for obtaining a caustic is presented. The specimen Sp is now a cracked medium and the light rays come from a laser (although this is not necessary) and form a focus at a distance z_i from the specimen. The ratio

$$\lambda_m = (z_0 + z_i)/z_i \tag{3.15}$$

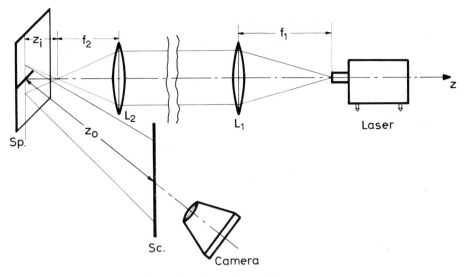

Figure 3.2. The experimental set-up.

is called the *magnification ratio* of the optical arrangement. This ratio can easily be seen to multiply the coordinates (x, y) of a point of a specimen in the formulas giving the coordinates (X, Y) of the corresponding point of the screen. For example, if $\lambda_m \neq 1$ (that is the light beam is not a parallel light beam when impinging on the specimen), equations (3.7) should be written as

$$X = \lambda_m x - 2z_0 \frac{\partial f(x, y)}{\partial x}, \quad Y = \lambda_m y - 2z_o \frac{\partial f(x, y)}{\partial y}. \tag{3.16}$$

Evidently, if the focus of the light beam lies behind the specimen, then z_i and λ_m take negative values.

3.3 The equations of caustics

In this section we will determine the equation of the caustic, as well as the equation of its initial curve, when a caustic is formed on a screen after a light beam impinges on an elastic specimen of thickness d under plane stress conditions. As explained in the previous section, the correspondence between the points $P(x, y)$ of the specimen and $R(X, Y)$ of the screen, at a distance z_0 from the specimen, is established by

$$X = \lambda_m x + z_0 \frac{\partial \Delta s}{\partial x}, \quad Y = \lambda_m y + z_0 \frac{\partial \Delta s}{\partial y}, \qquad (3.17)$$

where λ_m is the magnification ratio of the optical arrangement and Δs the increase of the optical path s of the rays of the light beam due to the loading of the specimen, both faces of which are assumed completely plane and parallel to each other before loading. In order that a caustic be formed on the screen, it is necessary that X takes a maximum or minimum value for $Y = $ constant, or, inversely, that Y reaches a maximum or minimum value for $X = $ constant. By using elementary algebra, we can see that these conditions are satisfied if the Jacobian determinant vanishes

$$J = \frac{\partial(X, Y)}{\partial(x, y)} = \begin{vmatrix} \partial X/\partial x & \partial X/\partial y \\ \partial Y/\partial x & \partial Y/\partial y \end{vmatrix} = 0, \qquad (3.18)$$

or, because of equations (3.17),

$$\begin{vmatrix} 1 + (z_0/\lambda_m)\partial^2 \Delta s/\partial x^2 & (z_0/\lambda_m)\partial^2 \Delta s/\partial x \partial y \\ (z_0/\lambda_m)\partial^2 \Delta s/\partial x \partial y & 1 + (z_0/\lambda_m)\partial^2 \Delta s/\partial y^2 \end{vmatrix} = 0 \qquad (3.19)$$

and, finally,

$$1 + \frac{z_0}{\lambda_m}\left(\frac{\partial^2 \Delta s}{\partial x^2} + \frac{\partial^2 \Delta s}{\partial y^2}\right) + \left(\frac{z_0}{\lambda_m}\right)^2 \left[\frac{\partial^2 \Delta s}{\partial x^2}\frac{\partial^2 \Delta s}{\partial y^2} - \left(\frac{\partial^2 \Delta s}{\partial x \partial y}\right)^2\right] = 0. \qquad (3.20)$$

In practically all cases, the fulfillment of this equation for a set of pairs (x, y) on the specimen means the formation of a caustic on the screen by their corresponding points (X, Y), these points determined by using equations (3.17) for those points (x, y) of the specimen which satisfy equation (3.20). Hence, we can call equation (3.18), or its equivalent equations (3.19) and (3.20), the *equation of the initial curve of the caustic*. Equations (3.17) can further be called *equations of the caustic*.

In this way, and under the assumptions made, that the light beam impinges almost normally (with an angle less than 10°) on the specimen and that its faces are plane and parallel before loading, we can determine the caustics formed on the screen, if such curves exist, by using equations (3.20) and (3.17). We need to know only the constants z_0 and λ_m of the optical set-up, as well as the increase of the optical path

$\Delta s = \Delta s(x, y)$ as a function of the coordinates x and y of the points $P(x, y)$ of the specimen. This is not very easy in all cases. Here we will consider the most important ones.

At first, in the case when the screen lies in front of the specimen and we consider the light rays reflected from the front surface of the specimen, Δs being denoted in this case by Δs_f, we have

$$\Delta s_f = -d\varepsilon_z, \tag{3.21}$$

where d is the thickness of the specimen, as already mentioned, and ε_z is the strain component in the normal direction to the plane of the specimen. In the special case of isotropic elastic materials, by taking into account Hooke's law, we have

$$\varepsilon_z = -\frac{\nu}{E}(\sigma_x + \sigma_y), \tag{3.22}$$

ν being the Poisson ratio and E the Young modulus of the material of the specimen. Then, since

$$\sigma_x + \sigma_y = 4\,\mathrm{Re}\,\Phi(z), \tag{3.23}$$

where $\Phi(z)$ is the complex potential of Muskhelishvili [1], we find

$$\Delta s_f = dc_f(\sigma_x + \sigma_y) = 4dc_f \mathrm{Re}\,\Phi(z), \tag{3.24}$$

where the constant c_f is given by

$$c_f = \nu/E. \tag{3.25}$$

Second, we consider the rays of the light beam, which impinge on the front surface of the specimen, are refracted on this surface, traverse the specimen, are reflected on this surface, traverse the specimen, are reflected on the rear surface of it and, next, traverse once more the specimen and are refracted once more on the front face of it before they leave the specimen and impinge on the screen. The increase of the optical path Δs_r for these rays (reflected rays) would be opposite to Δs_f if the refractive index n of the material of the specimen were equal to 1 and did not change with loading. But this is not the case; the refractive index n of the material of the specimen changes by (see e.g. [13])

$$\Delta n_1 = b_1 \varepsilon_1 + b_2 (\varepsilon_2 + \varepsilon_3), \tag{3.26}$$

$$\Delta n_2 = b_1 \varepsilon_2 + b_2 (\varepsilon_1 + \varepsilon_3)$$

(where b_1 and b_2 are optical constants) along the directions of principal strains ε_1 and ε_2 on the plane of the specimen, the direction of the principal strain ε_3 coinciding with the normal to the plane of the specimen (Oz-axis). We can also mention that equations (3.26) are valid for isotropic elastic materials. By taking into account equations (3.26), we can easily find for the reflected light rays from the rear face of the specimen [13]

$$\Delta s_{r1} = 2d[b_1 \varepsilon_1 + b_2 \varepsilon_2 + (b_2 + n - \tfrac{1}{2}) \varepsilon_3],$$

$$\Delta s_{r2} = 2d[b_1 \varepsilon_2 + b_2 \varepsilon_1 + (b_2 + n - \tfrac{1}{2}) \varepsilon_3] \tag{3.27}$$

after the products $\varepsilon_i \varepsilon_3$ ($i = 1, 2, 3$) have been neglected. Furthermore, by taking into account Hooke's law, we have [13]

$$\Delta s_{r1} = 2d(\alpha_r \sigma_1 + \beta_r \sigma_2), \quad \Delta s_{r2} = 2d(\alpha_r \sigma_2 + \beta_r \sigma_1), \tag{3.28}$$

where

$$\alpha_r = [b_1 - 2 \nu b_2 - \nu(n - \tfrac{1}{2})]/E, \quad \beta_r = [b_2 - \nu(b_1 + b_2) - \nu(n - \tfrac{1}{2})]/E, \tag{3.29}$$

or, finally,

$$\Delta s_{r1,2} = dc_r[(\sigma_1 + \sigma_2) \pm \xi_r (\sigma_1 - \sigma_2)], \tag{3.30}$$

where

$$c_r = (a_r + b_r)/2 \quad \xi_r = (\alpha_r - \beta_r)/(\alpha_r + \beta_r). \tag{3.31}$$

From these equations it is clear that when $\xi_r \neq 0$, or $b_1 \neq b_2$, we have two values for the increase of the optical path Δs_r, Δs_{r1} and Δs_{r2}, even in mechanically isotropic materials. This means that we can obtain two caustics from the light rays reflected on the rear face of the specimen.

This fact is not very pleasant in the method of caustics since simply the calculations will become more complicated, as will be seen later, with almost no additional information. On the contrary, in photoelasticity it is of vital importance that $\xi_r \neq 0$. Here we will prefer to consider optically isotropic (inert) materials, for which $\xi_r \simeq 0$, like plexiglas. For such materials, equations (3.30) are simplified as [13]

$$\Delta s_r = dc_r(\sigma_1 + \sigma_2) = 4dc_r \, \text{Re} \, \Phi(z). \tag{3.32}$$

This equation is completely analogous to equation (3.24) with only a change in the value of the constant c.

Third, we consider the case when the screen is placed behind the specimen. In this case, we take into account only the light rays which traverse only once the specimen refracted on both its surfaces and not reflected at all (traversing rays). For these rays it can be similarly found that the increases $\Delta s_{t1,2}$ along the principal stress and strain directions are given by [86]

$$\Delta s_{t1,2} = dc_t[(\sigma_1 + \sigma_2) \pm \xi_t(\sigma_1 - \sigma_2)], \tag{3.33}$$

where

$$c_t = (a_t + b_t)/2 \qquad \xi_t = (\alpha_t - \beta_t)/(\alpha_t + \beta_t) \tag{3.34}$$

with the constants α_t and β_t given now by [86]

$$\alpha_t = [b_1 - 2\nu b_2 - \nu(n-1)]/E, \qquad \beta_t = [b_2 - \nu(b_1 + b_2) - \nu(n-1)]/E. \tag{3.35}$$

If $\xi_t = 0$, then we have

$$\Delta s_t = dc_t(\sigma_1 + \sigma_2) = 4dc_t \, \text{Re} \, \Phi(z). \tag{3.36}$$

These equations are completely analogous to those derived in the case of the reflected light rays considered previously.

Now, confining ourselves to the case when $\xi_r = \xi_t = 0$, that is to the case of optically isotropic and mechanically isotropic materials, we have for all three cases considered

$$\Delta s = 4dc \, \text{Re} \, \Phi(z), \tag{3.37}$$

where c takes one of the values c_f, c_r or c_t. Then from equations (3.17) we conclude that

$$W = X + iY = \lambda_m[z + C\overline{\Phi'(z)}], \quad z = x + iy, \tag{3.38}$$

where $\overline{\Phi'(z)}$ denotes the complex conjugate of the derivative $\Phi'(z)$ of $\Phi(z)$ and the overall constant C is given by

$$C = 4z_0 dc/\lambda_m. \tag{3.39}$$

Evidently, to derive equation (3.38), we have taken into account that for a complex function $\Phi(z)$ of the variable z of the form

$$\Phi(z) = u(x, y) + iv(x, y), \tag{3.40}$$

the well-known Cauchy-Reimann relations

$$\frac{\partial u(x, y)}{\partial x} = \frac{\partial v(x, y)}{\partial y}, \quad \frac{\partial u(x, y)}{\partial y} = -\frac{\partial v(x, y)}{\partial x} \tag{3.41}$$

hold. In a similar manner and on the basis of equations (3.37) and (3.39–3.41), it can be seen that the equation of the initial curve of the caustic (3.20) can be written under the present assumptions as

$$|C\Phi''(z)| = 1, \tag{3.42}$$

where $\Phi''(z)$ denotes the second derivative of the complex potential $\Phi(z)$. The pair of equations

$$|C\Phi''(z)| = 1 \text{ and } W = \lambda_m[z + C\overline{\Phi'(z)}], \tag{3.43}$$

the first of which is the equation of the initial curve of the caustic on the specimen and the second the equation of the caustic itself on the screen, have played a most important rôle during the use of the method of caustics in a series of practical applications. Indeed, most plane elasticity problems were solved long ago with the help of the complex potentials $\Phi(z)$ and $\Psi(z)$ of Muskhelishvili [1] or, under appropriate conditions, through the complex potential $Z_I(z)$ of Westergaard, related to $\Phi(z)$ by the simple relation

$$Z_I(z) = 2\Phi(z). \tag{3.44}$$

Thus, the determination of the caustic by using equations (3.43) is very easy.

3.4 Properties of the caustics at crack tips

We consider a plane isotropic elastic medium Sp under generalized plane stress conditions containing a crack L with a tip O as shown in Figure 3.3. Of course, the medium Sp may be finite or infinite, contain holes and cracks (besides L) and be arbitrarily loaded. For convenience, we consider a Cartesian coordinate system Oxy with its origin O coinciding with the crack tip under consideration and its Ox-axis along the tangent of the crack at O and in the direction of extension of the crack. Under the reasonable assumptions that the external loadings applied along the edges $(+)$ and $(-)$ of the crack L (and probably being different) do not present singularities near O of order absolutely greater than $(-\frac{1}{2})$, the stress field near O presents the well-known (see e.g. [5]) inverse square root singularity, that is it tends to infinity like $r^{-1/2}$ as

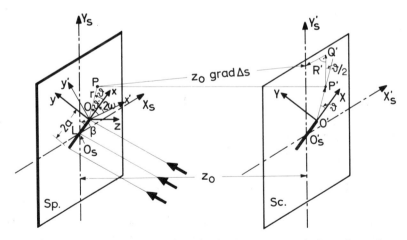

Figure 3.3. Geometry of the formation of a caustic from a cracked specimen Sp on a screen Sc.

$r \to 0$, where r is the polar radius about O, that is

$$z = x + iy = r \exp (i\vartheta). \tag{3.45}$$

Under these conditions, the stress intensity factor K at the crack tip O is determined by (see e.g. [5])

$$K = 2(2\pi)^{1/2}\lim_{z \to 0} [z^{1/2} \, \Phi(z)], \tag{3.46}$$

where $\Phi(z)$ is the complex potential of Muskhelishvili [1] for the problem under consideration. From equation (3.46) it is also clear that in the near vicinity of the crack tip O we can also write

$$\Phi(z) = K/[2(2\pi z)^{1/2}]. \tag{3.47}$$

Of course, in general, K is a complex quantity, that is

$$K = K_I - iK_{II}, \tag{3.48}$$

where K_I and K_{II} denote the mode I and mode II stress intensity factors at the crack tip O. We can also mention that recently in several publications the quantity

$$k = \pi^{-1/2}K \tag{3.49}$$

is defined as the stress intensity factor at a crack tip. In this chapter, we will not use at all this definition although it presents some advantages over the definition (3.46) of the stress intensity factor.

Now by taking into account equation (3.42) for the initial curve of a caustic and substituting in it $\Phi(z)$ by its expression given by equation (3.47), we find that this curve is a circle of radius

$$r = r_0 = \left[\frac{3}{8(2\pi)^{1/2}} |CK| \right]^{2/5} = 0.4677 \, |CK|^{2/5}. \tag{3.50}$$

This equation reveals that the initial curve of the caustic on the specimen depends only on the absolute value of the stress intensity factor K and on the overall constant C of the experimental arrangement.

Moreover, this curve is the circumference of a circle surrounding the crack tip *O*. Of course, it is evident that equation (3.50) is valid only up to the extent of validity of equation (3.47), or, otherwise, r_0 should result from equation (3.50) sufficiently small so that the initial curve of the caustic lies inside the near vicinity of the crack tip. If more accuracy in the determination of the initial curve of the caustic is required or the constant *C* of the experimental arrangement has a sufficiently large value, then equation (3.47) is inadequate and a more complete or the exact expression for $\Phi(z)$ should be used in equations (3.43). This will become clearer in the sequel.

We can finally mention that in reality the initial curve of the caustic on the specimen, as defined by equation (3.50), is a complete circumference of a circle but cut from the crack boundary *L*. This means that it starts from the (−) edge of *L*, corresponding to a polar angle $\vartheta = -\pi$, and it terminates on the (+) edge of *L*, corresponding to a polar angle $\vartheta = +\pi$.

Furthermore, equation (3.38) for the caustic on the screen takes, because of equation (3.47), the form

$$W = X + iY = \lambda_m \left\{ r_0 \exp(i\vartheta) - \frac{C\bar{K}}{4(2\pi)^{1/2}} r_0^{-3/2} \exp\left(\frac{3i\vartheta}{2}\right) \right\},$$

$$(3.51)$$

$-\pi \leqq \vartheta \leqq \pi$ and $\bar{K}$ is the complex conjugate of *K*.

By taking into account equation (3.50), we obtain

$$W = X + iY = \lambda_m r_0 \left\{ \exp(i\vartheta) + \frac{2\bar{\alpha}}{3} \exp\left(\frac{3i\vartheta}{2}\right) \right\}, \quad -\pi \leqq \vartheta \leqq \pi, \quad (3.52)$$

where

$$\alpha = -CK/|CK|. \tag{3.53}$$

and $\bar{\alpha}$ denotes the complex conjugate quantity of α. If we further write the stress intensity factor *K* under the polar form

$$K = |K| \exp(-i\omega), \tag{3.54}$$

that is

$$\omega = \tan^{-1}(K_{II}/K_I), \tag{3.55}$$

equation (3.52) can be written as

$$W = X + iY = \lambda_m r_0 \left\{ \exp\left(i\vartheta\right) + \frac{2\varepsilon}{3} \exp\left[i\left(\frac{3\vartheta}{2} + \omega\right)\right]\right\}, \quad -\pi \leqq \vartheta \leqq \pi,$$

$$(3.56)$$

where

$$\varepsilon = -C/|C| = -\text{sign } C. \tag{3.57}$$

Finally, by putting

$$\mu = K_{II}/K_I = \tan \omega, \tag{3.58}$$

we have from equation (3.56)

$$W = X + iY = \lambda_m r_0 \left\{ \exp\left(i\vartheta\right) + \frac{2\varepsilon}{3} \frac{1 + i\mu}{(1 + \mu^2)^{1/2}} \exp\left(\frac{3i\vartheta}{2}\right)\right\}. \tag{3.59}$$

We can also easily obtain the parametric equations of the caustic in real form. Thus, from equation (3.56) we have

$$X = \lambda_m r_0 \left\{ \cos \vartheta + \frac{2\varepsilon}{3} \cos\left(\frac{3\vartheta}{2} + \omega\right)\right\},$$

$$\qquad\qquad\qquad\qquad\qquad -\pi \leqq \vartheta \leqq \pi. \tag{3.60}$$

$$Y = \lambda_m r_0 \left\{ \sin \vartheta + \frac{2\varepsilon}{3} \sin\left(\frac{3\vartheta}{2} + \omega\right)\right\},$$

Similarly, from equation (3.59) we find:

$$X = \lambda_m r_0 \left\{ \cos \vartheta + \frac{2\varepsilon}{3}(1 + \mu^2)^{-1/2}\left[\cos\frac{3\vartheta}{2} - \mu \sin\frac{3\vartheta}{2}\right]\right\},$$

$$\qquad\qquad\qquad\qquad\qquad -\pi \leqq \vartheta \leqq \pi.$$

$$Y = \lambda_m r_0 \left\{ \sin \vartheta + \frac{2\varepsilon}{3}(1 + \mu^2)^{-1/2}\left[\sin\frac{3\vartheta}{2} + \mu \cos\frac{3\vartheta}{2}\right]\right\}, \tag{3.61}$$

On the basis of the above equations, we can draw these caustics and see their form. Since, as is clear from equation (3.57), ε is restricted to the values $\varepsilon = \pm 1$, only two forms for these caustics are possible for one value of μ. In Figure 3.4 we see the forms of the caustics for $\mu = 0, 0.25,$ 1 and ∞, drawn on the basis of equations (3.61) for $\varepsilon = 1$. In the same

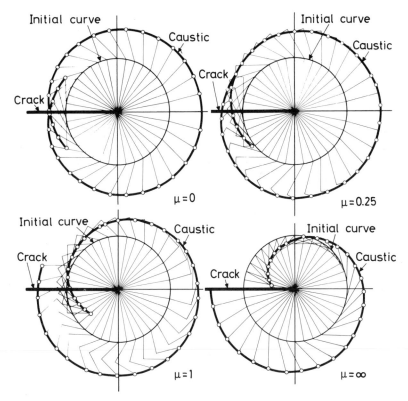

Figure 3.4. Variation of the theoretical form of the caustic about a crack tip with the ratio
μ for $\varepsilon = +1$.

figure the circular initial curve of the caustic (under the assumption that
$\lambda_m = 1$), as well as the way of formation of the caustic from its initial
curve are shown. Analogous curves can also be drawn for $\varepsilon = -1$. From
the above equations of the caustic, it is clear that in the case when
$\varepsilon = -1$, we can assume that $\varepsilon = +1$ and use the interval $[\pi, 3\pi]$ for the
parameter ϑ. Thus, if we assume that ε takes both values $\varepsilon = \pm 1$, then
the caustic is a closed curve and ϑ varies in a 4π interval. In Figure 3.5
such a complete caustic is shown for $\mu = 0.50$.

One more interesting remark is the following: If we rotate the coordinate system Oxy in Figure 3.3 by (-2ω) so that it coincides with the
$Ox'y'$ coordinate system (Figure 3.3), where

$$z' = x' + iy' = r \exp(i\vartheta'), \quad \vartheta = \vartheta' - 2\omega, \tag{3.62}$$

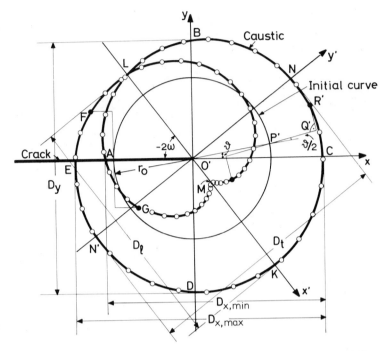

Figure 3.5. Theoretical form of the caustic formed at a crack tip, for $\mu = 0.50$, and its characteristic dimensions.

we can write equation (3.56) as

$$W' = X' + iY' = \lambda_m r_0 \left\{ \exp\left(i\vartheta'\right) + \frac{2\varepsilon}{3} \exp\left(\frac{3i\vartheta'}{2}\right) \right\},$$

$$-\pi + 2\omega \leq \vartheta' \leq \pi + 2\omega, \tag{3.63}$$

or in real form

$$X' = \lambda_m r_0 \left\{ \cos \vartheta' + \frac{2\varepsilon}{3} \cos \frac{3\vartheta'}{2} \right\},$$
$$Y' = \lambda_m r_0 \left\{ \sin \vartheta' + \frac{2\varepsilon}{3} \sin \frac{3\vartheta'}{2} \right\}, \qquad -\pi + 2\omega \leq \vartheta' \leq \pi + 2\omega. \tag{3.64}$$

In this form the equations of the caustic seem to be independent of μ and to result from equations (3.61) for $\mu = 0$. Of course, this is not true since the value of μ, or better ω, enters into the interval $(-\pi + 2\omega) \leqq \vartheta' \leqq \pi + 2\omega$ for the parameter ϑ. But, if we consider the complete form of the caustic, for both values $\varepsilon = +1$ and -1, this is completely independent of μ and has the same shape for every crack and specimen geometry and loading conditions. The interval for ϑ' can be considered coinciding with $[-2\pi, 2\pi]$ in this case.

Now, we will investigate some more interesting properties of the caustic formed at a crack tip, as seen in Figure 3.5 in its complete form for $\varepsilon = \pm 1$. In this case we have from equations (3.64)

$$X' = \lambda_m r_0 \left\{ \cos \vartheta' + \tfrac{2}{3} \cos \frac{3\vartheta'}{2} \right\}$$
$$-2\pi \leqq \vartheta' \leqq 2\pi. \qquad (3.65)$$
$$Y' = \lambda_m r_0 \left\{ \sin \vartheta' + \tfrac{2}{3} \sin \frac{3\vartheta'}{2} \right\}$$

From these equations we observe at first that the caustic is a symmetric curve about the Ox'-axis. Of course, in practice we do not see the caustic in its complete form obtained when ϑ' is permitted to vary in a 4π-interval, but we see only a part of it obtained when ϑ' varies in a 2π-interval. Then, the symmetry of the caustic is completely clear only when $\omega = 0$, that is when the Ox'- and Ox-axes coincide. When $\omega \neq 0$, the symmetry of the caustic is only partly observed in experiments.

Now, we can also express equations (3.65) in polar form by setting

$$X' + iY' = \rho \exp (i\varphi). \qquad (3.66)$$

Then we find

$$\tan \varphi = \frac{3 \sin \vartheta' + 2 \sin (3\vartheta'/2)}{3 \cos \vartheta' + 2 \cos (3\vartheta'/2)} \qquad (3.67)$$

for the polar angle φ and

$$\rho = \lambda_m r_0 \left\{ \cos (\vartheta' - \varphi) + \tfrac{2}{3} \cos \left(\frac{3\vartheta'}{2} - \varphi \right) \right\} \qquad (3.68)$$

for the polar radius ρ. We can further easily see that ρ takes a maximum value $\rho_{\max}$ for $\vartheta' = \varphi = 0$. Then we have

$$\rho_{\max} = 5\lambda_m r_0/3, \quad \varphi = 0, \quad \vartheta' = 0. \qquad (3.69)$$

Similarly, ρ takes a minimum value $\rho_{\min}$ when $\vartheta' = \pm 2\pi$, that is

$$\rho_{\min} = \lambda_m r_0/3, \quad \varphi = 0, \quad \vartheta' = \pm 2\pi. \tag{3.70}$$

In this case a cusp-point is seen to be formed on the caustic (Figure 3.5). For these values of ρ, the caustic intersects the positive Ox'-axis. Finally, for $\cos(\vartheta'/2) = 1/4$, ρ takes the value ρ_{mid}

$$\rho_{\text{mid}} = 4\lambda_m r_0/3, \quad \varphi = \pi, \quad \vartheta' = \pm 2\cos^{-1}(1/4), \quad \text{or} \quad \vartheta = \pm 151°3'. \tag{3.71}$$

In this case the caustic intersects the negative Ox'-axis. Moreover, from equations (3.69) and (3.71) we see that the maximum diameter of the caustic along the Ox'-axis is

$$D_l = 3\lambda_m r_0. \tag{3.72}$$

Finally, the diameter D_t of the caustic in a direction parallel to the Oy'-axis can be found to be [19]

$$D_t = 3.1702\,\lambda_m r_0, \quad \vartheta' = \pm 2\pi/5 \quad \text{or} \quad \vartheta' = \pm 72°. \tag{3.73}$$

It can be mentioned that this diameter is the maximum diameter of the caustic in a direction parallel to the Oy'-axis but it does not lie on this axis (Figure 3.5).

Now we can proceed to the estimation of the complex stress intensity factor K from the geometry of the caustic formed at a crack tip. This can be achieved very easily since the whole geometry of the caustic depends on the constants λ_m and C of the experimental arrangement and on the value of K. To determine K, we have to determine both $|K|$ and ω or μ (equations (3.55), (3.58)). As regards ω, this can be determined easily since the axis Ox' of symmetry of the caustic forms an angle (-2ω) with the Ox-axis along the tangent of the crack at the crack tip (Figure 3.5). The direction of the Ox-axis is well-known and the direction of the Ox'-axis can be directly determined from the geometry of the caustic near its cusp-point at $\rho_{\min}$ (Figure 3.5). In general, we obtain on the screen, if this is placed in front of the specimen, two caustics, with different values of the constant C, one with $\varepsilon = +1$ and another with $\varepsilon = -1$. Although the different values that the constant C takes for these two caustics make these curves completely different in size and not forming a closed curve, yet it is completely possible to see the cusp-point already mentioned being present in the caustic corresponding to $\varepsilon = -1$. If the angle formed between the Ox-axis and the tangent of the caustic at this cusp-point is φ_0, then we have

$$\omega = \varphi_0/2 \tag{3.74}$$

and further

$$\mu = K_{II}/K_I = \tan(\varphi_0/2). \tag{3.75}$$

As regards the absolute value $|K|$ of the complex stress intensity factor K, this can be obtained by measuring the maximum diameter D_t of the caustic corresponding to $\varepsilon = +1$ along the direction Oy' normal to the Ox'-axis, which has been already determined from the symmetry of the caustic corresponding to $\varepsilon = -1$ about its cusp-point. Thus, by combining equations (3.50) and (3.73), we find

$$|K| = 6.68434 r_0^{5/2}/|C| = 0.373544 (D_t/\lambda_m)^{5/2}/|C|. \tag{3.76}$$

It is worth-mentioning that, because of the fact that ϑ' varies only in the interval determined by the inequalities in equations (3.64), the transverse diameter D_t of the external caustic (corresponding to $\varepsilon = +1$) can be determined directly only if

$$|\omega| \leq 3\pi/10, \quad \text{or} \quad |\omega| \leq 54°. \tag{3.77}$$

Otherwise, the maximum distance of the points of this caustic from the Ox'-axis, equal to $D_t/2$, has to be determined. The double of this distance gives D_t.

In another way of thinking, we can work only along the Ox and Oy-axes of the initial coordinate system Oxy to determine $|K|$ and μ. In this case, we use data from the caustic corresponding to $\varepsilon = +1$ only. By defining the diameters $D_{x,\max}$, $D_{x,\min}$ and D_y along the Ox- and the Oy-axes, as shown in Figure 3.5, we can determine the ratio μ, equation (3.58), by measuring only $D_{x,\max}$ and $D_{x,\min}$, through the ratio $(D_{x,\max} - D_{x,\min})/D_{x,\max}$. In Figure 3.6 the function

$$(D_{x,\max} - D_{x,\min})/D_{x,\max} = f(\mu) \tag{3.78}$$

was drawn on the basis of the equations of the caustic (for $\varepsilon = +1$)

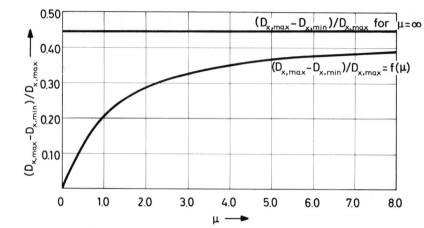

Figure 3.6. Variation of the ratio $(D_{x,\max} - D_{x,\min})/D_{x,\max}$ with the ratio μ.

already determined. From the curve of this figure we can obtain μ and further, if required, ω. As regards the absolute value of K, this can be determined from equation (3.50) or better from the first of equations (3.76), that is

$$|K| = 6.68434 r_0^{5/2}/|C|.$$
(3.79)

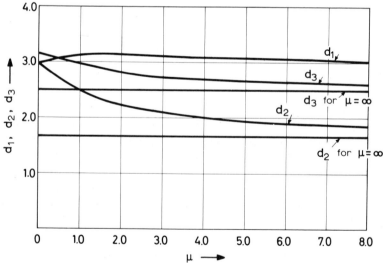

Figure 3.7. Variation of the quantities d_1, d_2 and d_3 with the ratio μ.

Hence, we need to determine r_0 from the geometry of the caustic. This can easily be made by using the curves of Figure 3.7, drawn on the basis of the equations of the caustic. In this figure the quantities

$$d_1 = D_{x,\max}/(\lambda_m r_0), \quad d_2 = D_{x,\min}/(\lambda_m r_0), \quad d_3 = D_y/(\lambda_m r_0) \tag{3.80}$$

are given as functions of the ratio μ, already determined as explained previously. Each one of the diameters $D_{x,\max}$, $D_{x,\min}$ and D_y can thus be used for the determination of r_0 and, further, $|K|$.

Finally, by using equations (3.48), (3.54) and (3.55), we can find also the following formulas for the determination of the mode I and mode II stress intensity factors, K_I and K_{II} respectively, at the crack tip under consideration

$$K_I = |K| \cos \omega, \quad K_{II} = |K| \sin \omega \tag{3.81}$$

or

$$K_I = |K|(1 + \mu^2)^{-1/2}, \quad K_{II} = |K|\mu(1 + \mu^2)^{-1/2}. \tag{3.82}$$

The method of caustics, as was developed for the experimental determination of stress intensity factors at crack tips inside mechanically and optically isotropic elastic media under generalized plane stress conditions, has been applied to a series of problems of simple cracks, branched cracks, arrays of cracks, interactions of cracks and boundaries etc. as will be seen in the sequel. In all cases, it was seen that, by using the method of caustics, it was possible to estimate quickly and accurately the values of the stress intensity factors at the crack tips, the accuracy being, in most cases, much better than 10%. Of course, in order to apply the simple formulas of this section, it is necessary that equation (3.47) hold with a sufficient accuracy in the neighborhood of the initial curve of the caustic on the specimen. This means that, if a crack tip lies very near another crack tip or a boundary, equation (3.47) does not constitute any more a sufficiently good approximation of the stress field in the vicinity of the crack tip. In this case, for a better accuracy we can take into account three terms in the expansion of $\Phi(z)$ near the crack tip for the determination of r_0 as will be seen in section 3.14. In more difficult cases, we can use the exact formula for $\Phi(z)$ in order to determine the form of the caustic and, further, the value of the stress intensity factor, if this is unknown, as will be seen later.

In most experiments plexiglas was found to be an appropriate material since it has the advantages that it is a mechanically and optically isotropic material and, moreover, it does not present a plastic zone at room temperature even in the close vicinity of the crack tip for sufficiently large loading intensities. In this way, plexiglas is an almost ideal material for the estimation of linear elastic stress intensity factors as was proposed in this section. Of course, it is evident that the values of the stress intensity factors for a cracked medium do not depend (except in exceptional cases) on the material of the medium; they depend only on the geometry and loading conditions of the medium.

As regards the mechanical constants of plexiglas, at room temperature, they were found to be approximately: $E = 3.2 \cdot 10^4$ kp/cm^2 and $\nu = 0.33$. Then from equation (3.25) we find that $c_f = 1.03 \cdot 10^{-5}$ cm^2/kp. As regards the constants c_r and c_t, they were found to be: $c_r = -3.22 \cdot 10^{-5}$ cm^2/kp and $c_t = -1.06$ cm^2/kp. In this way, we have

$$|c_f| : |c_r| : |c_t| = 1 : 3.12 : 1.03. \tag{3.83}$$

By taking further into account the equations of the caustic and its initial curve around a crack tip, equations (3.50) and (3.60) respectively, we can see that the following relations hold for all corresponding lengths l_f, l_r and l_t of the caustics and their initial curves from rays reflected from the front face of the specimen (f), the rear face of the specimen (r) and traversing the specimen (t) respectively

$$l_f : l_r : l_t = 1 : 1.58 : 1.01 \tag{3.84}$$

under the obvious assumption that the magnification ratio λ_m of the optical set-up, as well as the distance z_0 between the specimen and the screen are the same for all three caustics formed (the first two of which, (f) and (r), are observed when the screen is in front of the specimen and the third (t) when the screen is behind the specimen). In Figure 3.8, the relative shapes of these caustics and their initial curves are seen in the case of existence of only the mode I stress intensity factor K_I at a crack tip under the further assumptions that $\lambda_m = 1$ and $\varepsilon = 1$, that is the focus of the light beam, if this beam is not a parallel light beam, lies in front of the specimen (since, evidently, $K_I > 0$). In this figure the initial curves of the caustics are seen to be closed, circular curves (denoted by $(\times)$ for the rays reflected from the front face of the specimen (f), by $(\bigcirc)$ for the rays reflected from the rear face of the specimen (r), and by $(+)$ for the rays

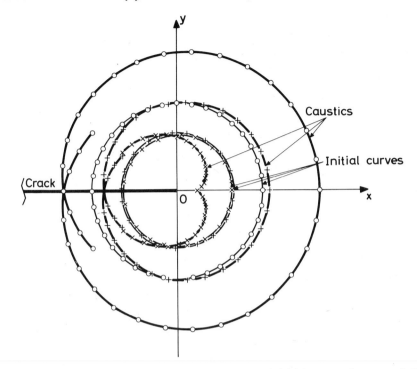

Figure 3.8. Relative shapes of the caustics, as well as their initial curves, about a crack tip in a plexiglas specimen.

traversing the specimen (t)). On the contrary, the corresponding caustics (denoted by the same symbols) are seen to be open curves as already explained ($-\pi \leqq \vartheta \leqq \pi$). Of course, only the caustics due to c_f and c_r are seen when the screen is in front of the specimen. Similarly, only the caustic due to c_t is seen when the screen is behind the specimen.

In general, since the caustic due to reflections of the light rays from the rear surface of the specimen (due to c_r) is the greatest one, because of the proportionalities (3.84), we prefer to place the screen in front of the specimen. At the same time, we avail ourselves of the fact that the caustic due to c_f is also formed. Since for $\varepsilon = -1$ this caustic (due to c_f) has a cusp-point, we can use it to determine the ratio $\mu = K_{II}/K_I$ as already explained. Next, the absolute value of K can be determined accurately from the caustic due to c_r. Of course, the second procedure described previously and based on measurements of the diameters of only one caustic (in our case of the one due to c_r, with $\varepsilon = 1$) is also applicable. In the remaining of this section only such caustics will be considered. Finally, it is preferable, although not necessary at all, that a monochromatic laser light beam be used. Such a beam has a greater intensity than an ordinary-light-source beam and can be concentrated in the near vicinity of the crack tip to produce a very clear caustic. Such a beam, because of its monochromatic character, produces, besides the caustic, an interference pattern around the caustic. But, because of the strong deviations of the light rays from their initial directions when reflected from the front or the rear face of the specimen in the vicinity of the crack tip (these deviations causing the creation of the caustic which *does not coincide with its initial curve*), as well as the fact that the two faces of the specimen are not absolutely parallel before loading (this being of little importance in the vicinity of the crack tip, where the gradients of the stress components are great), no easily exploited information can be gathered from this interference pattern either near or far from the crack tip. Besides this interference pattern, a diffraction pattern with secondary caustics appears in the cases where coherent light is used. These secondary caustics correspond to the parametric curves of equation (3.51) with $r_0 = $ constant (see, for instance, Figures 3.9 and 3.10).

In Figure 3.9 we see two typical experimental forms of the caustics formed at a straight crack tip when only the mode I stress intensity factor K_I is present (Figure 3.9(i)) or when both the mode I and the mode II stress intensity factors K_I and K_{II} are present (Figure 3.9(ii)). Of course, in both cases the screen was placed in front of the specimen. The values of the stress intensity factors can be directly determined from these caustics as described in the previous section.

A series of experiments were executed with tension cracked plexiglas specimens of length $l = 30$ cm, width $w = 16$ cm and thickness $d = 0.2$ cm. The optical constant c_r for these plexiglas specimens was found experimentally to be $c_r = -3.24 \cdot 10^{-5}$ cm^2/kp. Moreover, in the optical arrangement the screen was placed in front of the specimen at a distance $z_0 = 121$ cm whereas the focus of the light beam was formed also in front of the specimen and at a distance $z_i = 19.4$ cm from it. Thus the magnification ratio λ_m of the optical arrangement resulted from equation (3.15) to be $\lambda_m = 7.23$ and, further, the overall constant C was seen from equation (3.39) to be $C = -4.33 \cdot 10^{-4}$ cm^4/kp for the light rays reflected on the rear face of the specimen (corresponding to the optical constant c_r). As regards the tensile stress field on the specimen, it was of intensity $\sigma_m = 50$ kp/cm^2. The experimental determination of the stress intensity

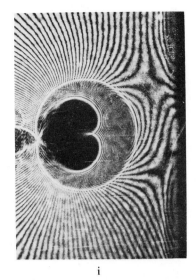

i

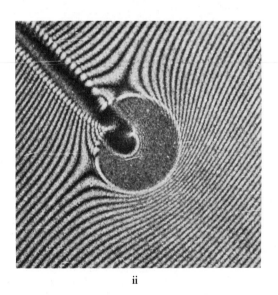

ii

Figure 3.9. Experimentally obtained caustics at the tip of a cracked tension plexiglas specimen with $\beta = 90°$ (i) and $\beta = 45°$ (ii).

216   *P. S. Theocaris*

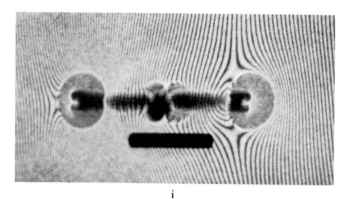

i

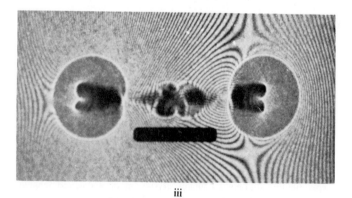

ii

iii

Figure 3.10. Variation of the experimental form of the caustics at the tips of a cracked tension plexiglas specimen, under mode *I* conditions, with loading.

factor K was made on the basis of equation (3.76) by measuring the transverse diameter D_t of the caustic. Then the absolute value of K results. Furthermore, the argument of K can also be estimated as proposed previously and thus both stress intensity factors K_I and K_{II} (with $K = K_I - iK_{II}$) be evaluated. In all experiments a straight crack of length $2a = 1$ cm lying in the central region of the specimen was used. Moreover, several orientations of the crack have been considered forming angles $\beta = 20° \div 90°$ with the tension axis. In all cases it was found that the discrepancies between the experimentally determined values of $K_{I,II}$ and the corresponding theoretical values, determined by [5]

$$K_I = 1.02\,\sigma_m(\pi a)^{1/2} \sin^2 \beta, \quad K_{II} = 1.02\,\sigma_m(\pi a)^{1/2} \sin \beta \cos \beta, \qquad (3.85)$$

did not exceed 10% and, generally, were smaller than 5%.

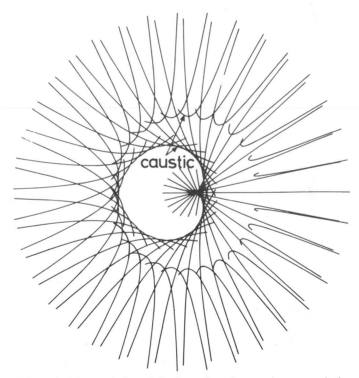

Figure 3.11. Geometrical interpretation of the formation of a caustic at a crack tip as an envelope of light rays.

Furthermore, in Figure 3.10 the increase of the size of the caustics at straight-crack tips under mode I loading conditions with the tensile loading intensity is presented. Of course, because of equations (3.50) and (3.60), the dimensions of the caustic do not vary proportionally to the stress intensity factor and to the tensile loading intensity (although the latter are proportional to one another).

Finally, in Figure 3.11 a characteristic geometrical interpretation of the formation of the caustic is presented. To draw this figure, we have considered the radial directions (at 10° intervals) on the specimen and have determined the corresponding curves on the screen, which are seen to be tangent to the caustic, or, otherwise, to form the caustic on the screen. The increase of light intensity in the neighborhood of the caustic (but only on one side of it) is also clear from Figure 3.11. These curves present the sections of the families of light rays in space, as they have been reflected from the vicinity of the initial curve on the specimen Sp, with the screen Sc and are obtained from equations (3.51) when ϑ is a constant and r_0 a variable. Similarly, the secondary caustics, clear from Figures 3.9 and 3.10, present the sections of the same rays with the screen Sc, but correspond to equation (3.51) with r_0 being a constant and ϑ varying in the interval $-\pi \leqq \vartheta \leqq \pi$.

3.5 The case of birefringent media

In the case of cracked birefringent media, although the caustics formed around a crack tip and due to reflections of the light rays from the front face of the specimen (due to c_f) remain unchanged, as was in the case of optically inert materials, the caustics due to reflections of the light rays on the rear face of the specimen (due to c_r), as well as the caustics due to the light rays traversing the specimen (due to c_t) change significantly because of the quantities ξ_r or ξ_t present in equations (3.30) and (3.33) respectively in this case. Moreover, as clear from these equations, both values $\pm\xi$ ($\pm\xi_r$ or $\pm\xi_t$) can be used. A consequence of this fact is that the caustics formed about a crack tip are double in this case.

For the derivation of the equations of the initial curve of the caustic and the caustic near a crack tip in the case of a medium made of a birefringent material, we take into account the asymptotic forms of the stress components near a crack tip, given by the Sneddon formulas, together with equations (3.30) and (3.33), as well as equations (3.17–3.20). In reference [21] this analysis was completely made for the case of existence of only the mode I stress intensity factor K_I at the crack tip.

From the results of this reference it is concluded that, for a given value of ξ, the initial curve of the caustic is not a circle (as in the case of optically inert materials) but its radius r varies with the polar angle ϑ as

$$r = r_0 = \left(\frac{3A}{8(2\pi)^{1/2}}|CK_I|\right)^{2/5} = 0.4677 A^{2/5}|CK_I|^{2/5}, \tag{3.86}$$

where

$$A = \left\{-\frac{\xi}{4}\sin\vartheta + \left[1 + \frac{\xi}{4}\left(7\sin\frac{\vartheta}{2} - \sin\frac{3\vartheta}{2}\right) + \frac{\xi^2}{32}(25 + 9\cos 2\vartheta)\right]^{1/2}\right\} \tag{3.87}$$

and C is given by equation (3.39). Evidently, if $\xi = 0$, then $A = 1$ and equation (3.86) coincides with equation (3.50). A similar remark is also valid for the equations of the caustic itself [21]

$$X = \lambda_m r_0\left\{A^{2/5}\cos\vartheta + \frac{2\varepsilon}{3}A^{-3/5}\left[\cos\frac{3\vartheta}{2} + \frac{3\xi}{4}\sin 2\vartheta\right]\right\},$$

$$Y = \lambda_m r_0\left\{A^{2/5}\sin\vartheta + \frac{2\varepsilon}{3}A^{-3/5}\left[\sin\frac{3\vartheta}{2} - \frac{\xi}{4}(1 + 3\cos 2\vartheta)\right]\right\}, \tag{3.88}$$

$$-\pi \leq \vartheta \leq \pi,$$

which, if $\xi = 0$, coincide with equations (3.60), of course with $\omega = 0$ since in this section only the mode I stress intensity factor K_I was assumed to be present. Moreover, the case of existence of the mode II stress intensity factor K_{II} can also be considered in an analogous manner and after simply an increase of the required calculations.

In references [21, 22] the geometrical properties of the caustic, determined by equations (3.86–3.88) were extensively investigated. In the same references the theoretically obtained forms of the caustics and their initial curves about a crack tip for several values of the constant ξ, as determined from equations (3.86–3.88) were presented with ϑ assumed varying in a 4π-interval in equations (3.88). This is not true, but in this way both cases $\varepsilon = \pm 1$ are considered. Moreover, if we use the value $-\xi$ instead of ξ, the caustics and their initial curves will take their symmetric forms about the crack axis. Here we present in Figure 3.12 the experimentally obtained caustics from a birefringent cracked medium (made of polycarbonate of bisphenol A, $|\xi| = 0.15$) as formed

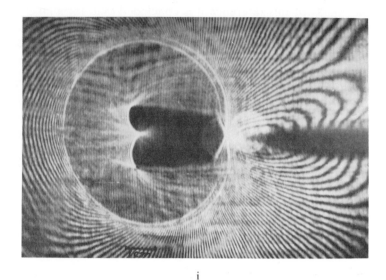

i

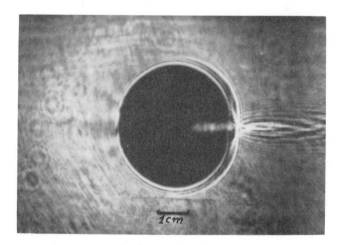

ii

(a)

Figure 3.12. Experimental (a) and theoretical (b) forms of the caustics obtained at a crack tip in a birefringent tension specimen (made of polycarbonate of bisphenol *A*) under mode *I* conditions from the rays reflected from the specimen (i) or traversing the specimen (ii).

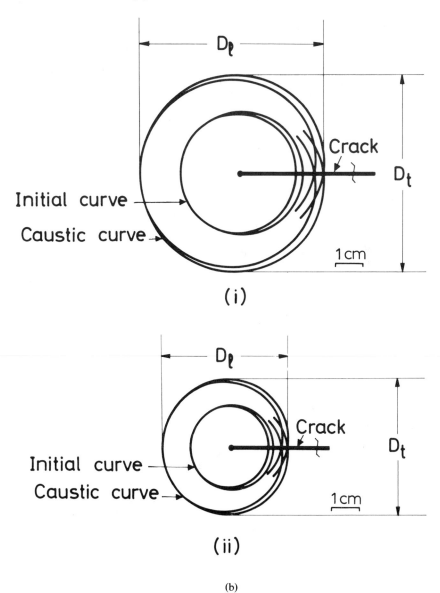

(b)

Figure 3.12. Continued.

when the screen lies in front (*i*) or behind (*ii*) the specimen together with their theoretical forms (for $\varepsilon = +1$). In Figure 3.12 we observe that the caustics due to $c_{r,t}$ (for c_r the external caustic) are double as expected.

Finally, for the determination of the stress intensity factor K_I from the transverse diameter D_t of the caustic (corresponding to $\varepsilon = +1$) or from the longitudinal diameter D_l of the caustic (corresponding also to $\varepsilon = +1$), as these are defined in Figure 3.12b, we can use the formula

$$K_I = \delta_{t,l}(|\xi|)(D_{t,l}/\lambda_m)^{5/2}/|C|. \tag{3.89}$$

The functions $\delta_{t,l}(|\xi|)$ were drawn in Figure 3.13. Moreover, equation (3.89) is completely analogous to equation (3.76), to which it reduces when the transverse diameter D_t is measured and $\xi = 0$ as is also clear from Figure 3.13.

As an application of the above theoretical results, we consider once more the experimentally obtained caustics of Figure 3.12a the first of which (*i*) corresponds to light rays reflected from the front and rear faces of an edge-cracked polycarbonate of-bisphenol-A tension specimen, whereas the second (*ii*) corresponds to light rays traversing the same specimen. The material properties of polycarbonate of bisphenol A were found to be $E = 28000 \text{ kp/cm}^2$, $\nu = 0.36$, $c_r = -2.20 \cdot 10^{-5} \text{ cm}^2/\text{kp}$,

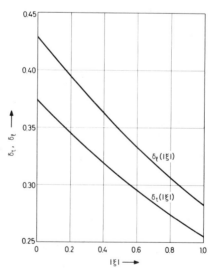

Figure 3.13. Variation of the functions $\delta_{t,l}(|\xi|)$ with $|\xi|$ in a birefringent medium.

$c_t = -1.40 \cdot 10^{-5}$ cm^2/kp and $|\xi| = 0.153$. Furthermore, the distances z_0 and z_i for the reflected light rays (the screen lying in this case in front of the specimen) were: $z_0 = 126$ cm and $z_i = 18$ cm. Hence, from equation (3.15) we deduce that $\lambda_m = 8.00$ in this case. Moreover, from the curves of Figure 3.13 (or, more accurately, through direct computations) we find that for $|\xi| = 0.153$ $\delta_t(|\xi|) = 0.3524$ and $\delta_l(|\xi|) = 0.4031$. As regards the corresponding diameters of the external caustic of Figure 3.12a(i) on the screen, they were found to be: $D_t = 6.1$ cm, and $D_l = 5.6$ cm. Under these conditions and with the help of equations (3.39) (with $d = 0.3$ cm for the present specimen and $c = c_r = -2.20 \cdot 10^{-5}$ cm^2/kp as already mentioned) and (3.89) the experimental value of the mode I stress intensity factor K_I at the crack tip can be calculated: $K_I = 215.39$ kpcm$^{-3/2}$ when the calculation is based on the transverse diameter D_t of the caustic or $K_I = 198.98$ kpcm$^{-3/2}$ when the calculation is based on the longitudinal diameter D_l of the caustic. By taking further into account that the length of the edge crack is $a = 1$ cm, the width of the specimen $w = 4.98$ cm and its thickness $d = 0.3$ cm and the tensile load $P = 160$ kp, we can evaluate the average stress component normal to the crack: $\sigma_m = P/(wd) = 107.1$ kp/cm^2 and further from the formula $K_I = 1.1215\, \sigma_m (\pi a)^{1/2}$ [6] the theoretical value of the mode I stress intensity factor K_I at the edge crack tip: $K_I = 212.83$ kp cm$^{-3/2}$. This value is clearly seen to be in very good agreement with the corresponding already mentioned experimental values for the same factor. Similarly, for the light rays traversing the specimen we have: $z_0 = 147$ cm, $z_i = 18$ cm, $\lambda_m = 9.17$, $c_t = -1.40 \cdot 10^{-5}$ cm^2/kp (as already mentioned), $D_t = 4.54$ cm and $D_l = 4.2$ cm, all other quantities of interest keeping their previous values. The experimentally obtained values of K_I were thus seen to be $K_I = 225.38$ kpcm$^{-3/2}$ (when the calculation was based on D_t) and $K_I = 199.80$ kpcm$^{-3/2}$ (when the calculation was based on D_l). These values are also in satisfactory agreement with the theoretical value $K_I = 212.83$ kpcm$^{-3/2}$, evidently being the same as in the previous case. Finally, it can be mentioned that the mean value of the experimental estimations of K_I is $K_I = 209.9$ kpcm$^{-3/2}$ and its discrepancy from the theoretical value $K_I = 212.83$ kpcm$^{-3/2}$ is less than 1.5%.

3.6 The case of anisotropic media

The method of caustics can also be used in the case of mechanically anisotropic media. This case was investigated in detail in reference [23]. By confining ourselves to the caustics formed from reflections of the

light rays from the front face of the specimen, in which case equation
(3.21) is valid, and taking into account the theory of anisotropic elasticity
for a medium under generalized plane stress conditions, the stress field
in such a medium characterized by two complex potentials $\Phi_1(z_1)$ and
$\Phi_2(z_2)$, where

$$z_1 = x + s_1 y, \quad z_2 = x + s_2 y, \tag{3.90}$$

s_1 and s_2 being characteristic quantities of the material of the anisotropic
medium under consideration, we can find, after a reasoning analogous to
that of section 3.3, that the equation of the initial curve of the caustic of
the specimen is given by [23]

$$\{1 - D \operatorname{Re}\left[\delta_1 \Phi_1''(z_1) + \delta_2 \Phi_2''(z_2)\right]\}\{1 - D \operatorname{Re}\left[\delta_1 s_1^2 \Phi_1''(z_1) + \delta_2 s_2^2 \Phi_2''(z_2)\right]\}$$
$$= \{D \operatorname{Re}\left[\delta_1 s_1 \Phi_1''(z_1) + \delta_2 s_2 \Phi_2''(z_2)\right]\}^2, \tag{3.91}$$

where

$$D = 2z_0 \, d/\lambda_m \tag{3.92}$$

and the constants δ_1 and δ_2 depend on the properties of the material of
the anisotropic medium, like s_1 and s_2. Moreover, the equations of the
caustic on the specimen can be seen to have the forms [23]

$$W_x = \lambda_m \{x - D \operatorname{Re}\left[\delta_1 \Phi_1'(z_1) + \delta_2 \Phi_2'(z_2)\right]\},$$
$$W_y = \lambda_m \{y - D \operatorname{Re}\left[\delta_1 s_1 \Phi_1'(z_1) + \delta_2 s_2 \Phi_2'(z_2)\right]\}. \tag{3.93}$$

Equations (3.91) and (3.93) correspond to equations (3.43), valid in the
case of isotropic elastic media.

On the basis of equations (3.91) and (3.93) the equations of the initial
curve of the caustic and the caustic itself about a crack tip in the special
case of an orthotropic elastic medium have been derived in reference
[23]. These equations are sufficiently more complicated than in the case
of isotropic elastic media and will not be given here. Here we give in
Figure 3.14a two typical forms of caustics about a crack tip in an
orthropic (polycarbonate of bisphenol A) elastic specimen, having being
submitted to an appropriate mechanical treatment [23], the direction of
the crack being along one (i) or the other (ii) of the two principal
directions of the orthotropic medium. These caustics have been obtained

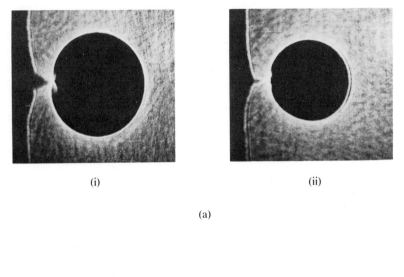

(i)   (ii)

(a)

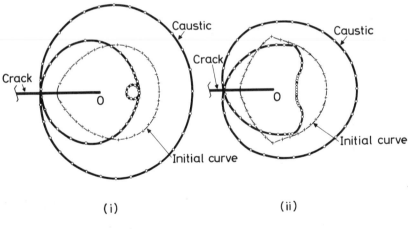

(i)   (ii)

(b)

Figure 3.14. Typical experimental (a) and theoretical (b) forms of the caustics (from reflected light rays) at a crack tip in a tension specimen made of an orthotropic material under mode *I* conditions.

from reflections of the light rays from the front face of the specimen, the focus of the light beam formed also behind the specimen. It is only in this case that equations (3.91) and (3.93) are valid. Finally, in Figure 3.14b the theoretical forms of the caustics corresponding to the experimentally obtained caustics of Figure 3.14a have been drawn, for the complete 4π-interval for the polar angle ϑ, together with their initial curves. A considerable similarity between the experimental and the theoretical forms of the caustics is observed from Figures 3.14a and 3.14b, although it is also seen that the shapes of the caustics in the present case, of an orthotropic elastic medium, are somewhat different than the standard shape of a caustic at a crack tip in an isotropic elastic medium.

Finally, on the basis of equations (3.90–3.93) it is possible to derive convenient formulas for the estimation of the stress intensity factors at crack tips as was made in section 4 in the case of isotropic elastic materials. This was made in reference [23] in the case of orthotropic elastic materials and the values of the stress intensity factors at the tips of the cracks of Figure 3.14 have been estimated on the basis of the material properties, the constants of the optical arrangement and, evidently, the dimensions of the caustics. Since orthotropic (and generally anisotropic) materials appear rather rarely in practice and also the whole procedure for the estimation of stress intensity factors at crack tips in such materials does not differ essentially from that presented in section 3.4 and applicable to the case of isotropic elastic materials, we will not give here the details, which can be found in reference [23].

3.7   Interacting crack problems

The method of caustics can further be applied to the determination of the stress intensity factors at crack tips approaching other crack tips or boundaries of the specimen. Several problems of interaction of cracks with other cracks or boundaries have been considered in references [24–30]. In these problems we can distinguish two special cases: In the first case the crack tip at which we want to determine the stress intensity factor does not lie very near another crack or boundary. In this case, the formulas derived in section 3.4 are applicable. This was the case in the crack problems considered in references [24–28]. On the other hand, in the second case the crack tip lies very near another crack tip (or boundary) and the shape of the caustic around this crack tip is influenced by the other crack (or the boundary). In this case the exact formulas for

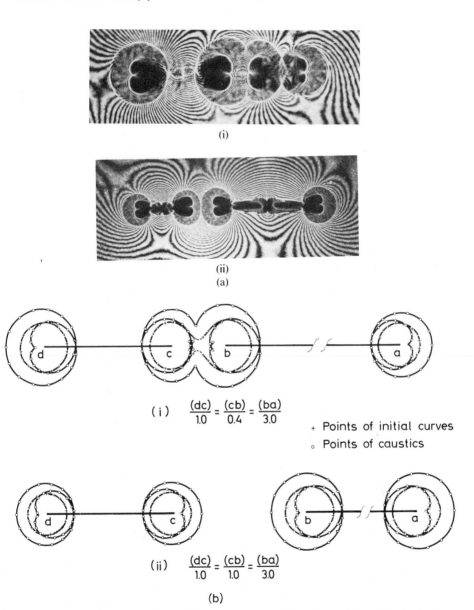

(i)

(ii)
(a)

(i) $\dfrac{(dc)}{1.0} = \dfrac{(cb)}{0.4} = \dfrac{(ba)}{3.0}$

+ Points of initial curves
∘ Points of caustics

(ii) $\dfrac{(dc)}{1.0} = \dfrac{(cb)}{1.0} = \dfrac{(ba)}{3.0}$

(b)

Figure 3.15. Typical forms of experimentally (a), as well as theoretically (b), obtained caustics at interacting crack tips in a tension specimen under mode I conditions.

P. S. Theocaris

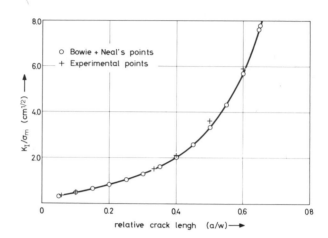

Figure 3.16. Comparison of experimentally and theoretically obtained reduced values K_I/σ_m of the stress intensity factor K_I at an edge-cracked, finite-width tension specimen under mode I conditions loaded by a tensile loading σ_m for various values of the ratio a/w.

the complex potential $\Phi(z)$ have to be used so that the theoretical forms of the caustics be in agreement with the corresponding experimental forms. This was made in references [29–30] for the cases of interacting straight cracks along a straight line.

Here we present in Figures 3.15 typical forms of experimentally obtained caustics at interacting cracks in plexiglas specimens. We can observe from Figure 3.15(i) that it is possible that only one caustic be formed around two interacting crack tips instead of two separate caustics as happens in Figure 3.15(ii).

As an application we present in Figure 3.16 the values of the mode I stress intensity factor K_I at the tip of an edge crack of length a in a tension specimen of width w reduced to the applied tensile stress σ_m (of direction normal to the direction of the crack). The experiments were made on plexiglas specimens and equation (3.76) has been used for the determination of the stress intensity factor K_I. In the same figure the theoretical values for K_I/σ_m are also presented as obtained from the paper by Bowie and Neal [87]. The agreement between the experimental and the theoretical results for K_I/σ_m is seen to be excellent. The interaction between the edge crack and the boundary of the specimen (when the crack tip approaches the boundary, that is a/w increases) is clearly seen from Figure 3.16.

3.8   Branched crack problems

The method of caustics for the determination of complex (mixed-mode) stress intensity factors can also be successfully applied to the determination of stress intensity factors at branched crack tips [31–33]. The method of section 3.4 for the determination of stress intensity factors remains applicable in this case, where it is not very easy to obtain the values of the stress intensity factors through theoretical techniques.

In Figure 3.17 we see typical forms of the caustics created at the tips of an asymmetrically branched crack. Furthermore, in Figure 3.18 a comparison between the theoretically and experimentally obtained values for the stress intensity factors K_{IA}, K_{IB}, K_{IIB}, K_{IC} and K_{IIC} at the tips A, B and C of the branched crack of this figure (under tensile loading normal to the main crack OA), reduced to the fundamental stress intensity factor $K_0 = \sigma(2a)^{1/2}$, are presented for several values of the ratio c/b of the lengths c and b of the crack branches OC and OB respectively under the assumption that $a/b = 4$ (where a is the length of the main branch OA), as well as for several values of the angle ϑ_c between the branch OC and the direction of extension of the branch OA. It can be observed from Figure 3.18 that there exists a good agreement between the experimental results for the stress intensity factors and the corresponding theoretical results, the latter obtained by using the Cauchy type singular integral equations method together with the Gauss-Legendre numerical integration rule [33].

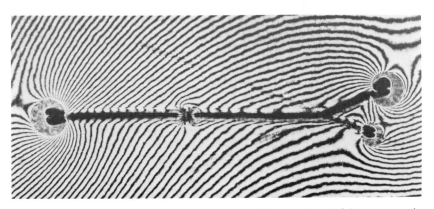

Figure 3.17. Typical forms of experimentally obtained (from reflected light rays) caustics at the tips of a branched crack in a plexiglas specimen under tension of direction normal to the main branch.

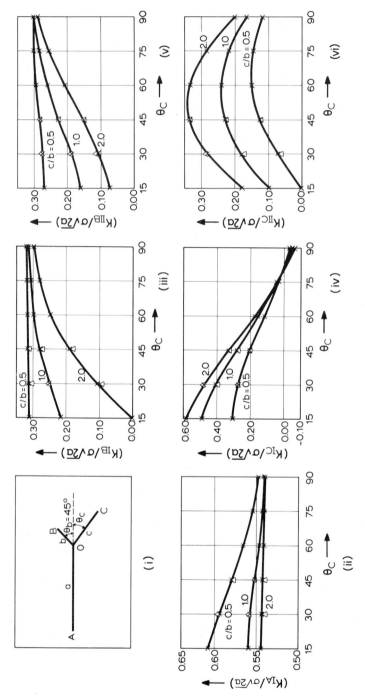

Figure 3.18. Geometry of the branched crack and variation of the reduced values of the stress intensity factors at the crack tips A, B and C with the angle ϑ_c under tensile loading normal to the branch OA (for $a/b = 4$) (The experimental values are denoted by the symbol $\triangle$ and the theoretical values by the symbol $\times$).

230

3.9 Interface crack problems

One further interesting application of the method of caustics is its application to crack problems with the crack lying along the interface of two plane isotropic elastic media, the elastic properties of which are described by the constants (μ_1, κ_1) and (μ_2, κ_2) [1]. The special cases when this crack is a straight crack along the straight boundary of two isotropic elastic half-planes or along the circular-arc-shaped boundary of an isotropic elastic inclusion inside an infinite isotropic elastic medium have been studied in references [34] and [35] respectively.

The results of these references remain valid even in the general case of a curvilinear crack of arbitrary shape along the interface, of arbitrary shape too, of two plane isotropic elastic media. This happens since the asymptotic behaviors for the complex potentials $\Phi_{1,2}(z)$ inside the two elastic media near a crack tip A are always of the form [35]

$$\Phi_1(z) = \frac{1-\beta}{2} K z^{-1/2-i\delta}, \quad \Phi_2(z) = \frac{1+\beta}{2} K z^{-1/2-i\delta}, \tag{3.94}$$

where K plays the rôle of a generalized stress intensity factor at the crack tip A and the constants β and δ are determined by

$$\beta = \frac{\mu_2(\kappa_1 - 1) - \mu_1(\kappa_2 - 1)}{\mu_2(\kappa_1 + 1) + \mu_1(\kappa_2 + 1)}, \tag{3.95}$$

$$\delta = \frac{1}{2\pi} \ln \frac{\mu_1 + \mu_2\kappa_1}{\mu_2 + \mu_1\kappa_2}. \tag{3.96}$$

By inserting the expressions (3.94) of the complex potentials $\Phi_{1,2}(z)$ into the equations (3.43) of the caustics, the formulas describing the geometry of the caustics formed at an interface crack tip A can be easily determined as was already made in references [34–35]. Of course, there are two such caustics near the crack tip A since two isotropic elastic media surround this point. Moreover, since all dimensions of these caustics depend on the generalized complex stress intensity factor K at the crack tip A, this factor can be directly determined experimentally following the developments of reference [35]. Evidently, in this case the formulas used are more complicated than the corresponding formulas valid for the case of a simple crack tip inside a homogeneous isotropic elastic medium.

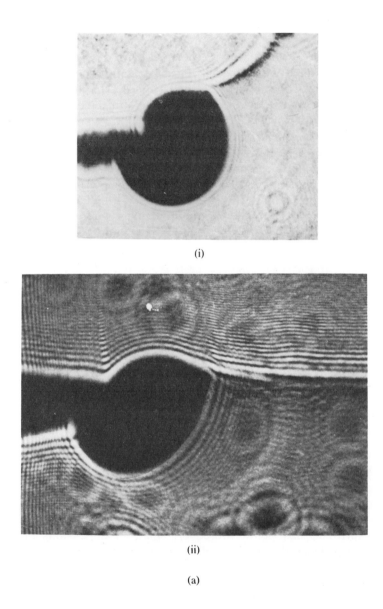

(i)

(ii)

(a)

Figure 3.19. Typical experimental (a) and theoretical (b) forms of caustics at the tip of a crack lying along a circular (i) or straight (ii) interface of two plane isotropic elastic media.

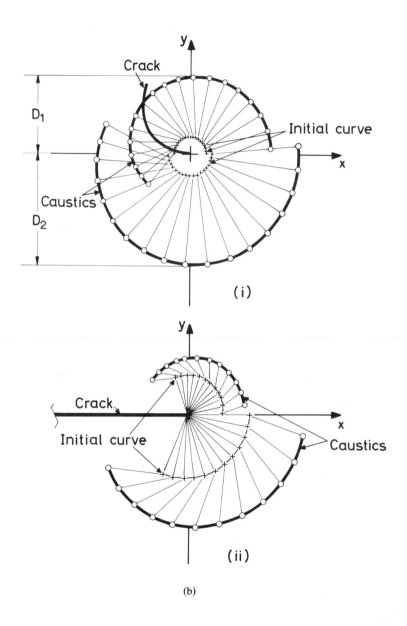

Figure 3.19. Continued.

In Figure 3.19(*i*) typical forms of the caustics formed at the crack tip *A* of a crack along the interface of a circular inclusion of a C-100-0-8 epoxy inside a matrix made of plasticized epoxy of the type C-100-40-8 are presented. These caustics have been seen to be in accordance with their expected theoretical forms in the cases under consideration shown also in Figure 3.19(*i*). Analogous theoretical and experimental typical forms of caustics for the case of a crack along a straight interface are shown in Figure 3.19(*ii*).

As an application, we can consider the experimentally obtained caustic of Figure 3.19a(*i*). The mechanical properties (Young's modulus and Poisson's ratio) of the material of the inclusion are: $E_1 = 33000\,\mathrm{kp/cm^2}$, $\nu_1 = 0.34$ whereas those of the matrix are: $E_2 = 18000\,\mathrm{kp/cm^2}$, $\nu_2 = 0.43$. The corresponding shear moduli, μ_1 and μ_2 respectively, result to be $\mu_1 = 12313\,\mathrm{kp/cm^2}$ and $\mu_2 = 6294\,\mathrm{kp/cm^2}$. Also the elastic constants κ_1 and κ_2, under generalized plane stress conditions, where the formula $\kappa = (3 - \nu)/(1 + \nu)$ holds, are found to be: $\kappa_1 = 1.9851$, $\kappa_2 = 1.7972$. Hence, the bi-elastic constants α and β [35] of the combination of these two materials are: $\alpha = -0.2997$ and $\beta = -0.0692$. Furthermore, the angle corresponding to the circular arc of the crack was $\vartheta = 45°$ and the specimen was loaded by a tensile loading $\sigma = 23.40\,\mathrm{kp/cm^2}$ of direction normal to the tangent of the circular crack at its middle-point. The experimental determination of the generalized stress intensity factor $K = K_I - iK_{II}$ of equations (3.94) was achieved by measuring the radii D_1 and D_2 of the caustics of Figure 3.19b(*i*). By use of the formulas and nomograms of reference [35], it was found that in the present case, when $D_1/D_2 = 0.60$, K resulted experimentally to be: $K = (10.02 - 3.69(i))\,\mathrm{kp/cm^2}$, the corresponding theoretical value being $K = (10.36 - 3.91(i))\,\mathrm{kp/cm^2}$. The agreement between the experimental and the theoretical value of K was thus seen to be very good.

3.10   *V*-notch problems

In the case of existence of a *V*-notch in a plane isotropic elastic medium, the stress components present a singularity at the *V*-notch tip of order λ, where $\lambda = -0.5$ when the *V*-notch degenerates into a crack and $\lambda \to 0$ when the *V*-notch degenerates into a straight boundary. In any case, the variation of the order of singularity λ of the stress field near the apex of a *V*-notch with the angle ϑ of the *V*-notch is well-known, evaluated for the first time by Williams [88].

This singularity at the *V*-notch apex causes the creation of a caustic.

The equations of this caustic can be easily found on the basis of the theoretical results of section 3.3 and, particularly, equations (3.43). Now in the neighborhood of the V-notch apex the complex potential $\Phi(z)$ of Muskhelishvili can be considered behaving like

$$\Phi(z) = Kz^\lambda, \quad -0.5 \leqq \lambda < 0, \tag{3.97}$$

where the constant K can be called the generalized stress intensity factor at the V-notch tip although it is not directly related to the stress intensity factor at a crack tip. By differentiating twice equation (3.97) with respect to z and replacing the values of $\Phi'(z)$ and $\Phi''(z)$ in equations (3.43), we can easily find that the initial curve of the caustic has a circular shape although it is not a complete circle but just a circular arc restricted by the boundaries of the V-notch. The radius of the initial curve is given by

$$r = r_0 = |CK\lambda(\lambda - 1)|^{1/(2-\lambda)}, \tag{3.98}$$

where C is given once more by equation (3.39). Moreover, the caustic on the screen is determined by

$$W = X + iY = \lambda_m[z + C\bar{K}\lambda\bar{z}^{\lambda-1}]. \tag{3.99a}$$

If we further take into account equation (3.98), we have

$$W = X + iY = \lambda_m r_0\left\{\exp(i\vartheta) + \frac{\varepsilon}{1-\lambda}\exp[i((1-\lambda)\vartheta + \omega)]\right\}, \tag{3.99b}$$

where the constants ε and ω are given by equations (3.57) and (3.55) once more. Evidently, for $\lambda = -\frac{1}{2}$, as happens at a crack tip, the form of the caustic at a V-notch tip coincides with the form (3.56) of the caustic at a crack tip determined in section 3.4.

As regards the experimental determination of the generalized stress intensity factor K_I at the V-notch tip, we can make use of the formula

$$K_I = \delta(\varphi)(D_t/\lambda_m)^{2-\lambda}/|C|, \tag{3.100}$$

completely analogous to equation (3.76), where $2(\pi - \varphi)$ is the angle of the V-notch (Figure 3.20), D_t the transverse diameter of the caustic and $\delta(\varphi)$ a function of the angle φ determined on the basis of equations (3.97–3.99) and presented in Figure 3.20 together with the order of the singularity $\lambda = \lambda(\varphi)$ [88]. We observe from Figure 3.20 that when $\varphi \to 90°$ $\lambda(\varphi) \to 0$, whereas when $\varphi \to 0°$ $\lambda(\varphi) \to -0.5$. Moreover, for $\varphi \to 0°$ $\delta(\varphi) \to 0.074511 = 0.373544/[2(2\pi)^{1/2}]$, where 0.373544 is the value used in equation (3.76) for crack problems.

In Figure 3.21 an experimentally obtained caustic about a V-notch tip in an isotropic elastic plexiglas specimen is presented together with the corresponding theoretical forms of the caustic and its initial curve. The specimen was a rectangular plexiglas specimen of width $w = 4$ cm, depth of the V-notch $a = 0.4$ cm, angle of the notch 270° ($\varphi = 45°$) and thickness of the specimen $d = 0.2$ cm. The tensile loading was $P = 80$ kp and the corresponding average tensile stress component resulted to be $\sigma_m = 100$ kp/cm². Furthermore, for the creation of the caustic, the diameter of which was found to be $D_t = 3.25$ cm, the screen was placed in front of the specimen and the caustic due to reflections on the rear face of the specimen (exterior caustic) has been used for the measurement of D_t. The distance of the focus of the light beam from the specimen was $z_i = 19.4$ cm and the distance of the screen from the specimen was $z_0 = 121$ cm. Thus from equation (3.15) the magnification ratio λ_m resulted to be $\lambda_m = 7.237$. Finally, the overall constant C of the experimental arrangement resulted from equation (3.39) to be $C = 4.33 \cdot 10^{-4}$ cm⁴/kp.

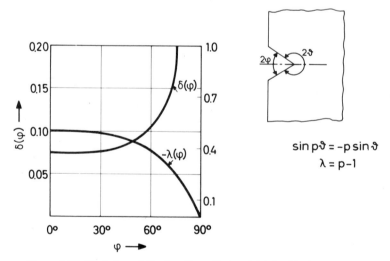

Figure 3.20. Variation of the functions $\delta(\varphi)$ and $\lambda(\varphi)$ with the angle φ.

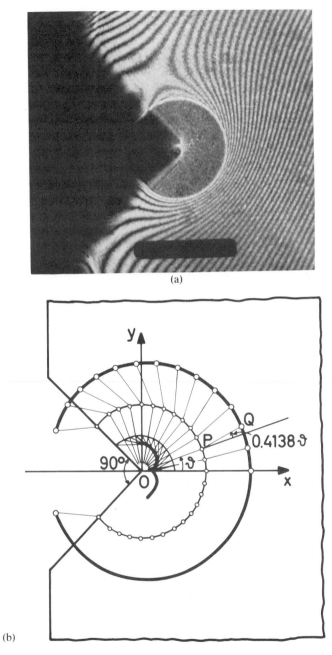

(a)

(b)

Figure 3.21. Experimental (a) and theoretical (b) form of the caustic at the tip of a V-notched tension specimen under mode I conditions ($\varphi = 45°$).

Under these circumstances, the experimental value for K_I was determined from equation (3.100) with the help of Figure 3.20 to be $K_I = 27.40 \, \text{kpcm}^{-3/2}$ ($\delta(\varphi) = 0.0848$, $\lambda(\varphi) = -0.4555$). The corresponding theoretical value for K_I was derived from the approximate formula of reference [89], the results of reference [90] taken also into account, and found to be $K_I = 28.22 \, \text{kpcm}^{-3/2}$. The difference between the theoretical and the experimental value for K_I was thus found to be 2.9%. This difference was due both to the experimental errors and to the approximate formulas used for the evaluation of the theoretical value for K_I.

Analogous results can also be obtained for the case of mixed-mode stress intensity factors. In this case the corresponding results of section 3.4 can be directly generalized to be applicable, under appropriate modifications, to the estimation of generalized stress intensity factors at V-notch tips. For example, in the mixed-mode case equation (3.100) holds for the absolute value of K.

3.11 Shell problems

The method of caustics can also be applied to cases when the cracked specimen is not a plane specimen under generalized plane stress conditions. Such is the case of shell problems, where, of course, the material of the shell is assumed to be isotropic and elastic. In this case, although the principles of the method of caustics, presented in section 3.2, remain valid, nevertheless, the formulas deduced in sections 3.3 and 3.4 are not valid any more. In shell problems one has to take into account that the surfaces of the shell are not plane both before and after loading; their curvatures have to be taken into account.

In references [39, 40] crack problems in cracked cylindrical shells under internal pressure [39] or under tension [40] have been considered. The shapes of the obtained caustics have been studied in detail in these references and satisfactory agreement between their theoretical and experimental forms was observed. Moreover, the method of determining the stress intensity factors at the crack tips from the corresponding caustics was presented. Unfortunately, the formulas for the caustics in cracked shell problems are more complicated than the corresponding formulas valid in generalized plane stress problems and they will not be repeated here. Here we present only a typical caustic in a cracked cylindrical shell under tension (the crack being normal to the tension direction) both in its theoretical and experimental form (Figure 3.22).

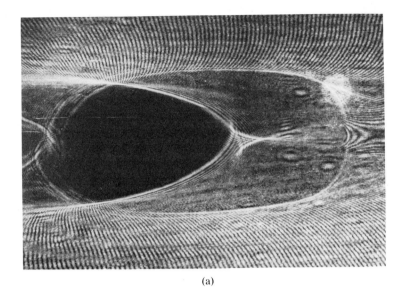

(a)

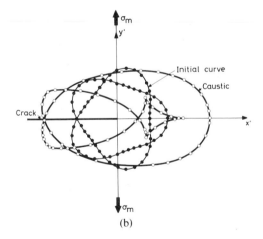

(b)

Figure 3.22. Experimental (a) and theoretical (b) form of the caustic at the tip of a cracked cylindrical shell under mode *I* tensile loading.

From this figure a satisfactory agreement between theoretical and experimental results can be observed.

In fact, in this case the radius of the cylindrical shell was $R = 10$ cm, its thickness $d = 0.232$ cm, the optical constants of the material of the specimen for light rays reflected from its rear face $c_r = -3.28 \cdot 10^{-5}$ cm^2/kp and $|\xi_r| = 0.153$, the tensile loading intensity $\sigma_m = 85.35$ kp/cm^2, the distance of the screen from the specimen (the screen lying in front of the specimen) $z_0 = 139.6$ cm and the magnification ratio of the optical set-up $\lambda_m = 4.825$. Moreover, the transverse diameter of the caustic on the screen was found to be $D_t = 5.4$ cm and the corresponding polar angle ϑ on the screen $|\vartheta| = 74°$. From these quantities and by using the procedure of determining the mode I stress intensity factor K_I at a crack tip described in reference [40], the experimental value of this factor was found to be: $K_I = 218.47$ kpcm$^{-3/2}$. The corresponding theoretical values of K_I can be found by using the method of Folias or the method of Erdogan and Kibler reported in reference [40]. The values of K_I determined by using these methods are: $K_I = 206.74$ kpcm$^{-3/2}$ and $K_I = 191.40$ kpcm$^{-3/2}$ respectively. The discrepancies between the theoretical values of K_I and its experimental value, reported previously, are thus seen to be small.

3.12   Plate problems

As a final application of the method of caustics to crack problems and to the determination of stress intensity factors at cracks tips, we consider the problem of a cracked thin isotropic elastic plate under bending. This problem has been recently considered in detail in reference [41], where both the classical (Kirchhoff's) and the more advanced (Reissner's) theories of bending of thin plates have been used.

In Figure 3.23 the standard theoretical form of the caustic formed about the crack tip (together with its initial curve) are presented under the assumptions that only the mode I stress intensity factor K_I is present, that the Reissner theory is valid and that the Poisson ratio ν of the isotropic elastic material is equal to $\nu = 0.34$. In the same figure the corresponding experimental form of the caustic is shown for comparison purposes. Of course, the screen was placed in front of the thin bent plate and at a distance z_0 from it.

By taking into account the theoretical developments of reference [41], we can see that the value of the stress intensity factor K_I at a crack tip

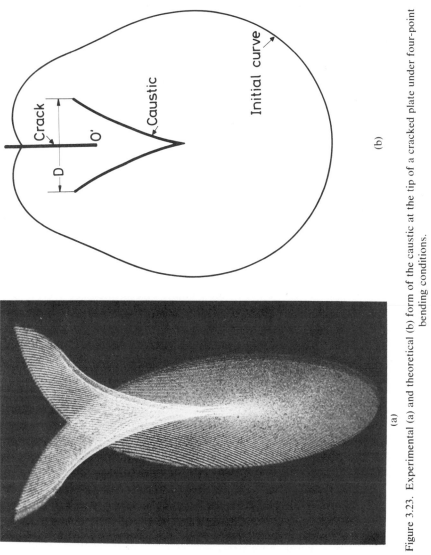

(a)

(b)

Figure 3.23. Experimental (a) and theoretical (b) form of the caustic at the tip of a cracked plate under four-point bending conditions.

in a symmetrically bent isotropic elastic thin plate can be determined by

$$K_I = \mu d(2D|\lambda_m|)^{1/2}/[8z_0(1 - \nu)], \qquad (3.101)$$

where μ is the shear modulus of the elastic material, d the thickness of the thin plate, λ_m' the magnification ratio of the optical set-up, determined once more by equation (3.15), and D the distance of the endpoints of the caustic on the screen as shown in Figure 3.23b.

In the case of the experimentally obtained caustic of Figure 3.23, the cracked thin plate was subjected to four-point bending of direction normal to the plane of the crack. This kind of bending produced a bending moment per unit width of the plate equal to $M = 0.374$ kp. Moreover, the material of the plate was plexiglas of thickness $d = 0.2$ cm with $\nu = 0.34$ and $\mu = 12300$ kp/cm^2 and the half-length of the crack was $a = 1.5$ cm. Also the distance z_0 of the screen from the specimen (the screen lying in front of the specimen) was $z_0 = 108$ cm whereas the distance z_i of the focus of the light beam impinging on the specimen (this focus formed behind the specimen) was $z_i = -20$ cm. Hence, from equation (3.15) the value of the magnification ratio λ_m resulted to be $\lambda_m = -4.4$. As regards the value of the distance D, it was found experimentally to be $D = 4.6$ cm. Then the value of the stress intensity factor was found from equation (3.101) to be equal to $K_I = 27.45$ kpcm$^{-3/2}$. On the other hand, the value of K_I can also be determined theoretically from the formula [41]

$$K_I = \frac{1 + \nu}{3 + \nu} \frac{6M}{d^2} a^{1/2}, \qquad (3.102)$$

which yields $K_I = 27.57$ kpcm$^{-3/2}$. The agreement between the experimental and the theoretical value of K_I was thus seen to be excellent.

3.13   Other applications

In the previous sections the method of caustics was applied to the determination of stress intensity factors at crack tips inside elastic media. Besides this category of problems, the method of caustics has been applied to several more problems of elasticity, viscoelasticity and plasticity either associated with cracks or not. Among them we can mention:

(i) Plane isotropic elasticity problems not associated with cracks but with circular holes [42] and elliptical holes [43]: Stress concentration factors and other quantities of interest similar to the stress intensity factors in cracks can be determined by the method of caustics in such problems.

(ii) Inclusion problems in plane elasticity [44–46]: Triangular inclusions, rectangular and other inclusions with cusp-points or not have been studied by the method of caustics.

(iii) Distributed loads along the boundary of a half-plane [47–52], where stress singularities of the order of one appear in the stress field. In this case both the caustics and pseudocaustics (that is the images of the boundary of the half-plane on the screen) formed are taken into account for the determination of the normal and shear loading distribution along the half-plane. Moreover, the method of pseudocaustics has been used as an experimental method for the solution of the general (first, second and mixed) fundamental problems of plane elasticity [51].

(iv) Crack problems in viscoelastic media [53–54]: In this case, equations analogous to those derived in the previous sections are seen to describe the shapes of the caustics. Nevertheless, in the case of viscoelastic media the magnitudes of the caustics formed change with time although their form remains almost unaltered.

(v) Crack problems in elastic-perfectly plastic media [55–58]. In this case the well-known Dugdale-Barrenblatt model is valid, either in its original form or properly modified, and has been used for the determination of the shapes of the caustics formed at crack-tips. The method of caustics has also been used in corrosion crack problems [59].

(vi) Dynamic crack problems [60–65]: The method of caustics has also been used for the experimental study of the initiation of cracks, acceleration, deccelaration and crack arrest phenomena and the accurate and complete determination of stress intensity factors at the tips of propagating cracks. Several phenomena in such a propagation, like bifurcation of a crack or interaction of a crack with an interface or another crack, have been studied in the dynamic case.

(vii) Elastic contact problems [66–68]: In this case, the method of caustics, together with the method of pseudocaustics, can be used for the determination of several quantities of interest and, under special circumstances, for the complete solution of the considered contact problem.

(viii) Elastic plate problems [69–71]: The method of caustics and pseudocaustics was also applied to several thin isotropic elastic plate problems, like circular, triangular and rectangular plate problems under uniform loading or loading by bending moments. A lot of information can be gathered in these problems from the shapes of the caustics and pseudocaustics formed.

(ix) Problems associated with reflectors [72–76]: All three cases of ellipsoidal, paraboloidal and hyperboloidal reflectors have been considered and the caustics formed have been studied in detail. Several interesting conclusions have been also drawn from the forms of these caustics.

(x) Three-dimensional elasticity problems [77]: Particular attention was paid to the problems of loading of an isotropic elastic half-space with a concentrated load, a uniformly distributed load over a circular region or a rigid circular punch. The results obtained for two-dimensional elasticity problems can easily be generalized to the three-dimensional case.

(xi) Finally, the method of caustics was applied to the direct experimental determination of the Poisson ratio ν of an isotropic elastic material and the stress-optical constants c_r and c_t of the same material [78–79]. Specimens with holes under out-of-plane bending and tension respectively have been used for the experimental determination of these quantities.

3.14 Discussion

The method of caustics constitutes an efficient experimental technique for the complete solution of a wide class of problems or the determination of various quantities of interest in these problems. Confining ourselves to the object of this chapter, that is the experimental determination of stress intensity factors at crack tips, we can compare the method of caustics, in the case of plane isotropic elastic media, with photoelasticity and interferometry. Such a comparison, made in reference [19], reveals that from the theoretical point of view the method of caustics is more accurate than both these methods in the general case of mixed-mode stress intensity factors. This is mainly due to the fact that the method of caustics ignores the constant part of the stress field near a crack tip because of the differentiations of the complex potential $\Phi(z)$ made in equation (3.43) of the caustics. This advantage of the method of caustics over both photoelasticity and interferometry seems not to have been sufficiently realized up to now.

Furthermore, in the special case of mode I stress intensity factors, the photoelastic method is sufficiently improved if we use it in its two-parameter variations, where the constant part of the stress field is taken into account as always achieved in the method of caustics. Thus, the first two terms in the asymptotic expansions of the stress components near the crack tip are taken into account. But, in this special case the method of caustics can also be improved so that all four first terms in the asymptotic expansions of the stress components near a crack tip are taken into account. To achieve this, we have just to determine the radius r_0 of the initial curve of the caustic, entering equation (3.79) for the determination of the mode I stress intensity factor K_I at a crack tip, not from equation (3.72) or equation (3.73) but from the following equation [19]

$$r_0 = 0.7236\,D_l - 0.3693\,D_t. \tag{3.103}$$

Thus both the longitudinal and the transverse diameter of the caustic are taken into account for the determination of K_I.

As regards the experimental point of view, it should be emphasized that the method of caustics provides a very clear and well-defined curve around the crack tip, the caustic, on which measurements can be made with a high accuracy. This is not the case in the (holographic) interferometric and especially in the photoelastic method. Moreover, in the method of caustics all information is gathered from the initial curve of the caustic, lying in the close vicinity of the crack tip. This is not the case with photoelasticity and (holographic) interferometry, where usually extrapolation techniques are used near the crack tip for the determination of the stress intensity factor (see e.g. [91]). This is also clearly seen in Figure 3.24, where we observe that the caustic is formed in the close vicinity of a crack tip even in an interferometric fringe pattern. This means that we cannot obtain information in the close vicinity of the crack tip from this pattern; only the method of caustics is capable to provide information in this region, which is of much interest in fracture mechanics.

Also with the magnification ratio λ_m, the caustic on the screen can have sufficient dimensions so that measurements on it can be easily made. Of course, it is accepted that triaxiality of the stress field near the

Figure 3.24. Interferometric pattern (isopachics) about a crack tip in a tension specimen under mode *I* conditions. The caustic appears as a black region in the detail of the pattern (courtesy of Dr. T. D. Dudderar).

crack tip, plasticity and other factors influencing the experimental determination of the stress intensity factor at a crack tip by the photo-elastic or the interferometric method have analogous influence on the experimental determination of this factor by the method of caustics. But this remark does not reduce the foregoing advantages of the method of caustics, which make it ideal for the experimental determination of stress intensity factors at crack tips. On the other hand, the definition of such a factor has nothing common either with triaxiality in the stress field or with plasticity.

The simple way in which an experiment by the method of caustics can be made for the experimental determination of the stress intensity factor at a crack tip, by using the simple formula (3.76), permits the wide use of this method in practical applications where cracked specimens are encountered. Of course, in several simple cases a stress intensity factor at a crack tip can be found in some paper or compendium on stress intensity factors obtained through theoretical considerations. In these cases, the determination of the same stress intensity factor by using the method of caustics can be used to confirm the available theoretical solutions, particularly on the question of how permissible is it to replace the real cracked specimen by the ideal specimen for which the theoretical stress intensity factors have been obtained. For example, the real specimen may be of finite dimensions or with geometrical imperfections, whereas the corresponding theoretical specimen may be considered infinite and without any geometrical imperfection.

But in the majority of practical cases under non-idealized geometry and boundary conditions, no values for the stress intensity factors are available in the literature. Evidently, even in these cases, these factors can be estimated approximately by using theoretical methods like the finite element method or the much more efficient Cauchy type singular integral equation method. But these methods require the use of a computer (and of computer time too . . .), as well as the existence or construction of an appropriate computer program. In such cases the method of caustics can be preferred over these theoretical techniques and provide in a short time and with almost no special experimental apparatuses the required value of the stress intensity factor. Evidently, if desired, the experimental results of the method of caustics can be confirmed by the foregoing theoretical techniques.

3.15 Conclusions

From the above developments it is concluded that:

(i) The method of caustics constitutes an efficient method of experimental determination of stress intensity factors at crack tips, without restrictions on geometry, material properties and loading conditions.

(ii) The accuracy of this method is sufficient (better than 5% in most cases) and, in any case, much better than the accuracy of relevant experimental techniques, like photoelasticity and interferometry.

(iii) The method of caustics constitutes the most appropriate technique for the determination of stress intensity factors in all practical

248

P. S. Theocaris

applications where no theoretical results are available in the literature.

References

[1] Muskhelishvili, N. I., *Some Basic Problems of the Mathematical Theory of Elasticity* (4th edition), P. Noordhoff, Groningen (1963).
[2] Westergaard, H. M., Bearing pressures and cracks, *Journal of Applied Mechanics*, 6, pp. A49–A53 (1939).
[3] Tada, H., Paris, P. C. and Irwin, G. R., *The Stress Analysis of Cracks Handbook*, Del Research Corporation, Hellertown, Pennsylvania (1973).
[4] Rooke, D. P. and Cartwright, D. J., *Compendium of Stress Intensity Factors*, Her Majesty's Stationery Office, London (1976).
[5] Paris, P. C. and Sih, G. C., Stress analysis of cracks, in: *Fracture Toughness Testing and Its Applications*, The American Society for Testing and Materials, Special Technical Publication 381, pp. 30–83 (1965).
[6] Sih, G. C. (editor), *Methods of Analysis and Solutions of Crack Problems* (*Mechanics of Fracture*, Vol. 1), Noordhoff, Leyden, The Netherlands (1973).
[7] Cartwright, D. G. and Rooke, D. P., Evaluation of stress intensity factors, *Journal of Strain Analysis*, 10, pp. 217–224 and 259–263 (1975).
[8] Ioakimidis, N. I., *General Methods for the Solution of Crack Problems in the Theory of Plane Elasticity*; Doctoral thesis at the National Technical University of Athens, Athens (1976) [Available from Univ. Micr. Int., order no. 76–21,056].
[9] Ioakimidis, N. I. and Theocaris, P. S., The second fundamental crack problem and the rigid line inclusion problem in plane elasticity, *Acta Mechanica*, 34, pp. 51–61 (1979).
[10] Ioakimidis, N. I. and Theocaris, P. S., The numerical evaluation of a class of generalized stress intensity factors by use of the Lobatto–Jacobi numerical integration rule, *International Journal of Fracture*, 14, pp. 469–484 (1978).
[11] Theocaris, P. S., Local yielding around a crack tip in plexiglas, *Journal of Applied Mechanics*, 37, pp. 409–415 (1970).
[12] Theocaris, P. S., A study of constrained zones in cracked plates by the optical method of caustics, *Proceedings of the Academy of Athens*, 46, pp. 116–130 (1971).
[13] Theocaris, P. S., Reflected shadow method for the study of constrained zones in cracked plates, *Applied Optics*, 10, pp. 2240–2247 (1971).
[14] Theocaris, P. S. and Joakimides, N., Some properties of generalized epicycloids applied to fracture mechanics, *Zeitschrift für angewandte Mathematik und Physik* (*ZAMP*), 22, pp. 876–890 (1971).
[15] Theocaris, P. S. and Gdoutos, E., An optical method for determining opening-mode and edge sliding-mode stress-intensity factors, *Journal of Applied Mechanics*, 39, pp. 91–97 (1972).
[16] Theocaris, P. S., Discussion on the paper: 'A new technique for viewing deformation zones at crack tips, by Hoeppner, D., Danford, V. and Pettit, D. E. (*Experimental Mechanics* 11, pp. 280–283 (1971))', *Experimental Mechanics*, 12, pp. 247–249 (1972).
[17] Theocaris, P. S., An optical method for the determination of constrained zones at crack-tips, Proceedings of the International Symposium of Experimental Mechanics, Waterloo, Canada, June 1972, pp. 511–530 (1972).

[18] Theocaris, P. S., Comments on previous discussion on the paper: 'The determination of mode *I* stress-intensity factors by holographic interferometry, by Dudderar, T. D. and Gorman, H. J. (*Experimental Mechanics*, 13, pp. 145-149 (1973))' *Experimental Mechanics* 15, pp. 150-152 (1975).

[19] Theocaris, P. S. and Ioakimidis, N. I., An improved method for the determination of mode *I* stress intensity factors by the experimental method of caustics, *Journal of Strain Analysis* 14 (1979), 14, pp. 111-118 (1980).

[20] Theocaris, P. S., Determination of crack-opening displacement by the method of caustics, *Journal of Strain Analysis*, 9, pp. 197-205 (1974).

[21] Theocaris, P. S., The reflected-shadow method for the study of the constrained zones in cracked birefringent media, *Journal of Strain Analysis*, 7, pp. 75-83 (1972).

[22] Theocaris, P. S. and Papadopoulos, G. A., Stress intensity factors from reflected caustics in birefringent plates with cracks, *Journal of Strain Analysis*, 16, pp. 29-36 (1981).

[23] Theocaris, P. S., Stress concentrations in anisotropic plates by the method of caustics, *Journal of Strain Analysis*, 11, pp. 154-160 (1976).

[24] Theocaris, P. S., The spread of plastic zones between symmetric edge cracks, *Israel Journal of Technology*, 8, pp. 367-373 (1970).

[25] Theocaris, P. S., The constrained zones in collinear asymmetric cracks by the method of caustics, Proceedings of the 7th All-Union Conference on Photoelasticity, Tallinn, U.S.S.R., November 1971, Vol. 2, pp. 221-236 (1971).

[26] Theocaris, P. S., Interaction of cracks with other cracks or boundaries, *Materialprüfung*, 13, pp. 264-269 (1971).

[27] Theocaris, P. S., Interaction of cracks with other cracks or boundaries, *International Journal of Fracture Mechanics*, 8, pp. 37-47 (1972).

[28] Theocaris, P. S., Interaction between collinear asymmetric cracks, *Journal of Strain Analysis*, 7, pp. 186-193 (1972).

[29] Theocaris, P. S., Constrained zones in a periodic array of collinear equal cracks, *International Journal of Mechanical Sciences*, 14, pp. 79-94 (1972).

[30] Theocaris, P. S., A theoretical consideration of the constrained zones in an array of interacting collinear and asymmetric cracks, *Acta Mechanica*, 17, pp. 169-189 (1973).

[31] Theocaris, P. S., Complex stress-intensity factors at bifurcated cracks, *Journal of the Mechanics and Physics of Solids*, 20, pp. 265-279 (1972).

[32] Theocaris, P. S. and Blonzou, C. H., Symmetric branching of cracks in PMMA (Plexiglas), *Materialprüfung*, 15, pp. 123-130 (1973).

[33] Theocaris, P. S., Asymmetric branching of cracks, *Journal of Applied Mechanics*, 44, pp. 611-618 (1977).

[34] Theocaris, P. S., Partly unbonded interfaces between dissimilar materials under normal and shear loading, *Acta Mechanica*, 24, pp. 99-115 (1976).

[35] Theocaris, P. S. and Stassinakis, C. A., Experimental solution of the problem of a curvilinear crack in bonded dissimilar materials, *International Journal of Fracture*, 13, pp. 13-26 (1977).

[36] Theocaris, P. S., Plastic strains at the roots of sharp notches in perspex, in: *Experimental Stress Analysis and Its Influence on Design*, Proceedings of the 4th International Conference on Stress Analysis, Cambridge, England, 1970, pp. 513-523 (1970).

[37] Theocaris, P. S., Stress and displacement singularities near corners, *Zeitschrift für angewandte Mathematik und Physik (ZAMP)*, 26, pp. 77-98 (1975).

250 P. S. Theocaris

[38] Theocaris, P. S., Singularities at the vertices of multiwedges, *Monographs of the Serbian Academy of Sciences and Arts*, 497, No. 11, pp. 1–38 (1976).
[39] Theocaris, P. S., Intensity factors of stationary cracks in cylindrical shells under internal pressure, Proceedings of the First International Conference on Structural Mechanics in Reactor Technology, Berlin 1972, Vol. 4, Part G, Paper G6/7, pp. 487–505 (1972).
[40] Theocaris, P. S. and Thireos, C. G., Stress intensity factors in cracked cylindrical shells under tension, *International Journal of Fracture*, 12, pp. 691–703 (1976).
[41] Theocaris, P. S., Symmetric bending of cracked plates studied by caustics, *International Journal of Mechanical Sciences*, 21, pp. 659–670 (1979).
[42] Theocaris, P. S., Optical stress rosette based on caustics, *Applied Optics*, 12, pp. 380–387 (1973).
[43] Theocaris, P. S., The evolution of caustics from holes to cracks, *International Journal of Solids and Structures*, 12, pp. 377–389 (1976).
[44] Theocaris, P. S. and Paipetis, S. A., Constrained zones at singular points of inclusion contours, *International Journal of Mechanical Sciences*, 18, pp. 581–587 (1976).
[45] Theocaris, P. S. and Paipetis, S. A., State of stress around inhomogeneities by the method of caustics, *Fibre Science and Technology*, 9, pp. 19–39 (1976).
[46] Theocaris, P. S. and Paipetis, S. A., Constrained zones at singular points of inclusion contours by the method of caustics, Proceedings of the XVII-th Polish Solid Mechanics Conference, Szczyrk, Poland, 3–9 September 1975, pp. 212–213 (1975).
[47] Theocaris, P. S., Stress-singularities at concentrated loads, *Experimental Mechanics*, 13, pp. 511–518 (1973).
[48] Theocaris, P. S., Stress singularities due to uniformly distributed loads along straight boundaries, *International Journal of Solids and Structures*, 9, pp. 655–670 (1973).
[49] Theocaris, P. S. and Razem, C., Deformed boundaries determined by the method of caustics, *Journal of Strain Analysis*, 12, pp. 223–232 (1977).
[50] Theocaris, P. S. and Razem, C., The end-values of distributed loads in half-planes by caustics, *Journal of Applied Mechanics*, 45, pp. 313–319 (1978).
[51] Theocaris, P. S., Ioakimidis, N. I. and Razem, C., The method of pseudocaustics for the experimental solution of simple elasticity problems, *International Journal of Mechanical Sciences*, 23, pp. 17–29 (1981).
[52] Theocaris, P. S. and Razem, C., Intensity, slope and curvature discontinuities in loading distributions at the contact of two plane bodies, *International Journal of Mechanical Sciences*, 21, pp. 339–353 (1979).
[53] Theocaris, P. S., The method of caustics for the study of cracked plates made of viscoelastic materials, Proceedings of the 3rd International Conference of Fracture, Munchen, Germany, Vol. VI, Paper No. 512, pp. 1–5 (1973).
[54] Theocaris, P. S., Method of caustics for the study of cracked plates made of visco-elastic materials, *International Journal of Mechanical Science*, 16, pp. 855–865 (1974).
[55] Theocaris, P. S., Stress intensity factors in yielding materials by the method of caustics, *International Journal of Fracture*, 9, pp. 185–197 (1973).
[56] Theocaris, P. S. and Gdoutos, E., Verification of the validity of the Dugdale-Barenblatt model by the method of caustics, *Engineering Fracture Mechanics*, 6, pp. 523–535 (1974).
[57] Theocaris, P. S. and Gdoutos, E. E., The modified Dugdale-Barenblatt model adapted to various fracture configurations in metals, *International Journal of Fracture*, 10, pp. 549–564 (1974).

[58] Theocaris, P. S. and Gdoutos, E. E., The size of plastic zones in cracked plates made of polycarbonate, *Experimental Mechanics*, 15, pp. 169–176 (1975).

[59] Theocaris, P. S. and Papadopoulos, G. A., Stress corrosion crack-growth in aluminum alloys, *Engineering Fracture Mechanics*, 9, pp. 781–794 (1977).

[60] Katsamanis, F., Raftopoulos, D. and Theocaris, P. S., The dependence of crack velocity on the critical stress in fracture, *Experimental Mechanics*, 17, pp. 128–132 (1977).

[61] Katsamanis, F., Raftopoulos, D. and Theocaris, P. S., Static and dynamic stress intensity factors by the method of transmitted caustics, *Journal of Engineering Materials and Technology*, 99, pp. 105–109 (1977).

[62] Theocaris, P. S., Dynamic behavior of polymers studied by reflected caustics, Proceedings of the First National Symposium of Tensometry, Iaşi, Roumania, 25–28 April 1977, pp. 207–221 (1977).

[63] Theocaris, P. S. and Katsamanis, F., Response of cracks to impact by caustics, *Engineering Fracture Mechanics*, 10, pp. 197–210 (1978).

[64] Theocaris, P. S., Dynamic propagation and arrest measurements by the method of caustics on overlapping skew-parallel cracks, *International Journal of Solids and Structures*, 14, pp. 639-653 (1978).

[65] Theocaris, P. S. and Milios, J., Dynamic crack propagation in composites, *International Journal of Fracture*, 16, pp. 31–51 (1980).

[66] Theocaris, P. S., The contact problem by the method of caustics, Proceedings of the Third Bulgarian National Congress of Theoretical and Applied Mechanics, Varna, September 13–16, 1977, Vol. 1, pp. 263–268 (1977).

[67] Theocaris, P. S. and Stassinakis, C. A., The elastic contact of two disks by the method of caustics, *Experimental Mechanics*, 18, pp. 409–415 (1978).

[68] Theocaris, P. S., Experimental study of plane elastic contact problems by the pseudocaustics method, *Journal of the Mechanics and Physics of Solids*, 27, No. 1, pp. 15–32 (1979).

[69] Theocaris, P. S. and Gdoutos, E. E., Experimental solution of flexed plates by the method of caustics, *Journal of Applied Mechanics*, 44, pp. 107–111 (1977).

[70] Theocaris, P. S., Caustics created from simply supported plates under uniform loading, *International Journal of Solids and Structures*, 13, pp. 1281–1291 (1977).

[71] Theocaris, P. S. and Ioakimidis, N. I., The equations of caustics for plate problems in bending by using complex potentials, *Theoretical and Applied Mechanics [Teoretichna i Prilozhna Mekhanika]*, 10, No. 2, pp. 48–51 (1979).

[72] Theocaris, P. S. and Gdoutos, E. E., Surface topography by caustics, *Applied Optics*, 15, pp. 1629–1638 (1976).

[73] Theocaris, P. S. and Gdoutos, E. E., Distance measuring based on caustics, *Applied Optics*, 16, pp. 722–728 (1977).

[74] Theocaris, P. S., Properties of caustics from conic reflectors. 1: Meridional rays, *Applied Optics*, 16, pp. 1705–1716 (1977).

[75] Theocaris, P. S., Basic properties of meridional and sagittal caustics from conic reflectors, *Transactions of the Academy of Athens*, 40, pp. 1–52 (1977).

[76] Theocaris, P. S., The topography of entirely or locally curved surfaces studied by caustics, Scientific publication of the National Technical University of Athens, Athens, pp. 1–67 (1977).

[77] Theocaris, P. S. and Ioakimidis, N. I., The method of caustics for the determination of normal loads acting on surfaces of elastic bodies, *Journal of Strain Analysis*, 15, No. 1, pp. 37–41 (1980).

[78] Theocaris, P. S. and Ioakimidis, N. I., On the determination of stress-optical constants by the method of reflected caustics, *Journal of Physics D: Applied Physics*, 12, pp. 497–504 (1979).

[79] Ioakimidis, N. I. and Theocaris, P. S., An optical method for the experimental determination of the Poisson ratio, *Materialprüfung*, 21, pp. 40–42 (1979).

[80] Theocaris, P. S., New methods based on geometric optics for the solution of fracture mechanics problems, *Technika Chronika*, 41, pp. 145–153 (1972).

[81] Theocaris, P. S., The method of caustics for the study of forms and deformations Report of the National Technical University of Athens, Athens, pp. 1–80 (1977).

[82] Theocaris, P. S., The method of caustics for the study of singular elastic fields, Proceedings of the 18th Polish Solid Mechanics Conference, Wisla-Jawornik, Poland, 7–14 September 1976, Polish Academy of Sciences, pp. 209–213 (1976).

[83] Theocaris, P. S., The study of singular stress fields by the method of caustics (Studiul cîmpurilor de tensiuni singulare cu ajutorul causticelor), in: *Analiza Experimentalá a Tensiunilor* (Mocanu, D. R., editor) (*Experimental Analysis of Stresses*), Vol. II, Chap. VI, pp. 567–637 (1977).

[84] Theocaris, P. S., The method of caustics—A powerful experimental method for the study of singular stress fields (Metoda Kaustyk—Nową metodą doświadczalną badania osobliwości pola naprężenia), *Engineering Transactions* (*Rozprawy Inżynierskie*), 26, pp. 131–178 (1978).

[85] Theocaris, P. S., The method of caustics applied to elasticity problems, in: *Developments in Stress Analysis* (G. Holister, editor), Applied Science Publishers, England, chapt. 2, pp. 27–63 (1979).

[86] Theocaris, P. S., Dependence of stress-optical coefficients on the mechanical and optical properties of polymers, *Journal of Strain Analysis*, 8, pp. 267–276 (1973).

[87] Bowie, O. L. and Neal, D. M., Single edge crack in rectangular tensile sheet, *Journal of Applied Mechanics*, 32, pp. 708–709 (1965).

[88] Williams, M. L., Stress singularities resulting from various boundary conditions in angular corners of plates in extension, *Journal of Applied Mechanics*, 19, pp. 526–528 (1952).

[89] N. I. Ioakimidis and P. S. Theocaris, A note on stress intensity factors for single edge V-notched plates in tension, *Engineering Fracture Mechanics*, 10, pp. 685–686 (1978).

[90] P. S. Theocaris and N. I. Ioakimidis, The V-notched elastic half-plane problem, *Acta Mechanica*, 32, pp. 125–140 (1979).

[91] Dudderar, T. D. and Gorman, H. J., The determination of mode I stress-intensity factors by holographic interferometry, *Experimental Mechanics*, 13, pp. 145–149 (1973).

G. Villarreal and G. C. Sih

4 | Three-dimensional photoelasticity: stress distribution around a through thickness crack

4.1 Introduction

The presence of a flaw or mechanical imperfection such as crack or a sharp notch in a stressed body produces a high elevation of stress in the surrounding region of the material. The redistribution of stresses and strains around the crack has been the subject of many past discussions both experimentally and analytically. Much of the work, however, has been limited to two-dimensional stress analyses because of the complexity of three-dimensional elasticity problems.

Post [1] introduced the application of photoelasticity to two-dimensional cracked plates. His work consisted of investigating the stress distribution around an edge crack subjected to tension. He used the interferometric isopachic technique for measuring the secondary principal stress sum $(\sigma_1 + \sigma_2)$ and the isochromatic pattern for the difference $(\sigma_1 - \sigma_2)$. The need for an experimental study of the three-dimensional character of the crack tip stress state was pointed out. Fessler and Mansell [2] studied the stress distribution in a thick plate with an edge crack using the frozen stress technique. No significant variation of stresses through the thickness was found. More recently, solutions of approximate plate theories have been obtained for plates with through cracks under bending and extension [3–6]. The problem of the bending of cracked plates has been solved by Hartranft and Sih [7] who used Reissner's plate bending theory. Smith and Smith [8] carried out a three-dimensional, frozen stress, photoelastic investigation to verify the Hartranft–Sih solution. It was concluded that the Hartranft–Sih solution agrees remarkably well with experiments, especially after precatastrophic crack growth has taken place. Using Irwin's method (suggested in a discussion of [9]) of obtaining stress-intensity factors from photoelastic data, Smith and Smith [10] performed experiments on other cracked specimens. These involved specimen configurations such as finite strips under remote tensile loading. They found that the values of the stress-intensity factors are dependent on whether data are taken very close to the crack tip or not since material imperfections in the crack tip region can have a strong influence on the results. Sih [11] has pointed out that in any analytical or experimental work, a core region surrounding the crack tip should always be eliminated from consideration on physical grounds. A discussion of this can be found in the work of Kipp and Sih [12]. An experimental study of the crack tip stress variation through the plate thickness in a bending problem was made by Mullinix and Smith [13] who verified Sih's [4] solution. Using the stress freezing and slicing technique the parameters which are involved in stress-intensity factor expressions were evaluated for a plate with finite thickness. It was found that the function $f_1^*(z)$ proposed by Sih [4] to describe the stress variation through the thickness agrees very well with the experimental data for thin to moderately thick plates.

It is the purpose of this investigation to experimentally test one of the assumptions of a recent modified version of the theory of generalized plane stress [14]. The form postulated by the theory for the stress variation through the thickness of a plate containing a crack will be compared with that obtained by three-dimensional photoelastic analysis. Specimens covering the range from thin to moderately thick plates were examined by the frozen stress technique. The experimentally measured transverse variation of the in-plane stress components σ_x and σ_y was in excellent agreement with that postulated by the theory.

4.2   Hartranft–Sih plate theory

Figure 4.1 considers a through crack of length 2a in a plate whose boundaries are sufficiently far away from the crack such that they do not affect the local stresses. The surfaces of the plate at $z = \pm h/2$ are free of tractions such that

$$\sigma_z = \tau_{xz} = \tau_{yz} = 0 \quad \text{for } z = \pm h/2 \tag{4.1}$$

In order to reduce the equations of elasticity to two-dimensions, the

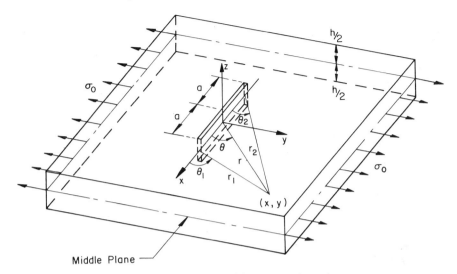

Figure 4.1. Plate containing a through crack.

following assumptions on the stress state are made in [14]:

$$(\sigma_x, \sigma_y, \tau_{xy}) = \frac{4}{h^2} f''(\zeta)(S_x, S_y, T_{xy}) \tag{4.2a}$$

$$(\tau_{xz}, \tau_{yz}) = -\frac{2}{h} f'(\zeta)(Z_x, Z_y) \tag{4.2b}$$

$$\sigma_z = f(\zeta)Z_z \tag{4.2c}$$

in which S_x, S_y, T_{xy}, Z_x, Z_y and Z_z are independent of z. The functions $f(\zeta)$, $f'(\zeta)$, and $f''(\zeta)$ depend on $\zeta = 2z/h$ only, and prescribe the transverse variation of the stresses.

The approximate differential equations and boundary conditions for the functions S_x, S_y, etc., are determined by requiring the complementary energy of the system to be as small as possible with the admitted equilibrium state of stress of equations (4.2). This enables the reduction of the three-dimensional equations of elasticity to a system of equations involving only the two variables x and y. Since the stresses in the z direction must satisfy the free surface conditions stated in equations (4.1), it follows that

$$f(\pm 1) = f'(\pm 1) = 0 \tag{4.3}$$

Applying the variational principle of the complementary energy of the system, a set of differential equations for the three generalized displacements and the six stress functions, i.e., u_x, u_y, u_z, S_x, S_y, etc., is obtained:

$$\alpha^4[1 + \beta^2/(1 - \nu^2)]\nabla^4 u_z - 2\alpha^2\nabla^2 u_z + u_z = 0 \tag{4.4a}$$

$$Z_x - \alpha^2\nabla^2 Z_x = \frac{1}{\alpha^2}\frac{\partial N}{\partial x} \tag{4.4b}$$

$$Z_y - \alpha^2\nabla^2 Z_y = \frac{1}{\alpha^2}\frac{\partial N}{\partial y} \tag{4.4c}$$

with the implicit relationship

$$\alpha^4(1 + \beta^2\kappa^2)\left(\frac{\partial Z_x}{\partial x} + \frac{\partial Z_y}{\partial y}\right) = (1 - \nu)D[(1 - \nu)u_z + \nu\alpha^2\nabla^2 u_z] \tag{4.5}$$

in which N stands for

$$N = \frac{D}{1 + \beta^2\kappa^2}[(1 - \nu)(u_z - \alpha^2\nabla^2 u_z) + \alpha^2\beta^2\kappa^2\nabla^2 u_z] \tag{4.6}$$

The parameter κ is equal to $[(1 - \nu)/(1 + \nu)]^{1/2}$ equations are

$$u_x = -\frac{\partial u_z}{\partial x} + \frac{2\alpha^2}{(1 - \nu)D}Z_x \tag{4.7a}$$

$$u_y = -\frac{\partial u_z}{\partial y} + \frac{2\alpha^2}{(1 - \nu)D}Z_y \tag{4.7b}$$

for the displacements and

$$S_x = -D\left(\frac{\partial^2 u_z}{\partial x^2} + \nu\frac{\partial^2 u_z}{\partial y^2}\right) + \frac{\alpha^2}{1 - \nu}\left[(2 - \nu)\frac{\partial Z_x}{\partial x} + \nu\frac{\partial Z_y}{\partial y}\right] \tag{4.8a}$$

$$S_y = -D\left(\frac{\partial^2 u_z}{\partial y^2} + \nu\frac{\partial^2 u_z}{\partial x^2}\right) + \frac{\alpha^2}{1 - \nu}\left[(2 - \nu)\frac{\partial Z_y}{\partial y} + \nu\frac{\partial Z_x}{\partial x}\right] \tag{4.8b}$$

$$T_{xy} = -D(1 - \nu)\frac{\partial^2 u_z}{\partial x\partial y} + \alpha^2\left(\frac{\partial Z_x}{\partial y} + \frac{\partial Z_y}{\partial x}\right) \tag{4.8c}$$

for the stress functions. In equations (4.7) and (4.8), $D = Eh^3/12(1 - v^2)$ is the flexural rigidity of the plate. The parameters α and β are given by

$$\alpha^2 = \frac{h^2}{6} I_2, \quad \beta^2 = \frac{3I_1}{2I_2^2} - 1 \tag{4.9}$$

in which I_1 and I_2 represent the integrals

$$I_1 = \int_{-1}^{1} [f(\zeta)]^2 \, d\zeta, \quad I_2 = \int_{-1}^{1} [f'(\zeta)]^2 \, d\zeta \tag{4.10}$$

The function $f(\zeta)$ is normalized so that

$$\int_{-1}^{1} [f''(\zeta)]^2 \, d\zeta = \tfrac{3}{2} \tag{4.11}$$

Let the surfaces of the crack be pressurized such that the plate is stretched symmetrically about xz and yz planes. Hence it is sufficient to formulate the problem in the quarter plane $x > 0$ and $y > 0$. The mixed boundary conditions on the edge $y = 0$ are

$$u_y(x, 0) = 0, \quad \text{for } x > a \tag{4.12a}$$

$$S_y(x, 0) = -S_0 P(x), \quad \text{for } x < a \tag{4.12b}$$

$$T_{xy}(x, 0) = Z_y(x, 0) = 0, \quad \text{for all } x \tag{4.12c}$$

Moreover, the displacements and stresses are required to be finite as $(x, y) \to \infty$.

The method of integral transforms was used to solve equations (4.4) together with (4.12) leading to the generalized displacement expressions

$$u_x(x, y) = \frac{2}{\pi D} \int_0^\infty \alpha^2 \Big\{ 2\alpha^2 \beta m s^2 \exp(-msy)$$

$$- \frac{1}{1 - v} \operatorname{Im}[(1 + i\beta\kappa)(q_0/p_0) \exp(-p_0 sy)] \Big\} sA(s) \sin(sx) \, ds \tag{4.13a}$$

$$u_y(x, y) = \frac{2}{\pi D} \int_0^\infty \alpha^2 \Big\{ 2\alpha^2 \beta s^2 \exp(-msy)$$

$$- \frac{1}{1 - v} \operatorname{Im}[(1 + i\beta\kappa)q_0 \exp(-p_0 sy)] \Big\} sA(s) \cos(sx) \, ds \tag{4.13b}$$

$$u_z(x, y) = \frac{2}{\pi(1-\nu)D} \int_0^\infty \alpha^2 A(s) \operatorname{Im}[(1-i\beta\kappa)$$
$$\times (q_0/p_0) \exp(-p_0 sy)] \sin(sx) \, ds \qquad (4.13c)$$

from which the stress functions in equations (4.8) may be obtained. The quantities m, p_0 and q_0 appearing in equations (4.13) are given by

$$m = \left[1 + \frac{1}{(\alpha s)^2}\right]^{1/2} \qquad (4.14a)$$

$$q_0 = \sqrt{1-\nu^2}\,[1 + (1 + i\beta\kappa)\alpha^2 s^2] \qquad (4.14b)$$

$$p_0 = |p_0| \exp(-i\vartheta/2) \qquad (4.14c)$$

in which

$$|p_0| = \left[\frac{(1-\nu^2)m^4 + \beta^2}{1-\nu^2+\beta^2}\right]^{1/4} \qquad (4.15)$$

such that

$$\vartheta = \tan^{-1}\left[\frac{\beta\sqrt{1-\nu^2}}{1-\nu^2+(\beta^2+1-\nu^2)\alpha^2 s^2}\right], \quad 0 \le \vartheta < \pi/2 \qquad (4.16)$$

The foregoing expressions are derived on the basis that β is real and hence the inequality

$$I_1 > \tfrac{2}{3} I_2^2 \qquad (4.17)$$

The conditions specified in equations (4.12) lead to a system of dual integral equations which can be solved to yield

$$A(s) = -\frac{\pi(1-\nu^2+\beta^2)S_0 a}{(1-\nu^2)\alpha^2\beta(1+\beta^2)s^2} \left\{\Phi(1)J_1(sa) - \int_0^1 \frac{d}{d\xi}\left[\frac{\Phi(\xi)}{\sqrt{\xi}}\right]J_1(sa\xi)\,d\xi\right\} \qquad (4.18)$$

If the applied load on the crack is uniform with magnitude, say σ_0, then S_0 can be written as

$$S_0 = \frac{h^2}{6}\sigma_0 \int_{-1}^1 f''(\zeta)\,d\zeta \qquad (4.19)$$

The function $\Phi(\xi)$ satisfies the Fredholm integral equation

$$\Phi(\xi) + \int_0^1 F(\xi, \eta)\Phi(\eta)\,d\eta = \frac{2}{\pi}\sqrt{\xi}\int_0^\xi \frac{P(a, \eta)\,d\eta}{\sqrt{\xi^2 - \eta^2}}, \quad 0 < \xi < 1 \qquad (4.20)$$

whose kernel is

$$F(\xi, \eta) = \sqrt{\xi\eta}\int_0^\infty sg\left(\frac{s}{a}\right)J_0(\xi s)J_0(\eta s)\,ds, \quad 0 < \xi \leq 1; 0 < \eta \leq 1 \qquad (4.21)$$

The function $g(s) = 0(s^2)$ as $s \to \infty$ and takes the form

$$g(s) = \frac{2\sqrt{1 - \nu^2}\,(1 - \nu^2 + \beta^2)}{(1 - \nu^2)(1 + \beta^2)\alpha^2\beta s^2|p_0|}\left\{-2\alpha^4\beta\kappa ms^2|p_0|\right.$$
$$+ [(1 + \alpha^2 s^2)^2 - (\beta\kappa)^2(\alpha s)^4]\sin\frac{\vartheta}{2}$$
$$\left. + [2\kappa(1 + \alpha^2 s^2)\alpha^2\beta s^2]\cos\frac{\vartheta}{2}\right\} - 1 \qquad (4.22)$$

In the above equations, J_1 and J_0 are first and zero order Bessel functions of the first kind.

4.3 Triaxial crack border stress field

The singular character of the stresses near the crack region can be obtained from the first term within the braces in equation (4.18). The other term does not contribute to the singular behavior of the local stress field. By means of a Laurent series expansion and the use of some Bessel integral identities, the following stress field was obtained:

$$\sigma_x = \frac{k_p(z)\sqrt{a}}{\sqrt{2r_1}}\left(\cos\frac{\theta_1}{2} - \frac{1}{2}\sin\theta_1\sin\frac{3\theta_1}{2}\right) + O(r_1^0) \qquad (4.23a)$$

$$\sigma_y = \frac{k_p(z)\sqrt{a}}{\sqrt{2r_1}}\left(\cos\frac{\theta_1}{2} + \frac{1}{2}\sin\theta_1\sin\frac{3\theta_1}{2}\right) + O(r_1^0) \qquad (4.23b)$$

$$\tau_{xy} = \frac{k_p(z)\sqrt{a}}{\sqrt{2r_1}}\left(\frac{1}{2}\sin\theta_1\cos\frac{3\theta_1}{2}\right) + O(r_1^0) \qquad (4.23c)$$

$$\sigma_z = \frac{k_t(z)\sqrt{a}}{\sqrt{2r_1}}\cos\frac{\theta_1}{2} + O(r_1^0) \qquad\qquad (4.23d)$$

$$\tau_{xz} = \tau_{yz} = O(r_1^0) \quad \text{as } r_1 \to 0 \qquad\qquad (4.23e)$$

where r_1 and θ_1 are defined in Figure 4.1.
The stress intensity factors $k_p(z)$ and $k_t(z)$ are given by

$$k_p(z) = \frac{2}{3}\Phi(1)f''(\zeta)\sigma_0\sqrt{a}\int_{-1}^{1} f''(\zeta)\,\mathrm{d}\zeta \qquad\qquad (4.24a)$$

$$k_t(z) = -\frac{2}{3}\,\Phi(1)p^2f''(\zeta)\sigma_0\sqrt{a}\int_{-1}^{1} f''(\zeta)\,\mathrm{d}\zeta \qquad\qquad (4.24b)$$

Note that the stress intensity factors in equations (4.23) take the general form

$$k_1(z) = \Phi(1)\sigma(z)\sqrt{a}$$

Numerical values of $\Phi(1)$ as a function of h/a are displayed in Figure 4.2 for different values of p and $\nu = 0.3$. To be noted also is that the stress

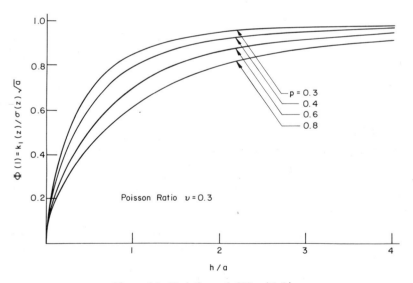

Figure 4.2. Variations of $\Phi(1)$ with h/a.

intensity factors k_p and k_t in equations (4.24) are not independent because the stress-state in the middle of the plate, $|z| < (1 - \varepsilon)h/2$, is required to satisfy the condition of plane strain [5]:

$$\sigma_z = \nu(\sigma_x + \sigma_y) \tag{4.25}$$

The above condition leads to the differential equation

$$f''(\zeta) + p^2 f(\zeta) = 0 \tag{4.26}$$

solving for the function $f(\zeta)$ that governs the stress distribution through the plate thickness. A solution of equation (4.26) is

$$f(\zeta) = A \cos p\zeta, \quad |\zeta| < 1 - \varepsilon \tag{4.27}$$

which together with the constraint given by equation (4.11) yields

$$A^2 = \frac{3}{2} \frac{1}{p^4} \left[1 + \frac{\sin 2p}{2p} \right]^{-1} \tag{4.28}$$

Equation (4.26) is valid outside a boundary layer of thickness ε next to the plate surface. Hartranft and Sih [14] have defined this thickness as

$$\epsilon \frac{h}{2} = \frac{h/2}{2 + (8h/a)} \tag{4.29}$$

Inside the boundary layer, $f(\zeta)$ is given in the following representation

$$\begin{aligned}
f(\zeta) = &- A(1 - u)^2[(1 + 2u) \cos p(1 - \varepsilon) \\
&- up\varepsilon \sin p(1 - \varepsilon)], \quad (1 - \varepsilon) < \zeta < 1
\end{aligned} \tag{4.30}$$

in which $u = (\zeta - 1 + \varepsilon)/E$. The restriction imposed on $f''(\zeta)$ in the boundary layer is that $f''(\zeta)$ and its derivatives must be continuous at $\zeta = 1 - \varepsilon$. It is found from equation (4.26) that

$$f''(\zeta) = Ap^2 \cos p\zeta, \quad 0 < \zeta < 1 \tag{4.31}$$

4.4   Experimental considerations: specimens and materials

The stresses in equations (4.23) refer to stresses applied on the crack surfaces. These equations must be modified when applied to the experimental specimen in Figure 4.1 in which the loads are on the plate edges at distances away from the crack being free from surface tractions. This can be accomplished by superimposing a uniform stress field σ_0, to the plate with no crack to equations (4.23). In this way, the complete stress field with a free crack surface is obtained. For a beam of light at normal incidence with the plane xy, the only stresses which produce photoelastic effects are those components which define the secondary principal stresses perpendicular to the z axis. Refer to page 333 of Vol. II [15]. The secondary principal stresses, σ_1 and σ_2, perpendicular to the z axis are the principal stresses related to the stress components σ_x, σ_y, τ_{xy} as

$$(\sigma_1, \sigma_2)_z = \frac{\sigma_x + \sigma_y}{2} \pm \sqrt{\left(\frac{\sigma_x - \sigma_y}{2}\right)^2 + \tau_{xy}^2} \tag{4.32}$$

The principal stress difference becomes

$$(\sigma_1 - \sigma_2)_z = 2\sqrt{\left(\frac{\sigma_x - \sigma_y}{2}\right)^2 + \tau_{xy}^2} \tag{4.33}$$

Neglecting the terms of $O(r_1^0)$ in equations (4.23) and substituting the local stresses into equation (4.33), it is found that

$$(\sigma_1 - \sigma_2)_z = \sigma_0 \left\{ \left[\frac{k(z)}{\sqrt{2r_1}} \sin \theta_1 + \sin \frac{3\theta_1}{2} \right]^2 + \left[\cos \frac{3\theta_1}{2} \right]^2 \right\}^{1/2} \tag{4.34}$$

where

$$k(z) = \frac{k_p(z)}{\sigma_0} \tag{4.35}$$

Using equations (4.24), the secondary principal stress difference given by (4.34) may also be solved for $f''(\zeta)$. That substitution gives

$$f''(\zeta) = \frac{3k(z)}{4A\Phi(1)\sqrt{a}\, p \sin p}. \tag{4.36}$$

A pair of measurements close to the crack tip for r_1 and θ_1 is approximately valid for the determination of $f''(\zeta)$.

Two different materials were selected for the photoelastic experiments, PLM-4B from Photoelastic, Inc., Malvern, Pennsylvania and CH-60 from Chapman Laboratories, West Chester, Pennsylvania. The mechanical and optical properties were given by the manufacturers. Both of these materials can be easily machined and have a low stress optic coefficient at the critical temperature. The six test specimens consisted of plates 30.5 cm long 10.2 cm wide with thicknesses between 1.25 cm and 3.64 cm. The crack length varied from 1.27 cm to 3.81 cm covering a range of h/a ratios of approximately 1.33 to 2.35. Table 4.1 gives the geometric dimensions and other pertinent parameters. The cracks were cut with a jig saw with a blade of 0.025 cm thick. A small circular hole is first drilled at the center of each plate so that the blade could be threaded through to produce a saw cut (representing the crack) with a radius of curvature of 0.0127 cm. The oven used to freeze the stresses in the cracked specimens was made by Chapman Laboratories and has a temperature-time control through a precut cam for the thermal cycle for each of the photoelastic materials under consideration. In addition, it has a polaroid glass window through which the photoelastic patterns can be observed during loading. A photoelastic fringe multiplication and shar-

TABLE 4.1

Specimen dimensions and applied tensile load

Test specimens	Thickness h		Crack length $2a$		Applied load P		Fringe optical constant	
	(in)	(cm)	(in)	(cm)	(lbs)	(N)	$\frac{lb}{in^2}$/fri	Pa cm/fri
1	0.492	1.250	0.500	1.270	42.0	187	2.610	4.57×10^4
2	0.491	1.247	0.750	1.905	42.0	187	2.610	4.57×10^4
3	0.987	2.507	0.750	1.905	41.0	182	1.430	2.50×10^4
4	0.992	2.520	1.060	2.692	31.0	138	1.450	2.54×10^4
5	0.994	2.525	1.500	3.810	36.0	160	1.430	2.50×10^4
6	1.430	3.637	1.250	3.175	52.0	231	2.250	3.94×10^4

Plate width for specimen no. 1, 2 and 6 is 4 in (10.16 cm) and 3.95 in (10.03 cm) for specimen no. 3, 4 and 5.

pening unit was used. This unit is located between the polarizer and analyzer such that a more accurate determination of the fringe pattern can be obtained.

4.5 Test procedure: frozen stress technique

Two types of end grips for applying the load were designed. In three of the specimens, four equally spaced holes of 0.63 cm diameter were drilled perpendicular to the plate surfaces at each end through which pins were inserted to carry the tensile load applied to the specimens as shown in Figure 4.3. In the other three specimens, a single hole of 0.63 cm diameter was drilled parallel to the plate surfaces and ends at each end to carry a loading pin. The latter setup permits certain freedom

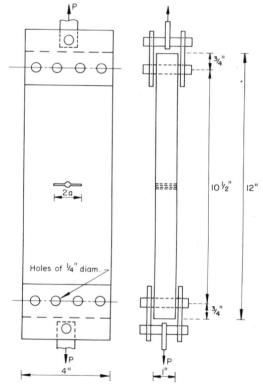

Figure 4.3. Geometry and grips of the photoelastic specimens.

of movement at the ends of the specimen at the critical temperature and achieves a more uniform distribution of the load at the ends. The specimens were supported at the top end, and loaded by dead weights suspended from the bottom. The loaded specimens were placed in an oven which has an automatic thermal cycle setting device. The temperature was raised from approximately 27°C to 138°C, at a rate of 8.3°C/hr. At the elevated temperature, dead weights were applied, but the load was kept slightly below the value that would produce fracture. The temperature was held constant for 6 hours, with the load held constant. The specimen was then cooled down to room temperature at a rate of 1.1°C/hr while the load was maintained.

In order to determine the stress variations through the plate thickness, the specimens with nominal thickness 1.25 cm, 2.5 cm, and 3.8 cm were cut into 3, 5, and 7 slices, respectively, parallel to the plane of the plate. Refer to Figure 4.6 for the precise location of the slices and subslices. These slices were cut with a band saw and the surfaces were finished with a milling machine to the desired thickness. Each slice was then hand-polished with a very fine sand paper. These parallel slices were used to determine the in-plane stress variation of σ_x, σ_y and τ_{xy} described in equations (4.23). Slices perpendicular to the plate surfaces were also obtained. These specimens were used for studying the plane strain equation $\sigma_z = \nu(\sigma_x + \sigma_y)$, a condition assumed in the theoretical problem to prevail in the interior region of the plate near the crack front. A portion of the specimen material between the crack and the loading holes was removed and used as a calibration test for determining the stress-optic constant. This calibration test was performed at the critical temperature. Figure 4.4 illustrates the directions of the planes along which the slices were taken.

The sliced specimens were then inserted into the polariscope assembly for determining the fringes. A schematic of the optical system is shown in Figure 4.5. In order to measure the fractional fringe orders quickly and accurately, the isochromatic fringe multiplication technique introduced by Post [16] was adopted. This method increases the optical path length by means of two partially reflecting mirrors slightly inclined toward each other. These mirrors are set between the quarter wave plates of an ordinary circular polariscope. The parallel beams of light that emerge from the mirrors are at different angles and are converted to different points in the focal plane of the field lens. Any one of those beams can then be observed by the eye or camera lens at the apex of the respective convergent light. Total isolation can be accomplished such

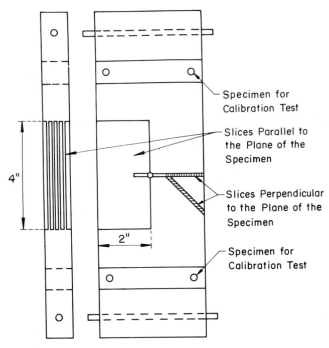

Figure 4.4. Primary slicing plan.

that the rays transversing through the specimen in a given number of times can be observed without interference from other rays. If the light passes through the specimen N times the effect is equivalent to that of a single light beam passing through N identical specimens all loaded the same way and hence the birefringence is multiplied by a factor N. The principle of this method is that the mirrors reflect part of the light and transmit part of the light that falls upon them. The fringe multiplications unit was used in conjunction with an immersion fluid to eliminate diffusing refraction caused by irregularities in the surfaces of the slices. The immersion fluid consists of mixing a low refraction index fluid with a high refraction index fluid following the method described by Post [16] until the refractive index of the mixed fluid matches that of the sliced material. Photographs of the fringe patterns using Kodak TRI-X Pan Film, and ILFORD-FP4 were taken in both light and dark fields and for multiplication factors from 1 to 11 as needed in the determination of the fringe pattern. Figures 4.7 to 4.9 show some typical fringe patterns obtained from the fringe multiplication method.

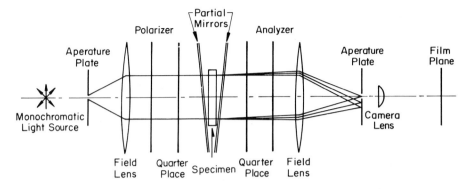

Figure 4.5. General polariscope assembly.

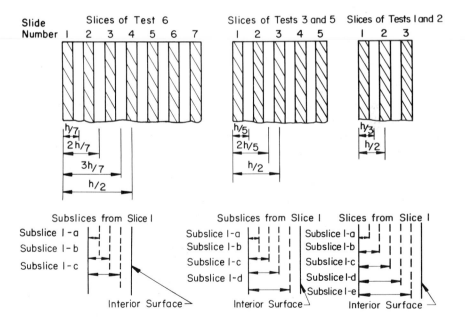

Figure 4.6. Slices and subslices location.

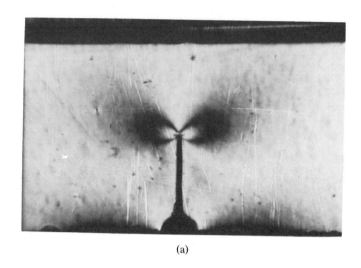

(a)

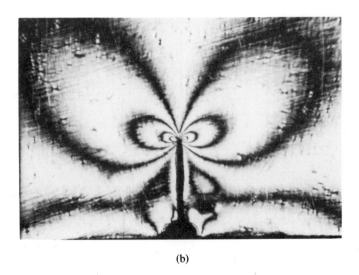

(b)

Figure 4.7. Isochromatic patterns from slice with $h = 0.5$ in (1.270 cm) and $2a = 0.75$ in (1.905 cm): (a) unmultiplied fringes and (b) fringes multiplied by three.

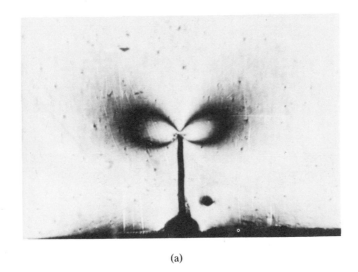

(a)

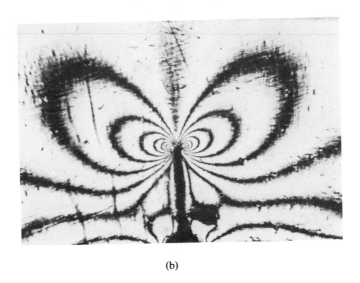

(b)

Figure 4.8. Isochromatic patterns for slice with $h = 0.5$ in (1.270 cm) and $2a = 0.75$ in (1.905 cm): (a) unmultiplied fringes and (b) fringes multiplied by five.

G. Villarreal and G. C. Sih

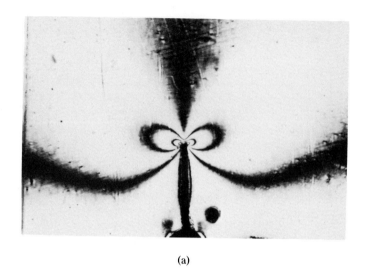

(a)

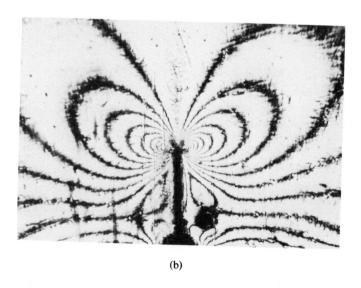

(b)

Figure 4.9. Isochromatic patterns for slice with $h = 0.5$ in (1.270 cm) and $2a = 0.75$ in (1.905 cm): (a) unmultiplied fringes and (b) fringes multiplied by seven.

Special attention was given to the slices adjacent to the plate surfaces. The thickness of these slices was reduced by milling the side opposite the original plate surface. They were analyzed photoelastically again as described earlier. This process of reducing the thickness of the exterior slices and reanalyzing them continued until these slices were approximately 0.102 cm thick. Further reduction in thickness was accomplished manually with sandpaper. The smallest thickness used was approximately 0.025 cm. Fewer fringes are observed for the thinner slices, primarily because the amount of retardation depends on the optical path length through the material. Higher-order fringe multiplications are required to obtain an optical path length equivalent to that of the thicker slices. This results in average stress in a small region rather than stress at a point. Thus, values of stress are less reliable, but trends observed remain valid.

4.6 Comparison of Hartranft–Sih theory with experiments

The objective of slicing is to measure the stress distribution ahead of the crack through the plate thickness. Subslicing as indicated in Figure 4.6 is carried out to gain information in the boundary layer so that the results can be used to compare with the predictions based on the Hartranft–Sih theory. To this end, a series of photographs were taken of each slice with different fringe multiplication factors in both the light and dark fields. A series of transparencies were prepared so that by projection the details of the fringes at the crack tip could be magnified. For each slice, the fringe order, n, was measured and recorded for a particular point with coordinates r_1 and θ_1 on one of the isochromatic loops. The values of r_1 in this investigation are between 0.07 cm and 0.15 cm.

Equation (4.34) can be rearranged in the form

$$\left(\frac{\sigma_1 - \sigma_2}{\sigma_0}\right)^2 = \frac{k^2(z)}{2r_1}\sin^2\theta_1 + \frac{2k(z)}{\sqrt{2r_1}}\sin\theta_1\sin\frac{3\theta_1}{2} + 1 \qquad (4.37)$$

By transposing the term $[(\sigma_1 - \sigma_2)/\sigma_0]^2$ to the right hand side, equation (4.37) becomes

$$A_1 k^2(z) + B_1 k(z) + C_1 = 0 \qquad (4.38)$$

which yields

$$k(z) = \frac{-B_1 \pm \sqrt{B_1^2 - 4A_1 C_1}}{2A_1} \tag{4.39}$$

The constants A_1 and B_1 stand for

$$A_1 = \frac{1}{2r_1} \sin^2 \theta_1 \tag{4.40a}$$

$$B_1 = \frac{1}{\sqrt{2r_1}} 2 \sin \theta_1 \sin \frac{3\theta_1}{2} \tag{4.40b}$$

$$C_1 = 1 - \left(\frac{\sigma_1 - \sigma_2}{\sigma_0}\right)^2 \tag{4.40c}$$

The secondary principal stress difference $(\sigma_1 - \sigma_2)$ is related to the fringe order through the stress optical law

$$\sigma_1 - \sigma_2 = \frac{nc}{t} \tag{4.41}$$

where t is the effective thickness of the slice and c is the stress-optic coefficient.

Once $k(z)$ is known from equation (4.39), $f''(\zeta)$ can be found from equation (4.36) where $\Phi(1)$ is already known from the curves in Figure 4.2 for different plate thickness to crack length ratios. The constant A is a function of p as given by equation (4.28). Thus, a value of p has to be assigned for the determination of $f''(\zeta)$.

Variation of in-plane stresses through plate thickness. Referring to equation (4.2a), it is seen that the z-dependence for σ_x, σ_y and τ_{xy} is governed by the function $f''(\zeta)$ where $\zeta = 2z/h$. Following the experimental procedure described earlier, $f''(\zeta)$ was found for $p = 0.3$, 0.4 and 0.6 and did not vary appreciably with p. The values of $f''(\zeta)$ for $p = 0.3$ differed approximately 1% from those for $p = 0.4$ and 6% for $p = 0.8$ and they are nearly the same in the interior or plane strain region. Within a boundary layer distance $\varepsilon h/2$, the experimental results in Figures 4.10 to 4.14 show that the in-plane stresses experience a rapid decrease in magnitude. This agrees with the general expectation that the two-dimensional plane strain solution holds well in the exterior of the plate

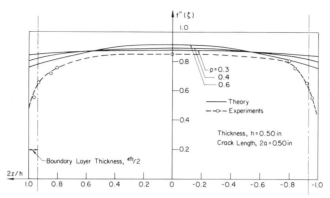

Figure 4.10. In-plane stress variation in z-direction for $h = 0.5$ in (1.27 cm) and $2a = 0.5$ in (1.27 cm).

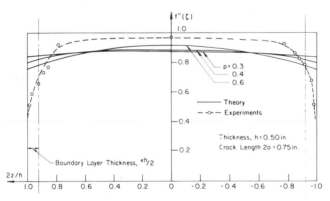

Figure 4.11. In-plane stress variation in z-direction for $h = 0.5$ in (1.27 cm) and $2a = 0.75$ in (1.91 cm).

except within a boundary next to the plate surface. The introduction of a boundary layer to be employed jointly with the interior solution obtained from the approximate plate theory was first advocated by Sih [5]. The theoretical results obtained from the Hartranft–Sih theory [14] given by the solid curves in Figures 4.10 to 4.14 are seen to agree well with experiments for $p = 0.3$ and 0.4 in the plane strain region. The plate thickness h was varied from 1.27 cm to 3.175 cm while the ratio $h/2a$ varied from 0.67 to 1.33. The difference between the experimental and theoretical results was less than 10%. Outside the region of plane strain, the difference is appreciable. This is to be expected since the ap-

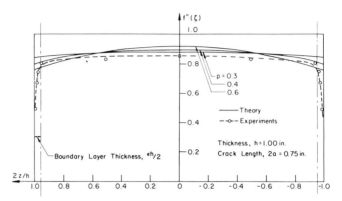

Figure 4.12. In-plane stress variation in z-direction for $h = 1.00$ in (2.54 cm) and $2a = 0.75$ in (1.905 cm).

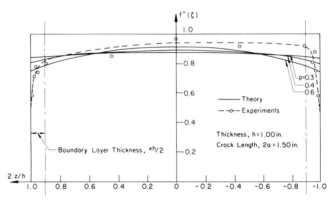

Figure 4.13. In-plane stress variation in z-direction for $h = 1.00$ in (2.54 cm) and $2a = 1.50$ in (3.81 cm).

proximate plate theory cannot yield accurate results in a layer of material near the plate surface where the free surface effect becomes significant.

Transverse normal stress. One of the distinct features of three-dimensional crack solution is the behavior of the transverse normal stress through the plate thickness. The function $f(\zeta)$ in equation (4.2c) that governs the behavior of σ_z through the plate thickness is given by equation (4.27).

Since $f(\zeta)$ and $f''(\zeta)$ obey the equation (4.26), $k_p(z)$ and $k_t(z)$ are

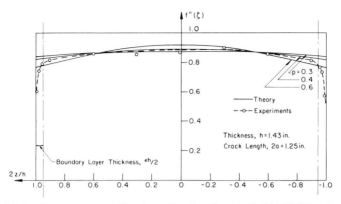

Figure 4.14. In-plane stress variation in z-direction for $h = 1.43$ in (3.632 cm) and $2a = 1.25$ in (3.175 cm).

related. Consider a slice of material normal to the plate surface in Figure 4.4 with $\theta_1 = 0$. This corresponds to a cross section of material in the plane of expected crack extension. The stresses which produce photo-elastic effects are σ_x, σ_z and τ_{xz}. By letting $\theta_1 = 0$ in equations (4.23), these stress components can be written as

$$\sigma_x = \frac{k_p(z)\sqrt{a}}{\sqrt{2r_1}} \tag{4.42a}$$

$$\sigma_z = \frac{k_t(z)\sqrt{a}}{\sqrt{2r_1}} \tag{4.42b}$$

$$\tau_{xy} = O(r_1^0) \tag{4.42c}$$

Hence, the plane strain condition, $\sigma_z = \nu(\sigma_x + \sigma_y)$, implies that

$$k_t(z) = \nu[2k_p(z)] \tag{4.43}$$

as $\sigma_x = \sigma_y$. Since the Poisson's ratio ν of the specimen is very nearly equal to 0.5 at the critical temperature, equation (4.43) gives $k_t(z) \approx k_p(z)$. In addition, because τ_{xz} is small in comparison with σ_x and σ_z as $r_1 \to 0$, it is obvious from

$$(\sigma_1 - \sigma_2)_y = 2\sqrt{\left(\frac{\sigma_x - \sigma_z}{2}\right)^2 + \tau_{xz}^2} \tag{4.44}$$

that $(\sigma_1 - \sigma_2)_y = \sigma_x - \sigma_z$ where $(\sigma_1)_y$ and $(\sigma_2)_y$ are principal stresses in a plane normal to the y-axis.

Under the above considerations, equations (4.42) gives

$$(\sigma_1 - \sigma_2)_y = 0 \tag{4.45}$$

except in the boundary layers. The photoelastic observations made on the slices perpendicular to the mid-plane of the specimen confirms the above relation. That is, no fringe pattern was observed except on the edges of the slice where the photoelastic pattern is affected by specimen boundaries. Measurements near these boundaries were not possible. The photoelastic observation based on the frozen stress technique showed that $\sigma_z = \nu(\sigma_x + \sigma_y)$ is indeed satisfied in the interior region of the plate as predicted by Sih [5].

The transverse variation of the stress intensity factor can now be normalized by its value at the midplane of the plate. According to equations (4.43), equation (4.27) is equivalent to

$$k_p(z) = k_p(0) \cos\left(\frac{2p}{h} z\right) \tag{4.46}$$

Both the theoretical and experimental values of equation (4.46) are displayed graphically in Figures 4.15 to 4.19. The predicted values are given by solid lines for $p = 0.2$, 0.4, 0.6 and 0.8 while the photoelastic results are shown as dotted curves through experimentally located points. The theoretical curves for $p = 0.4$ are in good agreement with experiment in the plane strain region for the range of h and $2a$ values treated. The errors committed in evaluating $k_p(z)/k_p(0)$ is only about 5% or less. In the boundary layer, $\varepsilon h/2$, the experimental curves are considerably lower.

Boundary layer effect. Figures 4.15 to 4.19 also show that the magnitude of the stresses drops sharply in a layer of the material near the plate surface. This drop will be referred to as the boundary layer effect. This can be observed from the values of $f''(\zeta)$ and $k_p(z)/k_p(0)$ which tend to decrease as the plate surface is approached. Recall that the Hartranft–Sih theory [14] assumes a layer thickness $\varepsilon h/2$ outside of which the material is under plane strain. This is the boundary layer of thickness given by equation (4.29) as indicated on each one of the Figures 4.15 to 4.19. In general, the curves begin to drop rapidly inside the theoretically

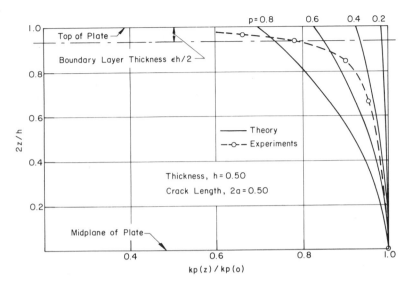

Figure 4.15. Normalized stress intensity factors along crack front for $h = 0.50$ in (1.27 cm) and $2a = 0.50$ in (1.27 cm).

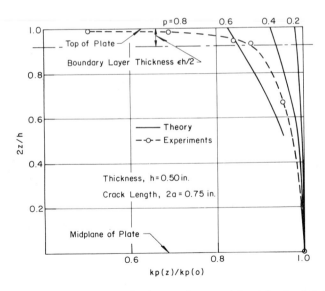

Figure 4.16. Normalized stress intensity factors along crack front for $h = 0.50$ in (1.27 cm) and $2a = 0.75$ in (1.91 cm).

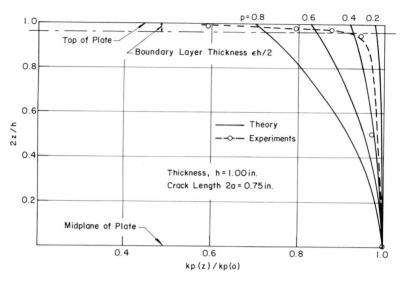

Figure 4.17. Normalized stress intensity factors along crack front for $h = 1.00$ in (2.54 cm) and $2a = 0.75$ in (1.905 cm).

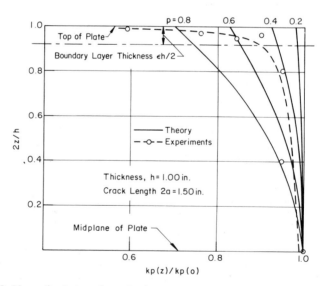

Figure 4.18. Normalized stress intensity factors along crack front for $h = 1.00$ in (2.54 cm) and $2a = 1.50$ in (3.81 cm).

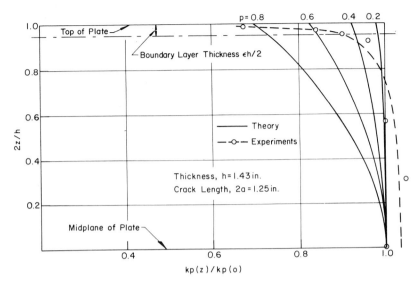

Figure 4.19. Normalized stress intensity factors along crack front for $h = 1.43$ in (3.632 cm) and $2a = 1.25$ (3.175 cm).

predicted boundary layer thickness. Hence, equation (4.29) serves as a good measure for determining the region within which surface effects are significant. Inside the layer, Hartranft and Sih [14] recommend $f(\zeta)$ to take the form given by equation (4.30).

References

[1] Post, D., 'Photoelastic Analysis for an Edge Crack in a Tensile Field', *Proceedings of the Society for Experimental Stress Analysis*, Vol. XII, No. 4, (1954).

[2] Fessler, H. and Mansell, D. O., 'Photoelastic Study of Stresses Near Cracks in Thick Plates', *Journal of Mechanical Engineering Sciences*, Vol. 4, No. 3, pp. 213–225, (1962).

[3] Knowles, J. K. and Wang, N. M., 'On the Bending of an Elastic Plate Containing a Crack', *Journal of Mathematics and Physics*, Vol. 39, pp. 223–236, (1961).

[4] Sih, G. C., 'Bending of a Cracked Plate with Arbitrary Stress Distribution Across the Thickness', *Journal of Engineering for Industry*, Trans. ASME, Series B, Vol. 92, No. 2, pp. 350–356, (February 1970).

[5] Sih, G. C., 'A Review of the Three-Dimensional Stress Problem for a Cracked Plate', *International Journal of Fracture Mechanics*, Vol. 7, No. 1, pp. 35–59, (March 1971).

[6] Smith, D. G. and Smith, C. W., 'Influence of Precatastrophic Extension and Other Effects on Local Stresses in Cracked Plates Under Bending Fields', *Experimental Mechanics*, Vol. 11, No. 9, pp. 394–401, (September 1971).

[7] Hartranft, R. J. and Sih, G. C., 'Effect of Plate Thickness on the Bending Stress Distribution Around Through Cracks', *Journal of Mathematics and Physics*, Vol. 47, No. 3, pp. 276–291, (1968).

[8] Smith, D. G. and Smith, C. W., A Photoelastic Evaluation of Closure and Other Effects Upon the Local Bending Stresses in Cracked Plates', *International Journal of Fracture Mechanics*, Vol. 6, No. 3, pp. 305–318, (1970).

[9] Wells, A. A. and Post, D., 'The Dynamic Stress Distribution Surrounding a Running Crack—A Photoelastic Analysis', *Proceedings of the Society for Experimental Stress Analysis*, Vol. XVI, pp. 69–92; discussion by G. R. Irwin, pp. 93–96, (1958).

[10] Smith, D. G. and Smith, C. W., 'Photoelastic Determination of Mixed Mode Stress-Intensity Factors', *Engineering Fracture Mechanics Journal*, Vol. 4, pp. 357–366, (1972).

[11] Sih, G. C., 'Application of Strain-Energy-Density Theory to Fundamental Fracture Problems', Institute of Fracture and Solid Mechanical Technical Report, Lehigh University, AFOSR-RT-73-1.

[12] Kipp, M. E. and Sih, G. C., 'The Strain Energy Density Failure Criterion Applied to Notched Elastic Solids', *International Journal of Solids and Structures*, Vol. 11, pp. 153–173, (1975).

[13] Mullinix, B. R. and Smith, C. W., 'Distributions of Local Stresses Across the Thickness of Cracked Plates Under Bending Fields', International Journal of Fracture Mechanics, Vol. 10, pp. 337–352, (1974).

[14] Hartranft, R. J. and Sih, G. C., 'An Approximate Three-Dimensional Theory of Plates with Application to Crack Problems', *International Journal of Engineering Science*, Vol. 8, pp. 711–729, (1970).

[15] Frocht, M. M., *Photoelasticity*, Vol. 2, Wiley, (1948).

[16] Post, D., 'Fringe Multiplication in Three-Dimensional Photoelasticity', *Journal of Strain Analysis*, Vol. 1, pp. 380–388, (1966).

J. Beinert and J. F. Kalthoff

5 | Experimental determination of dynamic stress intensity factors by shadow patterns

5.1 Introduction

The method of shadow patterns, which also is known under the name 'shadow spot method' or 'method of caustics', in numerous investigations has proven to be a powerful optical method to measure stress intensity factors in static and dynamic fracture mechanics problems. The technique yields high accuracy of measurement, but is still simple to apply. Besides the usual equipment for (high speed) photography, only a point light source is required. Merely a single, very exactly measurable quantity, a length, is to be evaluated. The method utilizes the information from the stress-strain-field within a small area very close to the crack tip. Any knowledge about external loading and geometric boundary conditions, therefore, is not necessary.

Thus, the method of shadow patterns appears to have greatest potential for dynamic fracture investigations. In static problems normally theoretically derived relations can be applied to determine stress intensity factors from external conditions (as load, crack length, crack opening displacement etc.). In dynamic problems, however, such relations are available only for special cases, since the theoretical description of dynamic fracture processes, in general, is more complicated. Statically deduced relations normally fail in dynamic problems [1, 2].

The method of shadow patterns was introduced in 1964 by Manogg [3, 4, 5], who performed a shadow optical analysis for a stationary crack under mode-I-loading. Later on, the method was further developed and its field of application extended by Theocaris [see his contribution to this volume] and by the authors and their coworkers [1, 6–10].

The shadow optical method for the measurement of stress intensity factors for both dynamically loaded stationary cracks and fast propagating cracks is based extensively on Manogg's analysis. Consequently, in the first of this article, the substantial characteristics and results of this analysis are summarized. Then its validity also for stationary cracks under dynamic loading is investigated. Furthermore, the shadow optical analysis for a propagating crack is carried out in analogy to Manogg's procedure. The results are compared to those of the static analysis. The experimental techniques in performing and evaluating a measurement are discussed in detail. Finally, the applicability of the method is illustrated by several examples.

5.2 Physical and mathematical principles of the method

Physical principle. The physical principle underlying the method of shadow patterns is illustrated in Figure 5.1. A specimen containing a crack which is loaded by a tensile stress σ is illuminated with light generated by a point light source. In the upper part of the figure a specimen of a transparent material is considered. The stress intensification in the region surrounding the crack tip causes a reduction of both the thickness of the specimen and the refractive index of the material. As a consequence, the area surrounding the crack tip acts similar to a divergent lens: the light transmitted through the specimen is deflected outwards. Therefore, in the shadow image of the crack on a plane at any distance z_0 behind the specimen, the crack tip appears surrounded by a dark spot. The spot is bounded by a bright light concentration, a caustic. The light is concentrated, because the angular deflection of the light beams decreases with increasing distance from the tip. The shadow pattern is shown schematically on the right side in the figure. In the lower part of Figure 5.1 a specimen of a non-transparent material with a mirrored surface is considered. It is demonstrated that by using the light reflected from the surface, in principle the same light configuration is given as for the transparent specimen. On a *virtual* image plane behind the specimen, therefore, qualitatively the same shadow pattern can be observed. The light deflections in the reflection case are caused only by the thickness changes of the specimen.

Because the magnitude of the light deflections is correlated with the magnitude of the stress intensification at the crack tip, the shadow pattern contains information about the stress-strain conditions in the vicinity of the crack tip and, in particular, about the stress intensity factor. Manogg [3] first theoretically calculated the shadow pattern and found quantitative correlations between the dimensions of the caustic and the stress intensity factor.

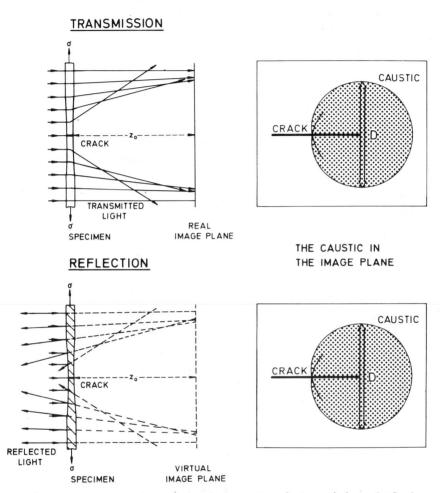

Figure 5.1. The principle of the method of shadow patterns for transmission and reflection. (from [9])

Principle of the calculation of caustics. The shadow optical analysis
consists of calculating the light deflections and resulting shape and size of
the shadow pattern for a given stress distribution with a given stress
intensity factor. The principles of Manogg's procedure [3] which apply
likewise for stationary and propagating cracks, will be sketched in the
following.

A light beam is considered which traverses the specimen at the point
$P(r, \varphi)$ in the object plane E as shown in Figure 5.2. The non-deflected
beam would pass the shadow image plane E' which is also called the
reference plane, at the point P_m defining the vector $\mathbf{r}_m$. Due to the light
deflection in P, however, the beam is displaced to the point $P'(x', y')$ by
a vector $\mathbf{w}$. $\mathbf{w}$ is a function of the co-ordinates r, φ of the point P. As can
be seen from the figure, the vector $\mathbf{r}'$ of the image point P' is given as

$$\mathbf{r}' = \mathbf{r}_m + \mathbf{w}(r, \varphi) \tag{5.1}$$

This general relation is also valid if light reflected at the mirrored surface
of the object is used. The shadow image on the plane E' then is virtual.
The shadow optical image is completely described by equation (5.1):

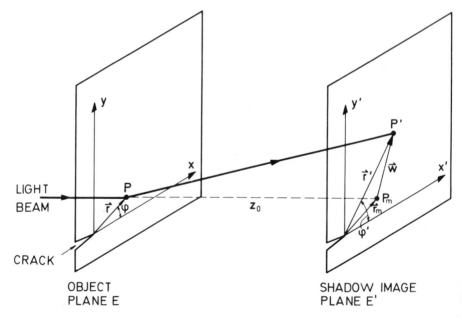

Figure 5.2. Geometrical conditions of the shadow optical analysis. (after [8])

For each point $P(r, \varphi)$ of the object, the corresponding image point $P'(x', y')$ of the shadow image is obtained.

After passing the object, the deflected light beams envelope a shadow space (see Figure 5.1). On the image plane which is a cross section of this space, they form a caustic. As an envelope, the caustic is a singular curve of the image equation (5.1). A necessary condition for the existence of such a singularity is that the Jacobian functional determinant J of equation (5.1) vanishes, i.e.

$$J = \frac{\partial x'}{\partial r} \frac{\partial y'}{\partial \varphi} - \frac{\partial x'}{\partial \varphi} \frac{\partial y'}{\partial r} = 0 \tag{5.2}$$

Those points P whose co-ordinates (r, φ) fulfill the condition (5.2) are imaged onto the caustic. The caustic itself then results by inserting these points (r, φ) into the image equation (5.1).

Thus, the main problem in analyzing a shadow pattern is finding for the given physical problem the vectors $\mathbf{r}_m$ and $\mathbf{w}$ in the image equation (5.1). The vector $\mathbf{r}_m$ is the projection of $\mathbf{r}$ onto the image plane (see Figure 5.2) and can easily be determined (see below). The vector $\mathbf{w}$, according to the theory of the eikonal, is given as

$$\mathbf{w}(r, \varphi) = z_0 \, \text{grad} \, \Delta s(r, \varphi) \tag{5.3}$$

where Δs is the change of the optical path length caused by the object and z_0 is the reference distance which is defined as the distance from the object plane E to the shadow image plane E' (see Figure 5.2). The path length change Δs (r, φ) in equation (5.3) is correlated to the stresses $\sigma(r, \varphi)$ in the object by elasto-optical relations. The stresses σ at each point P near the crack tip are given by fracture mechanics equations. These elasto-optical and fracture mechanics equations on which the shadow optical analysis is based are discussed in the following section.

5.3 Theoretical analysis of the shadow pattern after Manogg: the mode-I-loaded stationary crack

The basic elasto-optical equations

General case. As a first step in calculating the caustic, the path length change Δs in equation (5.3) will be determined as a function of the stresses. An object may be considered as being a transparent, plane parallel plate.

Due to a tensile load, its initial thickness d and its refractive index n are reduced to d-Δd and n-Δn, respectively. Then, for normally incident light, Δs is given as

$$\Delta s = (n - 1)\Delta d + d\Delta n. \tag{5.4}$$

The correlation between the change Δn of the refractive index and the principal stresses σ_1, σ_2 and σ_3 in the plate is described by Maxwell-Neumann's law [11]:

$$\Delta n_1 = A\sigma_1 + B(\sigma_2 + \sigma_3) \tag{5.5a}$$

$$\Delta n_2 = A\sigma_2 + B(\sigma_1 + \sigma_3). \tag{5.5b}$$

Δn_1 and Δn_2 are the refractive index changes for light polarized parallel to the principal stresses σ_1 and σ_2, respectively, and propagating in the direction of σ_3. A and B are elasto-optical material constants. Materials for which $A \neq B$ are called optically anisotropic. For those materials $\Delta n_1 \neq \Delta n_2$, i.e. they are birefringent (since the refractive index depends on the polarization direction of the light relative to the principal stresses). For optically isotropic materials $A = B$ and, consequently, $\Delta n_1 = \Delta n_2$, i.e. these materials are not birefringent.

Plane stress conditions. For plane stress conditions (this case was treated by Manogg) $\sigma_3 = 0$ and (due to Hooke's law)

$$\Delta d = -\frac{\nu}{E}(\sigma_1 + \sigma_2)d \tag{5.6}$$

where ν is Poisson's ratio and E is Young's modulus. Then, by using the equations (5.5), equation (5.4) becomes

$$\Delta s_1 = (a'\sigma_1 + b'\sigma_2)d \tag{5.7a}$$

$$\Delta s_2 = (a'\sigma_2 + b'\sigma_1)d \tag{5.7b}$$

with

$$a' = A - (n - 1)\frac{\nu}{E} \tag{5.8a}$$

$$b' = B - (n - 1)\frac{\nu}{E} \tag{5.8b}$$

where Δs_1 and Δs_2 are the path length changes for light polarized parallel to the principal stresses σ_1 and σ_2, respectively.

Plane strain conditions. For plane strain conditions $\Delta d = 0$ and $\sigma_3 = \nu(\sigma_1 + \sigma_2)$. Then equation (5.4), using the equations (5.5), becomes

$$\Delta s_1 = (a''\sigma_1 + b''\sigma_2)d \tag{5.9a}$$

$$\Delta s_2 = (a''\sigma_2 + b''\sigma_1)d \tag{5.9b}$$

with

$$a'' = A + \nu B = a' + \nu b' + (\nu + 1)(n - 1)\frac{\nu}{E} \tag{5.10a}$$

$$b'' = B + \nu B = (\nu + 1)\left[b' + (n - 1)\frac{\nu}{E}\right]. \tag{5.10b}$$

Non-transparent specimen. For the light reflected at the mirroring surface of the specimen, the optical path length change Δs is identical to the change of the geometrical path length, i.e.

$$\Delta s = 2\left(\frac{1}{2}\Delta d\right) = -\frac{\nu}{E}(\sigma_1 + \sigma_2)d, \tag{5.11}$$

if the thickness change is symmetrical about the mid-plane of the specimen, and plane stress conditions (see equation (5.6)) are given* (for plane strain conditions the thickness does not change and, therefore, no shadow optical effect exists).

Summary. The elasto-optical relations (5.7), (5.9) and (5.11) can be summarized to the more general form

$$\Delta s_1 = (a\sigma_1 + b\sigma_2)d \tag{5.12a}$$

$$\Delta s_2 = (a\sigma_2 + b\sigma_1)d \tag{5.12b}$$

which also can be written in terms of the principal stress sum and

---

* This result agrees with the calculations in [12], where also the case is considered that the. light after traversing a transparent specimen is reflected at the rear face.

difference:

$$\Delta s_{1,2} = cd[\sigma_1 + \sigma_2 \pm \lambda(\sigma_1 - \sigma_2)] \tag{5.13}$$

where the positive (negative) sign of λ relates to Δs_1 (Δs_2), and

$$c = \frac{a+b}{2} \tag{5.14a}$$

$$\lambda = \frac{a-b}{a+b} \tag{5.14b}$$

with

$a = a'$, $b = b'$ for plane stress conditions and transmitted light,
$a = a''$, $b = b''$ for plane strain conditions and transmitted light, and
$a = b = -\dfrac{\nu}{E}$ for plane stress conditions and reflected light.

The constants a', b' and a'', b'' are correlated to each other and to the constants A and B by the equations (5.10) and (5.8). The value of λ describes the amount of the optical anisotropy. For optically isotropic materials $a = b$ and, consequently, $\lambda = 0$.

It can be seen that for the shadow optical analysis, the cases of plane stress, plane strain, transmitted light and reflected light are basically the same and differ only by the values of the elasto-optical constants, c and λ, which are used in the basic equation (5.13). Numerical values for c and λ are given in Table 5.1 for the different cases considered and for different materials.

The basic fracture mechanics equations

For a crack under mode I loading conditions in linear elastic materials the stress distribution in the vicinity of the crack tip is given by Sneddon [13] and Williams [14]:

$$\sigma_x = \frac{K_I}{\sqrt{2\pi r}} \cos\frac{\varphi}{2}\left(1 - \sin\frac{\varphi}{2}\sin\frac{3\varphi}{2}\right) + a_2 \tag{5.15a}$$

$$\sigma_y = \frac{K_I}{\sqrt{2\pi r}} \cos\frac{\varphi}{2}\left(1 + \sin\frac{\varphi}{2}\sin\frac{3\varphi}{2}\right) \tag{5.15b}$$

$$\tau_{xy} = \frac{K_I}{\sqrt{2\pi r}} \cos\frac{\varphi}{2}\sin\frac{\varphi}{2}\cos\frac{3\varphi}{2} \tag{5.15c}$$

TABLE 5.1

Elastic and shadow optical constants for several materials

| Material | Elastic Constants | | Shadow Optical Constants | | | | | | | | |
| | Young's Modulus MN/m² | Poisson's Ratio | for Plane Stress | | | | k | for Plane Strain | | | |
			c m²/N	$\|\lambda\|$	f_o	f_i		c m²/N	$\|\lambda\|$	f_o	f_i
TRANSMISSION:											
Optically Anisotropic:											
Araldite B	3660*	0.392*	-0.970×10^{-10}	0.288	3.32	3.05	0.507	-0.580×10^{-10}	0.482	3.42	2.99
CR-39	2580	0.443	-1.200×10^{-10}	0.148	3.24	3.10	0.534	-0.560×10^{-10}	0.317	3.33	3.04
Plate Glass	73900	0.231	-0.027×10^{-10}	0.519	3.44	2.98	0.636	-0.017×10^{-10}	0.849	3.62	2.97
Homalite 100	4820*	0.310*	-0.920×10^{-10}	0.121	3.23	3.11	0.518	-0.767×10^{-10}	0.149	3.24	3.10
Optically Isotropic:											
PMMA	3240	0.350	-1.080×10^{-10}	~0	3.17	3.17	0.526	-0.750×10^{-10}	~0		3.17
REFLECTION:											
All materials	E	ν	$-\nu/E$	0	3.17	3.17	0.526	—	—	—	—

* dynamic values

where K_I is the stress intensity factor, r, φ are polar co-ordinates with the crack tip as origin (see Figure 5.2), σ_x, σ_y, τ_{xy} are the normal and shear stresses related to the rectangular co-ordinate system, and a_2 is a constant (its physical meaning is discussed below).

The evaluation relations

Utilizing the elasto-optical equations (5.13) and the fracture mechanics equations (5.15), equation (5.3) results. Then from equation (5.3), the image equation (5.1) is established. Following the guidelines given in section 5.2, the caustic is determined from equation (5.1) as a function of the stress intensity factor K_I.

The vector $\mathbf{r}_m$ (see Figure 5.2) is given as

$$\mathbf{r}_m = m\mathbf{r} \tag{5.16}$$

where m is a scale factor. When parallel incident light is used, $m = 1$, and for convergent (divergent) incident light, $m < 1$ ($m > 1$). With equations (5.13) and (5.3), equation (5.1) can then be written as

$$\mathbf{r}' = m\mathbf{r} + cdz_0 \, \text{grad}[\sigma_1 + \sigma_2 \pm \lambda(\sigma_1 - \sigma_2)]. \tag{5.17}$$

The principal stresses σ_1, σ_2 can be determined from the equations (5.15) using the relations

$$\sigma_1 + \sigma_2 = \sigma_x + \sigma_y \tag{5.18a}$$

$$\sigma_1 - \sigma_2 = [(\sigma_x - \sigma_y)^2 + 4\tau_{xy}^2]^{1/2}. \tag{5.18b}$$

For optically isotropic materials ($\lambda = 0$), the components x' and y' of $\mathbf{r}$ in the image equation (5.17) are then

$$x' = mr \cos \varphi - \frac{K_I}{\sqrt{2\pi}} \, cdz_0 r^{-3/2} \cos \frac{3}{2} \varphi \tag{5.19a}$$

$$y' = mr \sin \varphi - \frac{K_I}{\sqrt{2\pi}} \, cdz_0 r^{-3/2} \sin \frac{3}{2} \varphi \tag{5.19b}$$

with $-\pi \leq \varphi \leq +\pi$.

The determinant condition (5.2) leads to

$$r = \left(\frac{3}{2} \frac{K_I}{\sqrt{2\pi}} \frac{1}{m} \, cdz_0 \right)^{2/5} \equiv r_0 \tag{5.20}$$

where r is denoted as r_0 indicating that it is a constant independent of φ. The expression (5.20) represents a circle around the crack tip with radius r_0. This circle is imaged onto the caustic. The equation for the caustic, therefore, is obtained by replacing r in the image equations (5.19) by r_0:

$$x' = mr_0 \left(\cos \varphi + \frac{2}{3} \cos \frac{3}{2} \varphi \right) \qquad (5.21a)$$

$$y' = mr_0 \left(\sin \varphi + \frac{2}{3} \sin \frac{3}{2} \varphi \right) \qquad (5.21b)$$

with $-\pi \leqslant \varphi \leqslant +\pi$.

It can be seen that the caustic* has the form of a generalized epicycloid. The caustic is plotted in the upper part of Figure 5.3. The comparison with shadow patterns observed experimentally in both a transmission and a reflection arrangement shows good agreement.

The equations (5.21) show that only the size, but not the shape of the caustic depends on the value of r_0 (and, according to equation (5.20), on the value of K_I). From the equations (5.21) the maximum caustic diameter $2y'_{max} = D$ (see Figure 5.3) can be determined. D can be written as

$$D = mr_0 f \qquad (5.22)$$

where f is a constant. Inserting r_0 from equation (5.20) one obtains the wanted correlation between the experimentally measurable caustic diameter D and the stress intensity factor K_I:

$$K_I = \frac{2\sqrt{2\pi}}{3m^{3/2}f^{5/2}|c|dz_0} D^{5/2} \qquad (5.23)$$

For optically anisotropic materials ($\lambda \neq 0$) an analog calculation [3] shows that due to the term $\pm \lambda(\sigma_1 - \sigma_2)$ in equation (5.17), the caustic

---

* The equations (5.21) are valid for positive reference lengths z_0. For negative values of z_0, the signs of the second terms must be altered to minus. This yields another caustic line, which closes the open ends of the caustic shown in Figure 5.3 by a narrow, heart-shaped line, as is discussed in [15]. (Due to its smaller size this caustic yields a reduced accuracy if used for determining stress intensity factors). Negative reference lengths physically mean that the reference plane is positioned on the opposite side of the specimen than shown in Figure 5.1, i.e. the shadow image is virtual (real) in the transmission (reflection) arrangement.

THEORY	EXPERIMENT

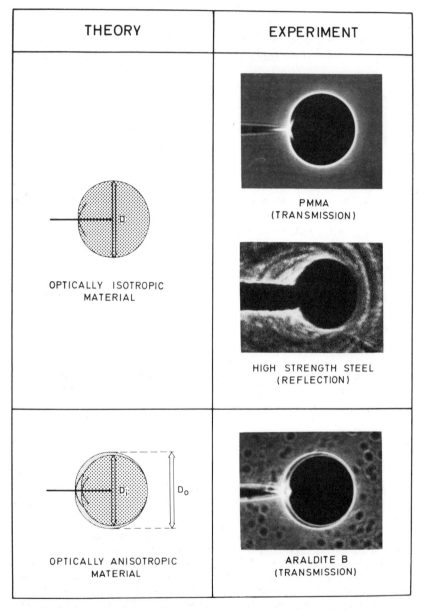

Figure 5.3. Theoretically calculated and experimentally observed shadow patterns.

splits into a double caustic with an outer and an inner branch. The magnitude of the split-up is determined by the value of λ. The calculated double caustic for Araldite B in comparison to the one experimentally observed is shown in the lower part of Figure 5.3. Two caustic diameters, D_o and D_i can be defined.

For optically anisotropic materials, the equations (5.22) and (5.23) after [3] are modified to*

$$D_{o,i} = mr_0 f_{o,i} \tag{5.24}$$

and

$$K_I = MD_{o,i}^{5/2}, \tag{5.25}$$

respectively, where

$$M = \frac{2\sqrt{2\pi}}{3m^{3/2}f_{o,i}^{5/2}|c|dz_0}. \tag{5.26}$$

The constants f_o and f_i apply when the caustic diameters D_o and D_i are considered. Numerical values of f_o and f_i are determined from the definition equation (5.24) by using the calculated caustic diameter D_o and D_i, respectively. The values of f_o and f_i depend only on the value of λ (defined in equation (5.14b)). In Figure 5.4 f_o and f_i are plotted as functions of λ. For $\lambda = 0$ it can be seen that $f_o = f_i = f = 3.17$. The constant c is calculated from equation (5.14a) using the elasto-optical constants of the material. Numerical values for f_o, f_i and c are given in Table 5.1 (after [3, 16, 17 and 18]) for different cases and different materials.

These values, especially those for c, are different for plane stress and plane strain conditions. Therefore, it is important to know what conditions prevail in an experiment under consideration. The term $(D_o - D_i)/D_o$ represents the magnitude of the caustic split-up for optically anisotropic materials. For Araldite B from equation (5.24) and Table 5.1

---

* Strictly speaking, the circle with radius r_0 for optically anisotropic materials is slightly deformed and split likewise in two branches. For small values of λ, however, and when the diameter D is oriented perpendicular to the crack as considered above, the deformation and the split-up can be neglected, and for r_0 the same value can be assumed as in the optically isotropic case [3].

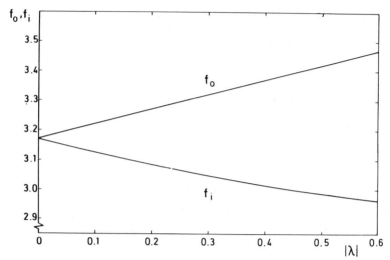

Figure 5.4. The shadow optical constants f_o and f_i as functions of the elasto-optical constant λ.

it can be seen that

$$\frac{D_o - D_i}{D_o} = \frac{f_o - f_i}{f_o} = \begin{cases} 0.079 \text{ for plane stress conditions, and} \\ 0.123 \text{ for plane strain conditions.} \end{cases} \qquad (5.27)$$

These values are significantly different. The splitting of the double caustic is larger for plane strain than for plane stress conditions. Thus, from an evaluation of the double caustic can be experimentally decided which state of stress exists.

The tip of the crack is usually not visible within the shadow pattern. But the crack tip can be localized numerically, since the shape of the caustic is constant. The distance between the crack tip and the inter-section of the caustic with the line defined by an imaginary extension of the crack is given as kD_0. Values for k are listed in Table 5.1.

The static shadow optical analysis of Manogg [3] is based on the equations (5.15) after Sneddon [13] and Williams [14]. These equations describe the stress distribution around the crack tip mathematically by forming an infinite series expanded in terms of $r^{-1/2}$. As the distance r increases relative to the specimen dimensions (or when the distance r is considered as being constant ($r = r_0$) and the distance from the crack tip to the specimen boundaries becomes smaller), then more higher order

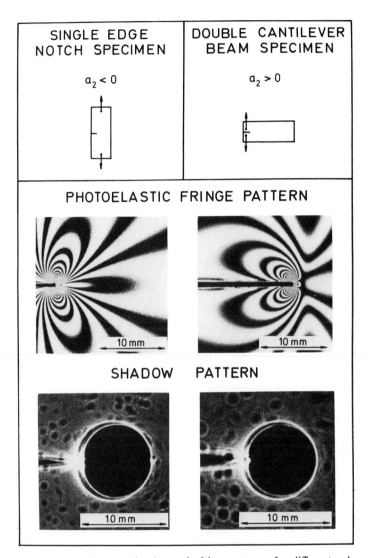

Figure 5.5. Shadow optical and isochromatic fringe patterns for different values of the second order term a_2 (The scales inserted in the photoelastic patterns relate to the specimen plane, the scales in the shadow patterns relate to the reference plane. The caustics are images of a circular line the radius of which in the specimen plane is $r_0 = 4.5$ mm) Araldite B.

terms must be taken into account. The second order term $a_2 r^0$ is taken into consideration in the equations (5.15). Since this term is constant, it is cancelled out in the shadow optical analysis due to the formation of the gradient in equation (5.17).* Physically this means that the influence of the specimen boundaries (in so far as represented by a_2) on the shadow pattern disappears. Thus, the shadow optical method in comparison with photoelastic methods for example, is much less sensitive to disturbances due to specimen boundaries. This is demonstrated in Figure 5.5: In a single edge notch specimen $a_2 < 0$ and the isochromatic fringes lean towards the crack propagation direction. In a double cantilever beam specimen $a_2 > 0$ and the fringes lean backwards. This behavior agrees with predictions in [19] and observations in [20]. The corresponding shadow patterns for both types of specimens, however, show no differences.

5.4 Validity of the analysis for the case of stationary cracks under dynamic loading

In order to verify whether the shadow optical analysis for the statically loaded stationary crack also is valid for the dynamically loaded stationary crack, the basic elasto-optical and fracture mechanics equations must be shown to be the same for both cases.

The basic elasto-optical equation (5.13), which expresses the linear correlation of path length changes and stresses, is also valid when the load is applied dynamically. (Non-linear effects may be excluded). The values of the constants c and λ, however, for many materials depend on the loading rate and, therefore, can be different for static and dynamic loading.

The basic fracture mechanics equations for a crack subjected to a time dependent load (elastic waves or impact) are given in [21]. The near field stress distribution for mode I loading conditions is given as

$$\sigma_x = \frac{K_I(t)}{\sqrt{2\pi r}} \cos \frac{\varphi}{2} \left(1 - \sin \frac{\varphi}{2} \sin \frac{3\varphi}{2} \right),$$ (5.28a)

$$\sigma_y = \frac{K_I(t)}{\sqrt{2\pi r}} \cos \frac{\varphi}{2} \left(1 + \sin \frac{\varphi}{2} \sin \frac{3\varphi}{2} \right),$$ (5.28b)

---

* For optically isotropic materials a_2 is cancelled out completely. For optically anisotropic materials, it can be shown that the remaining influence of a_2 on the shadow pattern is very small.

$$\tau_{xy} = \frac{K_I(t)}{\sqrt{2\pi r}} \cos \frac{\varphi}{2} \sin \frac{\varphi}{2} \cos \frac{3\varphi}{2}. \qquad (5.28c)$$

The dependence of the stresses on r and φ is the same as in the static case which is represented by the equations (5.15). But for the dynamic case, the stress intensity factor K_I is now a function of the time t. Consequently, if the same value of K_I is considered, the same caustic will result for both static and dynamic loading. Thus, the evaluation formulas for the static case can likewise be applied to dynamically loaded stationary cracks.*

5.5 The dynamic correction for the case of propagating cracks

Following the guideline of Manogg's analysis, the caustic for a crack propagating with constant velocity has been calculated [8, 10] using basic elasto-optical and fracture mechanics equations which are valid for this dynamic case. A comparison to the analysis for the stationary crack shows how the static evaluation relations must be modified in order to be applicable for propagating cracks.

The basic equations

The basic elasto-optical equation (5.13) can also be applied for propagating cracks when keeping in mind that the values of the constants c and λ could possibly change for strain rate sensitive materials. However, the basic fracture mechanics equations for propagating cracks have a different form. For a crack propagating with constant velocity v, the near field stress distribution is given in [24] and [25] as

$$\sigma_x = \frac{K}{\sqrt{2\pi}} f' \left[(1 + 2\alpha_1^2 - \alpha_2^2) \frac{\cos \frac{\varphi_1}{2}}{\sqrt{r_1}} - \frac{4\alpha_1 \alpha_2}{1 + \alpha_2^2} \frac{\cos \frac{\varphi_2}{2}}{\sqrt{r_2}} \right], \qquad (5.29a)$$

---

* Theocaris [see his contribution to this volume] has calculated the shadow pattern for a stationary crack subjected to superimposed tensile and shear loading (mode I + mode II). Since according to [21] for mode II loading conditions, the near field stress distribution for static and dynamic loading are also identical, the shadow optical evaluation formulas (see also [12, 22, 23, 6, 7]) for the mixed mode problem are valid likewise in both the static and the dynamic case.

$$\sigma_y = \frac{K_I}{\sqrt{2\pi}} f' \left[-(1+\alpha_2^2) \frac{\cos\frac{\varphi_1}{2}}{\sqrt{r_1}} + \frac{4\alpha_1\alpha_2}{1+\alpha_2^2} \frac{\cos\frac{\varphi_2}{2}}{\sqrt{r_2}} \right], \tag{5.29b}$$

$$\tau_{xy} = \frac{K_I}{\sqrt{2\pi}} f' \left[2\alpha_1 \left(\frac{\sin\frac{\varphi_1}{2}}{\sqrt{r_1}} - \frac{\sin\frac{\varphi_2}{2}}{\sqrt{r_2}} \right) \right]. \tag{5.29c}$$

x, y now represents a moving co-ordinate system with its origin in the crack tip. The co-ordinates φ_1, r_1 and φ_2, r_2 are connected with this system by the relations

$$x = r_1 \cos \varphi_1 = r_2 \cos \varphi_2 \tag{5.30a}$$

$$y = \frac{r_1}{\alpha_1} \sin \varphi_1 = \frac{r_2}{\alpha_2} \sin \varphi_2. \tag{5.30b}$$

Furthermore in the equations (5.29), α_1, α_2 and f' are defined by

$$\alpha_1 = \left(1 - \frac{v^2}{c_L^2} \right)^{1/2} \tag{5.31a}$$

$$\alpha_2 = \left(1 - \frac{v^2}{c_T^2} \right)^{1/2} \tag{5.31b}$$

$$f' = \frac{1 + \alpha_2^2}{4\alpha_1\alpha_2 - (1 + \alpha_2^2)^2} \tag{5.32}$$

where c_L is the longitudinal and c_T the transverse wave velocity.

The evaluation equations

Optically isotropic materials.   A computation analog to that which leads in the static case to the image equations (5.19), yields the corresponding equations for the dynamic case. For optically isotropic materials, these image equations, which give for each point $P(r, \varphi)$ in the specimen (see Figure (5.2)) the image point $P'(x', y')$ in the shadow image, are

$$x' = mr \cos \varphi - \frac{K_I}{\sqrt{2\pi}} cdz_0 r^{-3/2} F^{-1} G_1(\alpha_1, \varphi) \tag{5.33a}$$

$$y' = mr \sin \varphi - \frac{K_I}{\sqrt{2\pi}} cdz_0 r^{-3/2} F^{-1} G_2(\alpha_1, \varphi) \tag{5.33b}$$

where the abbreviations are used:

$$F = \frac{4\alpha_1\alpha_2 - (1 + \alpha_2^2)^2}{(\alpha_1^2 - \alpha_2^2)(1 + \alpha_2^2)} \tag{5.34}$$

$$G_1(\alpha_1, \varphi) = \frac{-1}{\sqrt{2}}(g^{1/2} + \cos \varphi)^{-1/2}(g^{-1/2} - g^{-1}\cos \varphi - 2g^{-3/2}\cos^2 \varphi) \tag{5.35a}$$

$$G_2(\alpha_1, \varphi) = \frac{1}{\sqrt{2}}(g^{1/2} + \cos \varphi)^{-1/2}(\alpha_1^2 g^{-1}\sin \varphi + \alpha_1^2 g^{-3/2}\sin 2\varphi) \tag{5.35b}$$

with

$$g = (\alpha_1^2 - 1)\sin^2 \varphi + 1. \tag{5.36}$$

The determinant condition (5.2) leads to a quadratic equation in $r^{-5/2}$, the solution of which has the form

$$r = \left[\frac{3}{2}\frac{K_I}{\sqrt{2\pi}}\frac{1}{m} cdz_0 F^{-1} H(\alpha_1, \varphi)\right]^{2/5} \equiv r_0^{dyn} \tag{5.37}$$

where r, in analogy to equation (5.20), is denoted as r_0^{dyn} and H is a complicated function dependent only on α_1 and φ [10]. A comparison with the analog relation (5.20) for the stationary crack shows that

$$r_0^{dyn} = F^{-2/5} H^{2/5}(\alpha_1, \varphi) r_0^{stat} \tag{5.38}$$

where r_0 from equation (5.20) is denoted here as r_0^{stat}.

Due to the definition (5.34), F is constant for a given crack velocity v. The function $H^{2/5}(\alpha_1, \varphi)$ has been computed numerically [8, 10] and is plotted in Figure 5.6 for steel ($c_L = 5940$ m/s). When crack velocities are considered within a range typical for fracture experiments (i.e. up to about 1200 m/s in steel), the value of $H^{2/5}$ for all angles φ nearly equals one. Thus in equation (5.38), $H^{2/5}$ can be approximated by one:

$$r_0^{dyn} = F^{-2/5} r_0^{stat}. \tag{5.39}$$

This means that the line which is imaged onto the caustic is practically a circle for the dynamic case also. The radius of this circle is enlarged in comparison to the static case by the factor $F^{-2/5}$.

Inserting r_0^{dyn} from equation (5.38) (and thus taking into account the

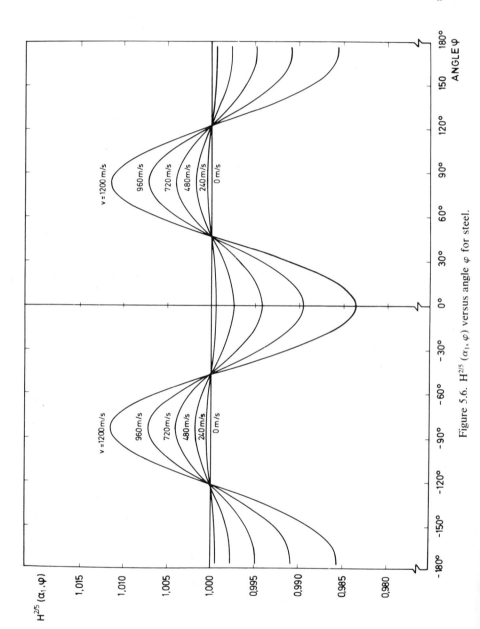

Figure 5.6. $H^{2/5}(\alpha_1, \varphi)$ versus angle φ for steel.

correct values of the function H) into the image equations (5.33) yields the expression for the dynamic caustic which can be written in the form

$$x' = mr_0^{dyn}(\cos \varphi + C_1(\alpha_1, \varphi)) \tag{5.40a}$$

$$y' = mr_0^{dyn}(\sin \varphi + C_2(\alpha_1, \varphi)) \tag{5.40b}$$

where C_1 and C_2 are functions of α_1 and φ, and influence the shape of the caustic. The size of the dynamic caustic depends in the same way on r_0^{dyn} (and thus on K_I) as it depends in the static case on r_0^{stat} (see equations (5.21)). In Figure 5.7 the caustic for steel ($c_L = 5940$ m/s, $c_T = 3230$ m/s) is plotted for three different crack velocities v with the stress intensity factor K_I being fixed. For comparison, the corresponding caustic for the stationary crack is also drawn. The shape of the caustic is approximately the same for all crack velocities and the size increases with increasing velocity. Within the velocity range of most practical interest, however, both the shape and the size of the dynamic caustic differ only slightly from that of the static one.

In order to study in more detail the influence of the functions C_1 and C_2 in the expressions (5.40) on the caustic shape, the caustic is plotted in Figure 5.8 in the normalized form $x' = \cos \varphi + C_1$, $y' = \sin \varphi + C_2$ for the crack velocities $v = 0$ and $v = 0.34c_L$. The latter value for steel ($v = 2000$ m/s) is in the range of the terminal crack velocity but for PMMA ($v = 950$ m/s) is considerably larger than the terminal crack velocity. The figure confirms that the shape of the dynamic caustic is nearly the same as that of the static caustic; for $v = 1000$ m/s in steel and $v = 475$ m/s in PMMA, the difference between the two caustics cannot be resolved in the scale of Figure 5.8.

Because the normalized dynamic and static caustics practically coincide the φ-dependent terms in the equation (5.40) can be replaced by those of the equations (5.21). Then, in analogy to equation (5.22) and using equation (5.39) the caustic diameter $2y'_{max} = D$ can be written as

$$D = mr_0^{dyn}f = F^{-2/5}mr_0^{stat}f \tag{5.41}$$

where f takes the same value as in equation (5.22) for the static case. Inserting r_0^{stat} from equation (5.20), one obtains the evaluation equation for the dynamic caustic for optically isotropic materials

$$K_I = FMD^{5/2} \tag{5.42}$$

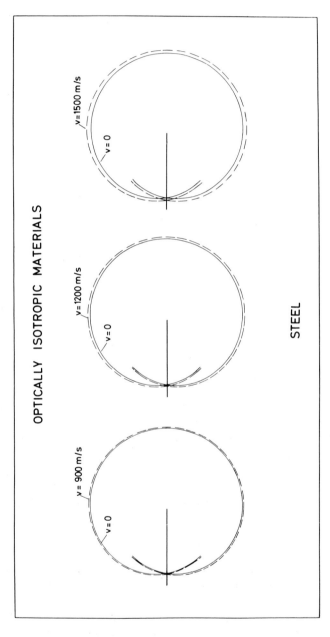

Figure 5.7. Caustics for propagating cracks at different crack velocities. Optically isotropic material (steel in reflection).

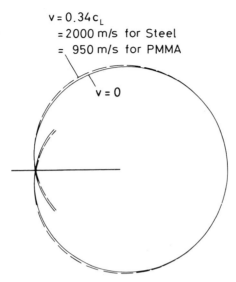

Figure 5.8. Normalized caustics for a propagating crack and for a stationary crack.

where M is defined by equation (5.26). A comparison with the corresponding static evaluation formula (5.25) shows that the static formula can also be used for the evaluation of the caustic for propagating cracks, when modified with the crack velocity dependent correction factor F defined by equation (5.34).*

The correction factor F is plotted as a function of the crack velocity v in Figure 5.9 for steel ($\nu = 0.29$) and PMMA ($\nu = 0.35$). Plotted versus the normalized crack velocity v/c_R, the curve depends only on Poisson's ratio ν of the material. This dependence is very slight, and in most practical cases, only one curve may be used for all materials. For v/c_R increasing from 0 to 1, the value of F decreases from 1 to 0. For crack velocities v up to 290 m/s in PMMA and up to 870 m/s in steel, the value of F varies only between 1 and 0.97. Thus, in most cases the dynamic correction factor can be neglected.

Optically anisotropic materials. For optically anisotropic materials ($\lambda \neq 0$), the term $\pm\lambda(\sigma_1 - \sigma_2)$ in equation (5.17) does not vanish and as a

---

* The error caused by the approximations used in equations (5.39) ($H^{2/5} = 1$) and in equation (5.41) (same shape for static and dynamic caustic) is discussed in [10].

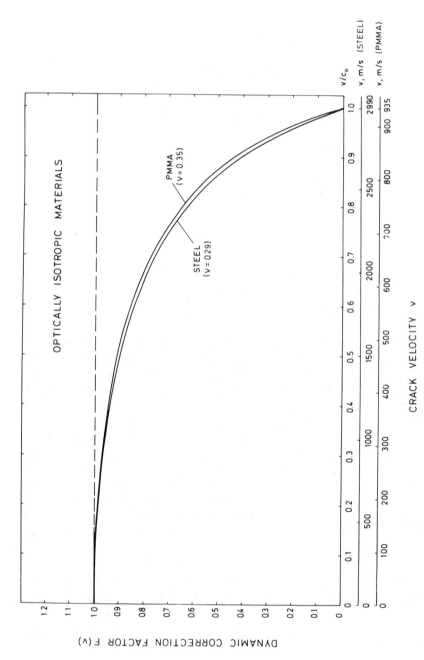

Figure 5.9. Dynamic correction factor F as a function of the crack velocity v.

consequence, the caustic splits up into a double caustic. The image equations (5.33) must be supplemented by an additional term containing the factor $\pm\lambda$. Because the image equation is then very complicated, the determinant equation (5.2) has not been solved analytically for the anisotropic case. As can be shown, [3, 10], however, the physical meaning of the condition (5.2) is equivalent to finding those light beams for which the length $|\mathbf{r}'|$ of the vector $\mathbf{r}'$ (see Figure 5.2) is a minimum when the length $|\mathbf{r}|$ of the vector $\mathbf{r}$ varies. The light beams which fulfill this condition form the caustic in the reference plane. The minimization problem, however, can also be solved numerically [10]. Caustics computed in this way for Araldite B ($c_L = 2500$ m/s, $c_T = 1060$ m/s) are shown in Figure 5.10 for three different crack velocities v (broken lines) in comparison to the caustic for the stationary crack (full lines). Within the velocity range of most practical interest (i.e. up to about 400 m/s for Araldite B), the dynamic caustics differ only slightly from the static ones.

The evaluation equation for optically anisotropic materials can be established by modifying the respective equation (5.42) for optically isotropic media to

$$K_I = F'_{o,i}(v)F(v)MD_{o,i}^{5/2} \tag{5.43}$$

where $F'_{o,i}(v)$ is an additional correction factor. The subscripts o and i relate to the outer and inner branch of the double caustic respectively. Due to compatibility reasons, $F'_{o,i} = 1$ for $v = 0$ and for $\lambda = 0$. For a given value of K_I from equation (5.43) follows

$$\frac{F_{o,i}(v)F(v)}{F'_{o,i}(v = 0)F(v = 0)} = F'_{o,i}(v)F(v) = \frac{D_{o,i}^{5/2}(v = 0)}{D_{o,i}^{5/2}(v)}. \tag{5.44}$$

By numerically computing the right hand side of this equation, the correction factor $F'_{o,i}F$ has been determined as a function of the crack velocity v [10]. The result for Araldite B is given in Figure 5.11 up to $v = 800$ m/s. For comparison, the function $F(v)$ for Araldite B ($\nu = 0.39$) is also drawn. It can be seen that $F(v)$, given analytically by equation (5.34), represents a good approximation for the complete correction factor $F'_o(v)F(v)$ when the outer caustic is evaluated. Then, in most cases of practical interest, the evaluation equation (5.42) can be used for optically isotropic as well as for anisotropic materials.

The radius r_0^{dyn} can also be computed numerically. The result is, that in

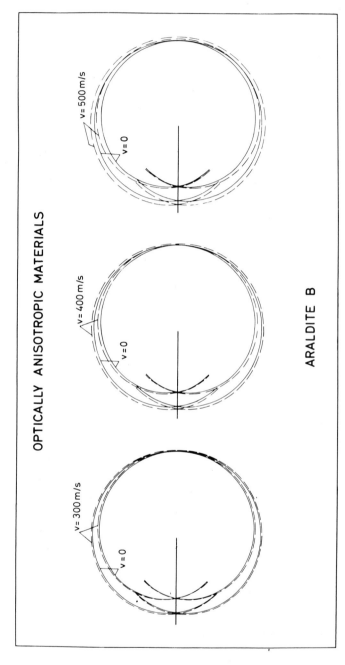

Figure 5.10. Caustics for propagating cracks at different crack velocities. Optically anisotropic material (Araldite B in transmission).

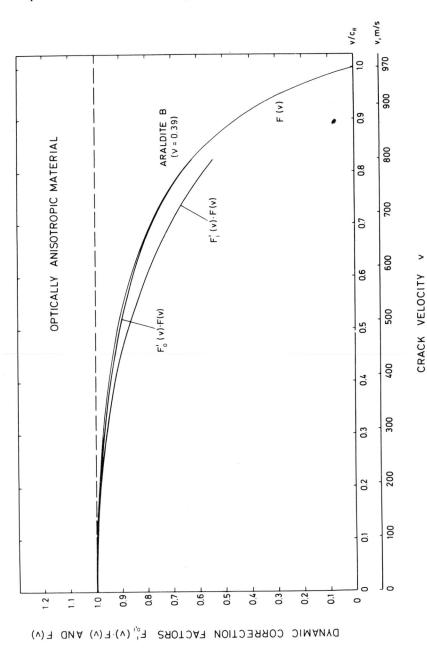

Figure 5.11. Dynamic correction factors $F_i'F$, $F_o'F$ and F as functions of the crack velocity v.

analogy to equation (5.41) r_0^{dyn} can be written*

$$r_0^{dyn} = \frac{D_{o,i}}{m f_{o,i}}. \tag{5.45}$$

The numerical computations indicate that the inner caustic diameter D_i increases stronger with increasing crack velocity than the outer diameter D_o. This can be seen also in Figure 5.10. Consequently, the magnitude of the caustic split-up in the dynamic case is slightly smaller than in the static case. Quantitatively, the static equation (5.27) for propagating cracks must be corrected to the form

$$\frac{D_o - D_i}{D_o} = \frac{F_i'^{2/5} f_o - F_o'^{2/5} f_i}{F_i'^{2/5} f_o} = 1 - \left(\frac{F_o'}{F_i'}\right)^{2/5} \frac{f_i}{f_o} \tag{5.46}$$

as can be seen by inserting $D_{o,i}(v = 0)$ from equation (5.25) into equation (5.44).

5.6 Experimental technique and evaluation procedure

In the previous sections the theoretical and mathematical foundation of the method of shadow patterns is described. In the following, emanating from practical experience, the technical and experimental details are given which must be observed when performing a shadow optical measurement, i.e. in conceiving the experiment, in designing the optical set up and in evaluating the data.

Concept of an experiment. For the conception as well as for the interpretation of a shadow optical measurement, it should be kept in mind that the caustic is the optical image of an approximately circular line of radius r_0 around the crack tip. An evaluation of the caustic, therefore, yields information on the stress-strain conditions along this circular line. Its radius r_0 at most is a few millimeters. The quantitative value of r_0 being expected for an experiment can be estimated by inserting the expected value of the stress intensity factor K_I in equation (5.20). For example for a specimen of Araldite B of thickness $d = 10$ mm in a transmission arrangement with a reference distance $z_0 = 1.3$ m, using

---

* The footnote related to equation (5.24) also applies to the dynamic case.

convergent light with a magnification factor $m = 0.7$, for a stress intensity factor $K_I = 0.79 \, MN/m^{3/2}$ (which represents the initiation toughness K_{Ic} of the material) and assuming plane stress conditions (see below), substitution into equations (5.20) and (5.24) results in the outer caustic diameter $D_0 = 8.6 \, mm$ and $r_0 = 3.7 \, mm$. The radius r_0 can be varied for a given specimen and given load by variation of the reference distance z_0 (see equation (5.20)).

The shadow optical analysis from which the evaluation formulas are deduced is based on the nearfield stress distributions (5.15) and (5.29); respectively, for linear elastic materials. Consequently, the idealizing assumptions underlying these stress distributions should be realized in the experiment as far as possible. Thus, r_0 should be sufficiently larger than the size of the plastic zone, as well as larger than the dimensions of the crack front curvature. (These conditions are fulfilled in the above given example for Araldite *B*). Furthermore, the specimen boundaries should be sufficiently remote from the crack tip. This condition is not very severe, since the shadow pattern is relatively insensitive to disturbances due to boundaries. Empirically, it is mostly satisfactory when the distance from the crack tip to the specimen boundaries is larger than about 20 to 30 mm.

The shadow optical constants c and $f_{o,i}$ in the evaluation formulas are different for plane stress and plane strain conditions (see Table 5.1). Therefore, the prevailing state of stress must be known. The actual state of stress, however, depends on the distance from the crack tip: very close to the tip, plane strain conditions apply, and more remote from the tip, plane stress conditions prevail. There are two ways to decide, what state of stress must be taken into account for the respective shadow optical measurement. One way is to check the amount of the caustic split-up according to equation (5.27). The other way—more sensitive, and also applicable for optically isotropic materials, but also more complicated—is to perform a static calibration measurement where the value of the stress intensity factor K_I is known. The equations (5.25) and (5.26) can then be used to calculate from the measured shadow spot diameter the unknown value of $f_{o,i}^{5/2}c$. It can then be checked to which state of stress this value corresponds. Such experiments have been performed [26] with specimens from Araldite *B* of 10 mm thickness for different values K_I with varying r_0. The resulting value of $f_{o,i}^{5/2}c$ increased from the plane-strain-value to the plane-stress-value for r_0 increasing from zero to about 3 mm. Thus, for $r_0 \gtrsim 3 \, mm$ plane stress conditions can be assumed. For propagating cracks this value can be expected to be even smaller. For $0 < r_0 < 3$ mm intermediate conditions prevail and, therefore, this range should not be used for a measurement.

When choosing the specimen material for an experiment, different material parameters must be considered. The shadow optical *sensitivity*—i.e. the magnitude of the caustic diameter D for a given value of K_I —is described by the constant $f_{o,i}^{5/2}c$ (defined by equations (5.14a) and (5.24)). Large values of $f_{o,i}^{5/2}c$ correspond to high shadow optical sensitivity (see equation (5.25)). The amount of the optical anisotropy is given by the constant λ (defined in equation (5.14b)) which determines the magnitude of the splitting of the caustic into a double caustic. A large value of λ correlates with a large caustic split-up (see Figure 5.4). *Viscoelasticity* leads to a special shadow optical phenomenon. After a crack is started from an originally statically loaded notch, a "remanent" shadow spot can be observed to remain at the tip of the initial notch for a long time after the initiation event. The use of materials, the elasto-optical constants of which are *loading rate sensitive* should be avoided, since in dynamic experiments, in general, the effective loading rate is not known. Furthermore, it should be observed that the size of the *plastic zone* is small compared to the value of r_0.

Plate glass, which is a common model material in experimental fracture mechanics, exhibits high optical anisotropy, but only a very small shadow optical sensitivity (see λ and c in Table 5.1). *Polymethylmethacrylate* (*PMMA*) is approximately optically isotrop, i.e. only a single caustic appears. The sensitivity is very high. The viscoelasticity, however, causes remanent shadow spots which need from a few seconds up to some minutes for relaxation. The relaxation time increases with increasing loading time before crack initiation. For *Homalite 100* the optical anisotropy is small and, therefore, the double caustic generally cannot be resolved on a shadow optical record. The sensitivity is high. Viscoelastic effects are observed but to a much lesser extent than in PMMA. An excellent material for shadow optical measurements is the epoxy resin *Araldite B.** The optical anisotropy and the shadow optical sensitivity are large. The viscoelastic effects are negligible. The differences of the mechanical and optical properties for static and dynamic loading are exceedingly small [27, 18]. A comparison of the properties of Araldite *B* and Homalite 100 is given in [18].

For all materials mentioned, plastic zone effects are negligibly small.

---

* Araldite *B* plates of high quality for shadow optical measurements are manufactured by TIEDEMANN, D-8100 Garmisch-Partenkirchen, Fed. Rep. of Germany.

Plasticity, however, plays a role when metallic materials—in a reflection arrangement—are utilized. Since linear elastic behavior is required except in a small zone around the crack tip, only very brittle metals should be used. The radius r_p of the plastic zone should be small compared to the shadow optical radius r_0. When $R_{p0.2}$ is the yield strength, r_p is proportional to $(K_I/R_{p0.2})^2$, and r_0 is proportional to $K_I^{2/5}$ (see equation (5.20)). Thus, r_p increases stronger with increasing stress intensity factor K_I than r_0. Therefore, the value of $r_p/r_0 \propto K_I^{1.6}/R_{p0.2}^2$ for the material should be as small as possible. The authors obtained good results with the high-strength maraging *steel HFX 760* (X2 Ni Co Mo 18 9 5 $\cong$ 18 Ni grade 300, Stahlwerke Südwestfalen). After a heat treatment 480°C/4h/air, the yield strength $R_{p0.2} \approx$ 2100 MN/m^2 and K_{Ic} is in the range of 70 to 100 MN/m$^{3/2}$. Thus, the plastic zone radius is $\leqslant 1$ mm. In order to achieve a high quality mirrored surface, one side of the specimen should be specially prepared by grinding, lapping and final polishing.

Optical arrangement. The only specific equipment required for a shadow optical arrangement is a light source which can be considered as a point light source. This condition can be fulfilled by utilizing a light emitting area of small size, and likewise by positioning the source at a large distance from the object. The light must be neither monochromatic nor polarized.* When a laser is used, possibly disturbances of the shadow pattern arise because of diffraction effects which are especially marked due to the coherent light.

In Figure 5.12, three examples for a shadow optical transmission arrangement are shown, using divergent (top), parallel (middle) and convergent (bottom) incident light. In the first example (top) no lenses or mirrors are required. The shadow pattern is observed on a screen and can be photographed with a film attached directly on this screen.

The scale factor m defined in equation (5.16), in this arrangement is given as $m = (z_0 + z_1)/z_1 > 1$, where z_0 and z_1 are defined in the figure. In the second example (middle) parallel light is generated ($m = 1$) by use of a lens S_1. The light is focused into the objective L of a camera by a second lens S_2. This camera is focused onto the reference plane.

* When linearly polarized light is used for optically anisotropic materials, the light intensity distribution along the caustic lines is non-uniform. Depending on the polarization direction relative to the respective directions of the principal stresses, the light beam is deflected onto one of the two branches of the double caustic (see equations (5.5) and (5.13)).

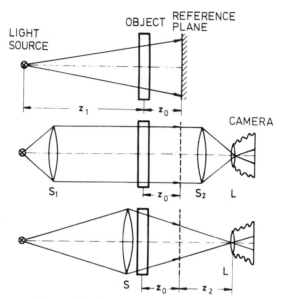

Figure 5.12. Shadow optical arrangements.

A real screen in the reference plane is not required. Therefore, this arrangement is in an analog way also applicable in a reflective arrangement for non-transparent materials (see Figure 5.1), where the reference plane is virtual. This is also valid for the third example (bottom), where only a single lens S is required which simultaneously generates convergent light and focuses it into the camera objective L. The scale factor $m = z_2/(z_0 + z_2) < 1$. In this arrangement, one obtains for increasing reference distance z_0 the strongest increase of the value of the radius r_0, i.e. one obtains also for small reference distances z_0 relatively large values of r_0. This can be shown by inserting m in equation (5.20) in terms of z_0. Consequently, the disturbances due to diffraction effects, which increase with increasing z_0, are lowest in a convergent arrangement. In a convergent arrangement, the caustic diameter D does not increase continually with increasing z_0, because the scale factor m then decreases. As can be calculated [3] by inserting m into equations (5.25) and (5.26), the diameter D takes a maximum value for the reference distance $z_0 = \frac{2}{5}(z_0 + z_2)$, where $z_0 + z_2$ is a constant for a given arrangement.

In order to determine the caustic diameter from a photographic record, during the experiment a scale should be mounted in the reference plane or more conveniently in the specimen plane. In the latter case, the scale is not in focus for the recording camera. However, if thin lines are used for the markings on the scale, the blurred image of these lines forms a symmetric pattern on the film and, therefore, the center of the markings, nevertheless, can be determined very accurately. Measured with a scale mounted in the specimen plane, the caustic diameter which exists in the reference plane appears too large (small) in a convergent (divergent) arrangement. The real caustic diameter is obtained, when the apparent diameter is multiplied by m.

Instead of the lenses S_1, S_2, S in the arrangements shown in Figure 5.12, concave mirrors with large focal lengths are usually used (see below). This allows a larger field of view, larger values and wider variations of z_0 and, because of the long distance between light source and specimen, a better approximation to a point light source.

In arrangements utilizing mirrors, however, astigmatism arises which acts in a manner such that different focal lengths must be taken into account for dimensions perpendicular and parallel to the plane which is defined by the optical axes. Therefore, small deformations of the shadow pattern may appear. These effects become smaller with decreasing angle between the optical axes. It has been verified by experiments that the effects due to astigmatism can be compensated for when the following precautions are observed.

Because of the astigmatism the image of the point light source is split into a line perpendicular to the plane of the optical axes, and into a second line parallel to this plane appearing at a slightly larger distance. The crack in the specimen should be oriented parallel to one of these two lines. The lens of the recording camera (L in Figure 5.12) should be positioned at the location of the light source image which is oriented parallel to the crack. The above described scale for measuring the caustic diameter should be oriented perpendicular to the crack.

For applications of the shadow optical method to dynamic fracture problems, high speed recording techniques must be employed. Figure 5.13 shows a Cranz Schardin high speed camera in both a shadow optical transmission and a reflection arrangement. These arrangements have been used for the applications given in the next Section. 24 sparks triggered in a predetermined time sequence are imaged by a concave mirror onto 24 objectives of a camera array. Each of the objectives corresponds to one of the sparks. For the sake of simplicity, the number of sparks and lenses in the figure is reduced to two. The diameter of the concave mirror is 0.3 m and its focal length is 2.5 m. The focal length of

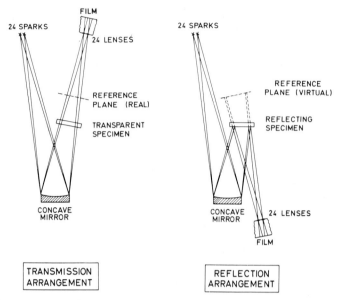

Figure 5.13. Shadow optical arrangements in combination with a Cranz-Schardin high speed camera.

the camera lenses is 500 mm. A film of at least 17 DIN (40 ASA) and a resolution of more than 150 lines/mm is used. The transmission arrangement corresponds to the third example at the bottom of Figure 5.12. In the reflection arrangement, the specimen acts as a mirror. The camera is focused on a virtual reference plane behind the specimen.

Evaluation. An evaluation of the shadow pattern can only yield correct results when the caustic exhibits the correct shape as shown, e.g., in Figure 5.3. A deformation of the caustic indicates that the assumptions underlying the above given analysis are not fulfilled.

The diameters D or D_o, D_i, respectively, defined in Figure 5.3, can be measured very accurately by aid of a microscope. Theoretically, the relevant caustic line should be defined by an abrupt transition from the dark inner space of the shadow pattern to the bright rim. This transition, however, is more or less smoothed due to diffraction effects and due to an imperfect point light source. As is confirmed by experiments [23], correct results are obtained when the line of maximum intensity within the bright rim is considered as the caustic line.

From the measured caustic diameters D or D_o, D_i, respectively, the stress intensity factor K_I and the radius r_0 can be determined. For optically anisotropic materials, a decision can be made concerning the prevailing state of stress. The evaluation formulas are summarized in the following.

The stress intensity factor K_I results from the formula

$$K_I = F'_{o,i}(v)F(v)MD_{o,i}^{5/2} \qquad \text{(equation (5.43))}$$

with

$$M = \frac{2\sqrt{2\pi}}{3m^{3/2}f_{o,i}^{5/2}|c|dz_0}, \qquad \text{(equation (5.26))}$$

$$F(v) = \frac{4\alpha_1\alpha_2 - (1+\alpha_2^2)^2}{(\alpha_1^2 - \alpha_2^2)(1+\alpha_2^2)}, \qquad \text{(equation (5.34))}$$

$$\alpha_1 = \left(1 - \frac{v^2}{c_L^2}\right)^{1/2}, \qquad \text{(equation (5.31a))}$$

$$\alpha_2 = \left(1 - \frac{v^2}{c_T^2}\right)^{1/2}, \qquad \text{(equation (5.31b))}$$

$F'_{o,i}(v)$ = additional dynamic correction factor for optically anisotropic materials. $F'_{o,i}(v)F(v)$ is given in Figure 5.11 for Araldite B.

m = scale factor = $\dfrac{\text{any length in the reference plane}}{\text{corresponding length in the specimen plane}}$;

(defined by eq. (5.16)),

$f_{o,i}$ = shadow optical constant (defined by equations (5.22) and (5.24), see Table 5.1 and Figure 5.4),

c = shadow optical constant (defined by equation (5.14a), see Table 5.1),

d = specimen thickness,

z_0 = reference distance = distance between specimen and reference plane,

v = crack propagation velocity,

c_L = longitudinal wave velocity,

c_T = transverse wave velocity.

The caustic is the image of a circle of radius r_0 around the crack tip in the specimen plane. The value of r_0 can be determined from the caustic diameter by the relation (see equations (5.22), (5.24), (5.41) and (5.45))

$$r_0 = \frac{D_{o,i}}{m f_{o,i}}.$$ (5.47)

By computing the ratio $(D_o - D_i)/D_o$ for optically anisotropic materials according to equation (5.46), it can be determined whether plane stress or plane strain conditions apply for the specific experiment.

In order to determine the crack length, the tip of the crack must be localized within the shadow pattern. The distance from the intersection of crack extension line and caustic to the crack tip is kD_o. Values for k are given in Table 5.1. From a series of shadow optical high speed records with known time sequence, a crack length versus time curve can be established for propagating cracks. By differentiating this curve, one obtains the crack velocity v for each record. From this velocity, the correction factors $F(v)$ and $F'_{o,i}(v) \cdot F(v)$ can be determined by equation (5.34) and Figure 5.11, respectively.

The accuracy of measurement for the shadow optical determination of stress intensity factors, is predominantly given by the accuracy of the values of the shadow optical constants c and $f_{o,i}$ for the material. In particular, for plastic materials these values very often depend on loading rate, temperature, humidity, batch and other parameters. The error due to those effects can be minimized by choosing a suitable material showing small variations of its properties (see discussion above). If necessary, the shadow optical constants must be measured[*] for the conditions prevailing in the experiment under consideration. Except for the shadow optical constants, all other parameters in the evaluation formulas are geometrical lengths which can be measured with very high accuracy. Errors which may arise due to diffraction effects, non point shaped light source, astigmatism and non-fulfilled assumptions of the shadow optical analysis (concerning near-field stress distribution, linear elastic behavior etc.) are discussed above. In practice, stress intensity factors can be determined with an accuracy of measurement of a few percent.

---

[*] For transparent materials, in an interferometric arrangement the path length change for a given load can be determined. On the basis of equations (5.12) the values of a and b are obtained, and from equations (5.14) and Figure 5.4 c, f_o and f_i are found. For non-transparent materials in a reflection arrangement $c = -\nu/E$ and $f_o = f_i = f = 3.17$ (see Table 5.1).

5.7 Applications

The usefulness and efficiency of applying the method of shadow patterns to dynamic fracture problems is demonstrated by two examples. Considered is the influence of dynamic effects on the determination of the dynamic fracture toughness values for both arresting cracks, K_{Ia}, and for dynamically loaded cracks, K_{Id}.

Dynamic stress intensity factors in notched bend specimens under impact loading

Instrumented impact tests (e.g., the Charpy test, the drop weight test) are used to measure the fracture toughness at high loading rates, K_{Id}. Usually loads measured at the tup of the striker are used to provide load-time or load-deflection curves. From these records critical fracture loads for the onset of crack propagation are obtained and, using a static stress intensity factor formula, K_{Id} values are determined.

There are, however, some difficulties inherent to this measuring and evaluation procedure [28, 29]. Firstly, since the load-time or load-deflection curves show characteristic oscillations, their interpretation is uncertain; secondly, the stresses acting in the specimen (which govern the fracture behavior) are inferred from load measurements at the tup without taking the influence of inertia effects into account. As a consequence, erroneous results are very often obtained for dynamic fracture toughness measurements.

In order to investigate the influence of dynamic effects on the determination of K_{Id}-values, therefore, the actual dynamic stress intensity factors were determined by direct measurements at the crack tip utilizing the method of shadow patterns. The results are compared to the load-time curves recorded simultaneously at the tup [2, 30].

The experimental setup is shown in Figure 5.14. Notched bend specimens with dimensions $550 \times 100 \times 10 \text{ mm}^3$ made from the transparent model material Araldite B were investigated. The specimens were struck by a drop weight at a velocity of $v_0 = 5$ m/s. Strain gages were glued close to the tip of the drop weight, thus permitting the recording of the load-time characteristics.

The shadow spots were photographed at successive times by a Cranz-Schardin high speed camera. A laser beam traversing the specimen close to its rear surface was interrupted by the falling drop weight and thereby gave a signal to trigger the high speed measuring devices. Ten out of 24 total photographs which are obtained for one experiment are shown in Figure 5.15. The first three pictures (up to the time of 55 μs) show the

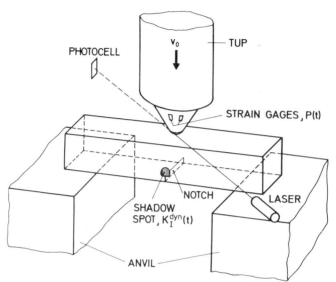

Figure 5.14. Experimental setup for the measurement of the dynamic fracture toughness in an impact test (schematically). (from [2])

increase of size of the shadow spot for the stationary crack as a function of time, the fourth photograph (75 μs) indicates that the crack became unstable (increasing crack length) and the subsequent photographs (up to 195 μs) show the behavior of the crack propagating through the specimen.

Figure 5.16 summarizes the experimental results of three cracks initiated under different conditions by using notch tips of different acuities. The load-time curves $P(t)$ (thin lines) are shown compared with the measured dynamic stress intensity factors $K_I^{dyn}(t)$ (data points). The dynamic stress intensity factors are given up to the time t_f at which crack initiation occurred. The K_I^{dyn} value at this time, therefore, represents the critical stress intensity factor $K_{Id,q}$ for the onset of crack propagation from a blunted notch. (For a sharp crack tip $K_{Id,q}$ would become equal to the fracture toughness K_{Id}).

By comparing the load-time curves $P(t)$ with the stress intensity factor curves $K_I^{dyn}(t)$ the following points emerge:

(1) The dynamic stress intensity factors K_I^{dyn} start increasing later (about 60 μs) than the loads P, as might be expected. This time shift is due to the different wave propagation paths between the point of impact and the positions were the respective signals are recorded.

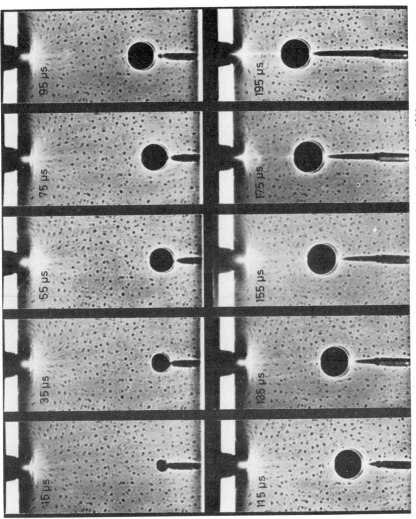

Figure 5.15. Shadow patterns for an impacted crack. (from [30])

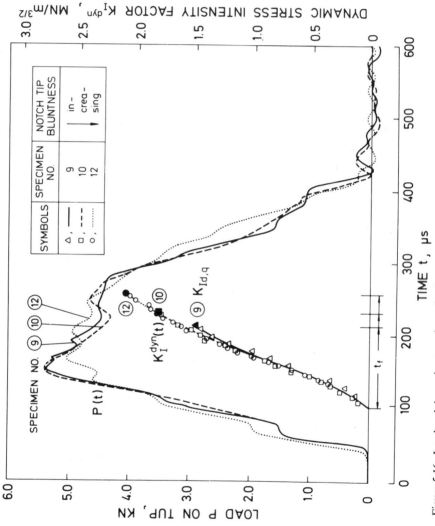

Figure 5.16. Load and dynamic stress intensity factor as functions of the time at an impact test. (from [2])

(2) The time at which maximum load is reached does not coincide with the time t_f at which the specimen actually starts breaking, i.e., the stress intensity factor is still increasing although the load is already decreasing.

(3) Although the conditions for crack initiation were changed for the three specimens, the $P(t)$ curves are practically the same, whereas the $K_I^{dyn}(t)$ curves show marked differences: Cracks initiated from blunter notches become unstable at later times t_f and larger $K_{Id,q}$ values than those from sharper notches. This last finding is in accordance with what is intuitively to be expected.

From these results it may be concluded that the load-time or load-deflection curves, measured at the tup and evaluated in the conventional way, do not provide full information on the actual loading conditions at the crack tip. Consequently the determination of dynamic fracture toughness values K_{Id} based on an estimate of the fracture load obtained from measurements at the tup and on an evaluation procedure using static bend formulas for the stress intensity factor leads to erroneous results. Rather a fully dynamic analysis, taking inertia effects into account, has to be applied. The whole system, consisting of tup, specimen, and anvil has to be treated as a complex time-dependent system.

Dynamic stress intensity factors for fast propagating and arresting cracks

Crack arrest toughness values K_{Ia} are evaluated using a static stress intensity factor analysis. According to this procedure which is advocated by Crosley and Ripling [31, 32], the process of crack arrest is treated mathematically identical to the process of crack initiation, with only the time scale being reversed. Dynamic effects (stress waves, structural vibrations) and their possible influence on the process of crack arrest are considered to be negligible.

This concept is contested by Hahn et al. [33, 34]. These authors analyzed the energetics of fracture arrest and concluded that, as the crack accelerates in the initial phase, kinetic energy is built up in the specimen which is subsequently available to contribute to the crack driving force during the arresting phase of crack propagation. Hahn et al. state that the crack arrest process is determined by the minimum fracture toughness K_{Im}, defined as the minimum value of the dynamic fracture toughness for propagating cracks, K_{ID}, as a function of crack velocity.

In order to investigate the influence of kinetic energy on the crack arrest process Kalthoff, Beinert, and Winkler [35, 1, 9] measured the actual dynamic stress intensity factors for arresting cracks utilizing the method of shadow patterns and compared the results to the corresponding static values. Crack arrest experiments were performed in wedge-loaded double cantilever beam specimens ($321 \times 127 \times 10$ mm^3) made from Araldite B. Notches with different root radii were used to achieve different initial crack velocities. For propagating and subsequently arresting cracks the shadow spots were photographed with a Cranz-Schardin high speed camera. Six out of 24 total photographs are reproduced in Figure 5.17. The momentary crack tip position is given with each photograph. It is recognized that with increasing crack length the shadow patterns decrease in size, thus indicating decreasing stress intensity factors for the advancing crack. The static stress intensity factors for the corresponding crack lengths were obtained using the conventional stress intensity factor formula from deflection measurements at the loading point.

Results for cracks, initiated at different values of the stress intensity factor at initiation (K_{Iq}) and thus propagating at different initial crack velocities, are summarized in Figure 5.18. Values of the dynamic stress intensity factor K_I^{dyn} (experimental points) are shown as a function of crack length a, together with the corresponding static stress intensity factor (K_I^{stat}) curve. Additionally, the measured velocities are given in the lower part of the diagram. The following characteristics of the crack arrest process can be deduced from these results:

At the beginning of the crack propagation phase the dynamic stress intensity factor K_I^{dyn} is smaller than the corresponding static value K_I^{stat}.

At the end of the crack propagation phase the dynamic stress intensity factor K_I^{dyn} is larger than the corresponding static value K_I^{stat}.

Only after arrest does the dynamic stress intensity factor K_I^{dyn} approach the static stress intensity factor at arrest K_{Ia}^{stat}.

Differences between the dynamic and the static stress intensity factor curves become smaller for cracks propagating at lower maximum velocities v_{max}, that is, the dynamic effects decrease with decreasing velocity as one might expect.

Similar experiments were performed in wedge-loaded double cantilever beam specimens made from high strength steel HFX 760 (Stahlwerke Südwestfalen), utilizing the shadow optical technique in reflection [9]. The steel specimens had the same dimension as the Araldite B specimens mentioned previously. A picture of a crack pro-

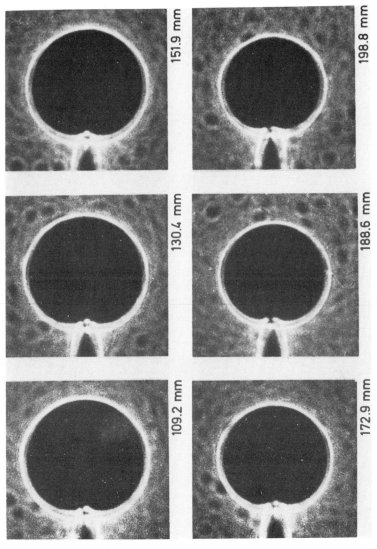

Figure 5.17. Shadow patterns for an arresting crack. (from [1])

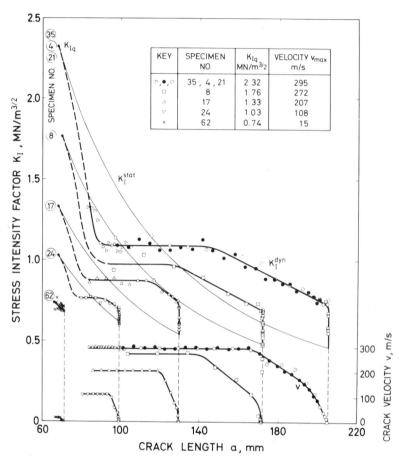

KEY	SPECIMEN NO.	K_{Iq} MN/m$^{3/2}$	VELOCITY v_{max} m/s
◁,●,○	35 , 4 , 21	2.32	295
□	8	1.76	272
△	17	1.33	207
▽	24	1.03	108
×	62	0.74	15

Figure 5.18. Stress intensity factors and crack propagation velocities for arresting cracks. (from [1])

pagating at a velocity of 1000 m/s is shown in Figure 5.19. In addition to the shadow spot elastic waves are visible which emanate from the tip of the propagating crack. The waves are reflected at the finite boundaries of the specimen and then interfere with the stress-strain field at the crack tip. Figure 5.19 gives visible evidence of the influence of dynamic effects on the crack arrest process. Quantitative data for propagating cracks in high strength steel specimens are shown in Figure 5.20 together with equivalent data for propagating cracks in Araldite B (taken from Figure 5.18). A dimensionless plot of the data was chosen for this comparison.

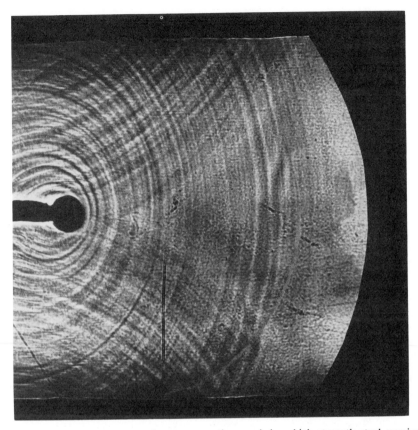

Figure 5.19. Shadow pattern for a propagating crack in a high strength steel specimen.
(from [9])

The (dynamic and static) stress intensity factors are normalized by the initiation stress intensity factor K_{Iq}, and the crack velocity v by the bar wave speed c_0. (In the high strength steel experiments the crack came to arrest only a few millimeters before the far edge of the specimen. The crack arrest process itself was not photographed).

The K_I^{dyn}-curves for the two materials exhibit a characteristic difference. Although similar in nature (K_I^{dyn} at the beginning of crack propagation is smaller and at the end is larger than K_I^{stat}), the K_I^{dyn}-values in the high strength steel specimen show large variations, whereas the data for the Araldite B specimens can be represented by a rather smooth

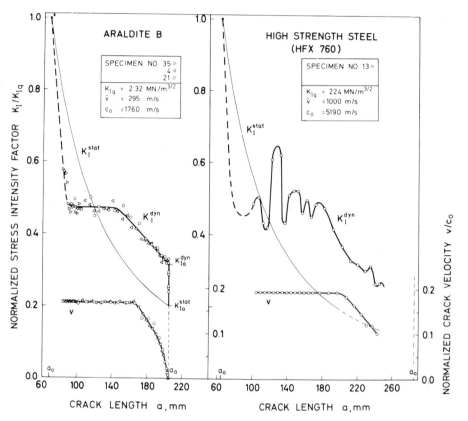

Figure 5.20. Stress intensity factors for arresting cracks in epoxy resin and high strength steel. (from [9])

curve. It is speculated that these large variations of the dynamic stress intensity factor values in the high strength steel specimen are due to higher frequency stress waves interacting with the crack. In Araldite B, very likely because of the larger attenuation, these waves are damped out to a larger extent and practically are not noticeable.

The behavior of the dynamic stress intensity factor in the post arrest phase is shown in Figure 5.21. The stress intensity factors, K_I^{dyn} and K_I^{stat}, and the crack length, a, are plotted as functions of time, t, for cracks in Araldite B and high strength steel specimens. It is recognized that the dynamic stress intensity factor oscillates around the value of the static stress intensity factor at arrest K_{Ia}^{stat}. In Araldite B a nearly harmonic

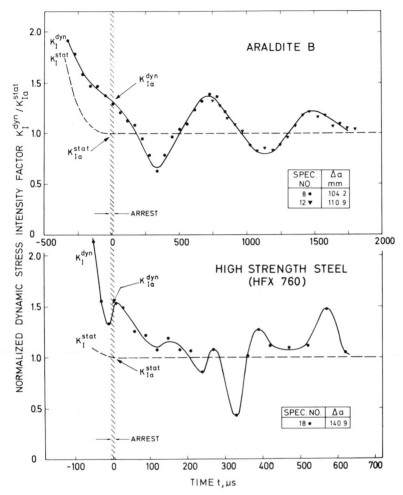

Figure 5.21. Oscillation of the dynamic stress intensity factor after arrest in epoxy resin and high strength steel. (after [9]).

oscillation with a damped amplitude is observed. For sufficiently large times after the crack arrest event, the dynamic stress intensity factor approaches the static value (see Figure 6 in [36]). Also in the high strength steel specimen an oscillation is observed which—due to the smaller attenuation of the elastic waves—shows higher frequency disturbances.

5.8   Concluding remarks

By the method of shadow patterns, stress intensity factors can be experimentally determined with high accuracy. The information which is evaluated in an experiment originates from a very small area surrounding the crack tip. This area can be localized very exactly. Either the light being transmitted through the specimen or the light being reflected from the specimen surface can be utilized. Thus, transparent and also nontransparent specimens can be investigated. The shadow pattern also yields information whether plane stress or plane strain conditions prevail. The method is largely insensitive to disturbances due to specimen boundaries.

For a stationary crack with a given value of the stress intensity factor, the shadow pattern is identical for both dynamically and statically applied load. In both cases, therefore, the same evaluation formulas are valid. For a propagating crack, the shadow pattern differs only very slightly from the one for the stationary crack, when the same value of the stress intensity factor is considered. Therefore, for a propagating crack the evaluation formulas for the stationary crack can be used after modification by a velocity dependent correction factor the value of which is very close to one.

From several applications, it has been shown that the method of shadow patterns is a powerful and efficient method which can be easily applied to investigate dynamic fracture problems. In combination with high speed techniques, dynamic stress intensity factors can be determined as functions of time, crack length, crack propagation velocity and other parameters. It may be expected that in the future the method will be developed further and its applicability will be extended.

References

[1] Kalthoff, J. F., Beinert, J. and Winkler, S., Measurements of Dynamic Stress Intensity Factors for Fast Running and Arresting Cracks in Double-Cantilever-Beam Specimens, *Fast Fracture and Crack Arrest*, The American Society for Testing and Materials, Philadelphia, Pa., U.S.A., ASTM STP 627, pp. 161–176 (1977).

[2] Kalthoff, J. F., Beinert, J. and Winkler, S., Dynamic Stress Intensity Factors for Arresting Cracks in DCB Specimens, *International Journal of Fracture*, 12, pp. 317–319 (1976).

[3] Manogg, P., Anwendung der Schattenoptik zur Untersuchung des Zerreißvorgangs von Platten, Dissertation, Freiburg, Germany (1964).

[4] Manogg, P., Die Lichtablenkung durch eine elastisch beanspruchte Platte und die Schattenfiguren von Kreis- und Rißkerbe, *Glastechnische Berichte*, 39, pp. 323–329 (1966).

[5] Manogg, P., Investigation of the Rupture of a Plexiglas Plate by Means of an Optical Method Involving High Speed Filming of the Shadows Originating Around Holes Drilling in the Plate, *International Journal of Fracture Mechanics*, 2, pp. 604–613 (1966).

[6] Beinert, J., Kalthoff, J. F., Seidelmann, U. and Soltész, U., Das schattenoptische Verfahren und seine Anwendung in der Bruchmechanik, VDI-Berichte Nr. 313, Verlag des Vereins Deutscher Ingenieure, Düsseldorf, Germany, pp. 15–25 (1977).

[7] Beinert, J., Experimental Methods in Dynamic Stress Analysis, *Modern Problems in Elastic Wave Propagation*, edited by J. Miklowitz and J. D. Achenbach, John Wiley, pp. 23–43 (1978).

[8] Beinert, J., Kalthoff, J. F. and Maier, M., Neuere Ergebnisse zur Anwendung des Schattenfleckverfahrens auf stehende und schnell-laufende Brüche, Preprints of the 6th International Conference on Experimental Stress Analysis, VDI-Berichte Nr. 313, Verlag des Vereins Deutscher Ingenieure, Düsseldorf, Germany, pp. 791–798 (1978).

[9] Kalthoff, J. F., Beinert, J., Winkler, S. and Klemm, W., Experimental Analysis of Dynamic Effects in Different Crack Arrest Test Specimens, *Crack Arrest Methodology and Applications*, The American Society for Testing and Materials, Philadelphia, Pa., U.S.A., ASTM STP 711, pp. 109–127 (1980).

[10] Maier, M., Beinert, J. and Kalthoff, J. F., Einfluß der Rißgeschwindigkeit bei der schattenoptischen Bestimmung dynamischer Spannungsintensitätsfaktoren für laufende Risse nach dem Kaustikenverfahren, Report Nr. W 1/79 of the Institut für Festkörpermechanik, Freiburg, Germany (1979).

[11] Föppl, L., and Mönch, E., *Praktische Spannungsoptik*, Springer-Verlag, Berlin, Heidelberg, New York (1972).

[12] Theocaris, P. S., Complex Stress-Intensity Factors at Bifurcated Cracks, *Journal of the Mechanics and Physics of Solids*, 20, pp. 265–279 (1972).

[13] Sneddon, I. N., The Distribution of Stress in the Neighbourhood of a Crack in an Elastic Solid, Proceedings of the Physical Society, London, 187, pp. 229–260 (1946).

[14] Williams, M. L., On the Stress Distribution at the Base of a Stationary Crack, *Journal of Applied Mechanics*, 24, pp. 109–114 (1957).

[15] Theocaris, P. S., Interaction of Cracks With Other Cracks or Boundaries, *Materialprüfung*, 13, pp. 264–269 (1971).

[16] Soltész, U., Experimentelle Untersuchungen zur stoßangeregten Wellenausbreitung in streifenförmigen Platten, Dissertation, Karlsruhe, Germany (1973).

[17] Blauel, J. G. and Kerkhof, F., Optische Verfahren und geeignete Materialien zur Analyse dynamischer Spannungszustände, Report No. 2/67 of the Ernst-Mach-Institut, Freiburg, Germany (1967).

[18] Metcalf, J. T. and Kobayashi, T., Comparison of Crack Behavior in Homalite 100 and Araldite B, *Crack Arrest Methodology and Applications*. The American Society for Testing and Materials, Philadelphia, Pa., U.S.A., ASTM STP 711, pp. 128–145 (1980).

[19] Cotterell, B., Notes on the Paths and Stability of Cracks, *International Journal of Fracture Mechanics*, 2, pp. 526–533 (1966).

[20] Kalthoff, J. F., Theoretische und experimentelle Untersuchungen zur Ausbreitungsrichtung gegabelter Risse, Report No. 7/72 of the Institut für Festkörpermechanik, Freiburg, Germany (1972).

330   J. Beinert and J. F. Kalthoff

[21] Sih, G. C., *Handbook of Stress Intensity Factors*, Institute of Fracture and Solid Mechanics, Lehigh University, Bethlehem, Pa., U.S.A. (1973).
[22] Theocaris, P. S. and Gdoutos, E., An Optical Method for Determining Opening-Mode and Edge Sliding-Mode Stress-Intensity Factors, *Journal of Applied Mechanics*, 39, pp. 91–97 (1972).
[23] Seidelmann, U., Anwendung des schattenoptischen Kaustikenverfahrens zur Bestimmung bruchmechanischer Kennwerte bei überlagerter Normal- und Scherbeanspruchung, Report No. 2/76 of the Institut für Festkörpermechanik, Freiburg, Germany (1976).
[24] Rice, J. R., Mathematical Analysis in the Mechanics of Fracture, *Fracture*, edited by H. Liebowitz, Academic Press, New York-London, pp. 235–238 (1968).
[25] Gross, D., Beiträge zu den dynamischen Problemen der Bruchmechanik, *Habilitationsschrift*, Stuttgart, Germany (1974).
[26] Beinert, J. and Soltész, U., unpublished results.
[27] Clark, A. B. J. and Sanford, R. J., A Comparison of Static and Dynamic Properties of Photoelastic Materials, *Experimental Mechanics*, 3, pp. 148–151 (1963).
[28] *Impact Testing of Metals*, The American Society for Testing and Materials, Philadelphia. Pa., U.S.A., ASTM STP 466 (1970).
[29] *Instrumented Impact Testing*, The American Society for Testing and Materials, Philadelphia, Pa., U.S.A., ASTM STP 563 (1974).
[30] Kalthoff, J. F., Beinert, J. and Winkler, S., Dynamische Effekte beim instrumentierten Kerbschlagbiegetest, 9. Sitzung des Arbeitskreises Bruchvorgänge im Deutschen Verband für Materialprüfung, DVM-Verlag, Berlin (1977).
[31] Crosley, P. B. and Ripling, E. J., Crack Arrest Toughness of Pressure Vessel Steels, *Nuclear Engineering and Design*, 17, pp. 32–45 (1971).
[32] Crosley, P. B. and Ripling, E. J., Characteristics of a Run-Arrest Segment of Crack Extension, *Fast Fracture and Crack Arrest*, The American Society for Testing and Materials, Philadelphia. Pa., U.S.A., ASTM STP 627, pp. 203–227 (1977).
[33] Hahn, G. T., Gehlen, P. C., Hoagland, R. G., Kanninen, M. F., Rosenfield, A. R. et al., *Critical Experiments, Measurements and Analyses to Establish a Crack Arrest Methodology for Nuclear Pressure Vessel Steels*, BMI-1939, 1959, 1995, Battelle Columbus Laboratories, Columbus, Ohio, U.S.A. August 1975, October 1976, May 1978.
[34] Hoagland, R. G., Rosenfield, A. R., Gehlen, P. C. and Hahn, G. T., A Crack Arrest Measuring Procedure for K_{IM}, K_{ID} and K_{Ia} Properties, *Fast Fracture and Crack Arrest*, The American Society for Testing and Materials, Philadelphia, Pa., U.S.A., ASTM STP 627, pp. 177–202 (1977).
[35] Kalthoff, J. F., Winkler, S. and Beinert, J., Dynamic Stress Intensity Factors for Arresting Cracks in DCB Specimens, *International Journal of Fracture*, 12, pp. 317–319 (1976).
[36] Kalthoff, J. F., Beinert, J. and Winkler, S., Influence of Dynamic Effects on Crack Arrest, EPRI RP 1022-1, First Semi-Annual Progress Report, Report V9/78, Institut für Festkörpermechanik, Freiburg, Germany (1978).

E. Sommer

6 | Experimental determination of stress intensity factor by COD measurements

6.1 Introduction

In transparent materials a simple interference optical technique can be used to measure very precisely the opening displacement of cracks in dependence of its local position. With an appropriate experimental arrangement, interference of light from opposing crack surfaces produces the well known Newton's fringes which are lines of equal crack opening displacement. Since the opening displacement is closely related to the stress intensity factor K, a K-calibration can then be established [1, 2]. This technique allows a local K-determination along the front of a crack even in three dimensional cases whereas most other experimental methods based for example on stress or compliance measurements yield K-values in an integral manner. The technique can be applied to transparent materials only, therefore its application is restricted. But some transparent plastics which exhibit elastic behavior may be considered as model materials so that the conclusions drawn may be generalized.

6.2 Principle of the interference optical technique

When parallel monochromatic light is reflected from the two opposing crack surfaces, a system of interference fringes will be produced due to a change in the phase difference of the lightwaves. The interference fringes are lines of equal distance between the two crack surfaces [3].

This may be seen in more detail from Figure 1. An idealized slot filled with a certain medium may be considered. If the wave length of the monochromatic light and the index of refraction of the medium in the slot and those of the surrounding material are λ_1 and n_1 and λ_2 and n_2,

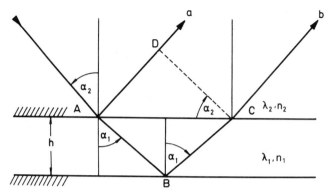

Figure 6.1 Principle of interference

respectively, the phase difference between the two interfering rays *a* and *b* is given by:

$$\phi = \frac{2\pi}{\lambda_1}(\overline{AB} + \overline{BC}) - \frac{2\pi}{\lambda_2}\overline{AD} = \frac{2\pi}{\lambda}2hn_1\cos\alpha_1 \qquad (6.1)$$

The last part of the equation results from geometrical considerations and the use of the law of refraction: $\sin\alpha_2/\sin\alpha_1 = n_1/n_2$; whereas $\lambda = n_i\,\lambda_i$ is the wave length in vacuum.

An additional phase change of π takes place for the reflection at the interface of a dense to a rare medium. Extinction of light (a fringe) occurs if the complete phase change is an odd multiple of π. The condition for interference of light, therefore, results in the equation:

$$\phi = \frac{2\pi}{\lambda}2hn_1\cos\alpha_1 + \pi = m\pi \qquad (m = 1, 2, 3, \ldots) \qquad (6.2)$$

or

$$\frac{2\pi}{\lambda}2hn_1\cos\alpha_1 = (m - 1)\pi = 2q\pi \qquad \left(q = \frac{m-1}{2} = 0, 1, 2, \ldots\right) \qquad (6.3)$$

This interference condition may be rewritten as:

$$h = \frac{\lambda q}{2\,n_1\cos\alpha_1} \qquad (q = 0, 1, 2, 3, \ldots) \qquad (6.4)$$

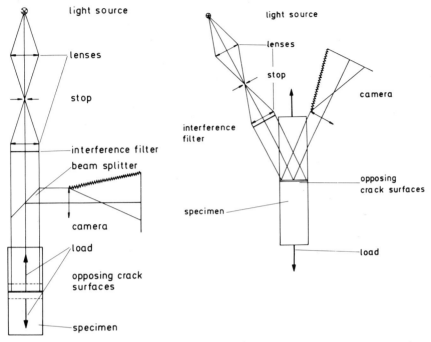

Figure 6.2. Schematic diagram of the experimental arrangement (a) incidence of light perpendicular to crack surfaces and (b) oblique incidence.

To make use of this interference condition two types of experimental arrangements are described in Figures 6.2a and 6.2b. In both cases parallel monochromatic light is produced in the same way. A spectral lamp is taken as a light source which is focused with the aid of a first lens in the plane where the stop is located. The second lens forms the parallel light when its distance from the stop is exactly the focal length. The interference filter ensures that the spectral width of the wave length used remains small.

Figure 6.2a shows the case where the incidence of light is perpendicular to the two opposing crack surfaces, i.e. $\cos \alpha_1 = 1$. For the idealized situation of a slot with two parallel surfaces filled with air ($n_1 \approx 1$), the camera would image a picture which is dark or bright depending whether the distance h is an odd or even multiple of the wave length. Since the real crack surfaces are wedge-like inclined, a whole system of interference fringes are formed.

It is clear then, if n_1 and α_1 are prescribed, the change in the

Figure 6.3. Typical interference fringe pattern for white light; crack opening displacement in a glass specimen [4].

separation distance h is given in terms of the change of q, where q is the fringe number. For the case of a crack, the opening displacement at various distances from the crack tip may be obtained by equation (6.4) where q is the number of light-interference fringes counted from the crack tip.

When only white light is used the interference pattern shows the typical distribution of spectral colors due to the wave length dependent interference condition. Figure 6.3 shows an example of this kind.

For some specimen- and crack-geometries an oblique incidence of light to the crack surfaces is more helpful. An experimental arrangement of this kind requires the additional determination of the angles of light incidence (Figure 6.2b). To relate the angle $\cos \alpha_1$ in the interference condition to the experimentally determined angles of light incidence (β_0^i) and refraction (β_0^e), the following geometrical consideration will be helpful (Figure 6.4). Because it cannot be assumed that the crack surface is always perpendicular to the specimen surface, a small additional angle $\pm \varepsilon$ will be included. From the angles of incidence β_0^i and refraction β_0^e the angles β_2^i and β_2^e can be calculated by means of the law of refraction $\sin \beta_0 / \sin \beta_2 = n_2 / n0$ ($n_2 \approx 1$).

Using the following relations (see Figure 6.4)

$$\beta_2^i - \varepsilon + \alpha_2 = 90° \qquad (6.5a)$$

$$\beta_2^e + \varepsilon + \alpha_2 = 90° \qquad (6.5b)$$

one obtains

$$\sin \alpha_2 = \cos \tfrac{1}{2}(\beta_2^i + \beta_2^e). \qquad (6.6)$$

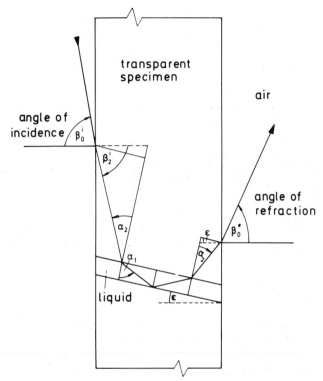

Figure 6.4. Geometrical angle relations.

The angle α_2 is again related to α_1 by the law of refraction $\sin \alpha_2 / \sin \alpha_1 = n_1/n_2$ as follows:

$$\cos \alpha_1 = \left(1 - \left(\frac{n_2}{n_1}\right)^2 \sin^2 \alpha_2\right)^{1/2} \tag{6.7}$$

This result is inserted into equation (6.4) for calculating h.

6.3 Determination of stress intensity factor from crack opening displacement

Analytical background The opening displacement of a crack in an elastic medium is governed by the stress intensity factor K. Therefore K can be calculated from the experimentally determined crack opening displacement $2v = h$ (equation (6.4)). The displacement $2v$ for the two dimensional case from fracture mechanics analysis for pure mode I and plane strain condition is given as [5]:

$$2v = \frac{4K}{E} \left(\frac{2r}{\pi}\right)^{1/2} (1 - \nu^2) + \text{higher order terms} \qquad (6.8)$$

where E is Young's modulus, ν Poisson's ratio and r the radial distance measured from the crack tip.

For an accurate calculation of the stress intensity factor from the scattering experimental data $(2v, r)$, the expected linear relationship between the square of the opening displacement and the distance coordinate is helpful. From a least square fit of the kind

$$4v^2 = a_0 + a_1 r + a_2 r^2 + \ldots \ldots \qquad (6.9)$$

the stress intensity factor K results as:

$$K = \frac{E}{(1 - \nu^2)} \left(\frac{\pi}{32} a_1\right)^{1/2} \qquad (6.10)$$

In three dimensional cases the direct evaluation of the stress intensity factor distribution along the crack contour is in principle possible, but difficult. The method can best be used to verify numerical finite element calculations of the opening displacement for various interesting crack configurations under complex geometry and loading conditions.

Conditions for the applicability of the interference technique. The application of the optical technique is restricted to transparent materials like glasses, glassy polymers or epoxy resins. Its resolution depends upon some limiting factors:

The contrast and the spacing of the fringes are important because the absolute number of a fringe has to be known for the exact determination of the crack opening displacement. Therefore the spacing of the fringes close to the crack tip should not be too small. In order to guarantee a high contrast and also a high number of fringes, the light source of the optical system must emit monochromatic light and the material of the specimens investigated must show good optical quality. The smoothness of the natural fracture surfaces should be as perfect as possible.

The best position of the fringe of the first order is where it is influenced neither by the plastically deformed crack tip region nor by the higher order terms.

K-Determination in crack-line loaded glass plates. The fracture tests have been conducted on glass plates of the sizes $(L \times W \times B : 250 \times 150 \times 6 \text{ mm}^3$ and $250 \times 200 \times 6 \text{ mm}^3)$.

The material was not especially annealed. A notch was cut at the center of the 250 mm side of each plate, and a crack was initiated at the notch root with the aid of thermal stresses. The level of crack-line loads (applied by means of high strength steel wires) were chosen to produce a very slow rate of crack propagation (approximately 10^{-5} to 10^{-3} mm/sec) to ensure good measuring accuracy.

The position of the interference fringes produced by monochromatic Na-light $(\lambda = 0.598 \text{ m}\mu)$ as well as the crack length were measured with a traveling microscope on the center line of the crack surfaces. The resolution of the microscope is $2.5 \cdot 10^{-3}$ mm.

The crack was filled with either purified water (H_2O; $n_1 = 1.33$) or diiodomethane (CH_2I_2: $n_1 = 1.74$). Because of the higher index of refraction of diiodomethane, more fringes per unit length can be evaluated (see equation (6.4)) and greater accuracy can be achieved. The improved accuracy can be attained only if the index of refraction would remain constant over the time period necessary to perform one experiment. The latter condition is better fulfilled with water since the indices of refraction of aqueous solutions of fairly high concentration are nearly equal to that of pure water. On the other hand, water has a very strong influence on crack growth in glass. For the same load, the crack velocity is increased by several orders of magnitude due to stress corrosion effects [6].

Since the goal is to calculate the stress intensity factor K from the crack opening displacement measurements, it is necessary to prove first that the experimental behavior of the crack opening agrees with analytically or numerically calculated values [7]. Due to the influence of real material properties as well as the simplicity of the idealized crack model, a deviation could be expected.

A representative experimental curve is shown in Figure 6.5. The square of the displacement, $2v$, is plotted vs the distance from the crack tip r. Although it is possible to measure the crack opening displacement accurately down to the order of one-half the wave length of light, the smallest corresponding position from the crack tip r is still three orders of magnitude larger. The experimental data can be reduced by a correction factor $(1 + \frac{5}{6} r/a)^2$ which represents the second order term for this load application according to equation (6.9). In agreement with equation (6.8) a straight line (dotted), then, can be constructed.

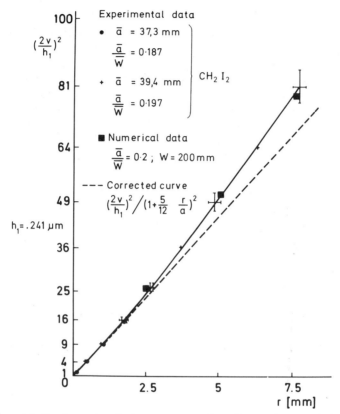

Figure 6.5. Crack opening displacement squared vs. distance from crack tip.

In most experiments, however, the measured displacements are larger than that predicted by numerical computation.

As mentioned earlier, the investigated plates are not annealed. Therefore, residual stresses may exist and influence the results. In order to evaluate the K-values due to external forces only the residual stresses must be removed or known. The superposition of such stress fields is described in the next section.

Influence of residual stresses.   To investigate the influence of residual stresses, the distribution of which will be explained in more detail in the following section, compact tension specimens ($125 \times 120 \times 20$ mm^3) were used. The transparent plastic material found to be suitable as a model material, Araldit B (an epoxy resin), shows almost no creep behavior under load in the range considered. Its elastic constants are $E = 3240 \, N/mm^2$ and $\nu = 0.3$ Care was taken in precracking specimens in order to produce smooth fracture surfaces and sharp crack tips.

A crack was propagated by means of an impact loaded wedge from a saw cut which served as a starter notch. A compressive stress acting perpendicular to the propagation direction and parallel to the crack plane helped to stop the running crack in the required range $0.45 < a/W < 0.55$ where $1.2 \, W =$ width of the specimen.

Such a prepared specimen was adjusted in the optical device described in the previous section (Figure 6.2a). In order to determine the opening displacement from the interference fringe system at different load steps photographs were taken. Two typical examples for load $P_{ext} = ON$ where the fracture surfaces are partially opened due to residual tensile forces and for $P_{ext} = 145.5N$ where the crack is completely open are shown in Figure 6.6. In accordance with equation (6.4), the photographs can be evaluated point for point when the wave length of the monochromatic light $\lambda = 0.546 \, \mu m$ and the exact fringe number q are known. A continuous but slow increase of load allows the increasing number of fringes to be followed at selected markings on the scale photographed through the same optical system.

In this way the curvature of the complete fracture surface as a function of load may be obtained. Examples are shown in the next figures. The square of the opening displacement in the center of the specimen ($z = 0$) is plotted as a function of the distance coordinate r for several steps of load P_{ext} in Figure 6.7a. A similar plot for varying z but constant load P_{ext} is given in Figure 6.7b. The stress intensity factor K_{total} can be determined from these measurements and equation (6.4) when the elastic constants E and ν are known.

In accordance with equations (6.4) and (6.8), the ratio

$$q/r^{1/2} \, \alpha K_{total}(P, z) = K_{ext}(P, z) + K_{int}(z) \qquad (6.11)$$

is expected to be a linear function of P_{ext} for constant values of the thickness parameter z. Since $K_{internal}$ is independent of the external load the slopes of the straight lines will show the variation of K_{ext} with the thickness z and their intercepts with the ordinate the variation of K_{int} with z. Three examples are plotted in Figure 6.8.

A photoelastic examination was made of the stress distribution in the (y, z) plane on a strip of material cut out immediately ahead of the crack

340

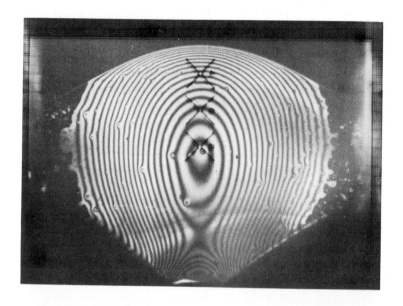

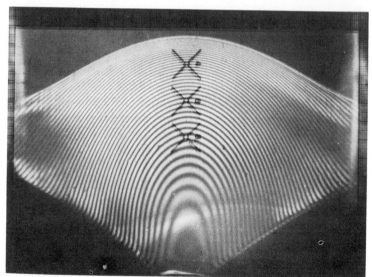

Figure 6.6. Interference pictures of a crack in a transparent specimen under two loading conditions: (a) $P_{external} = ON$ and (b) $P_{external} = 145.5N$

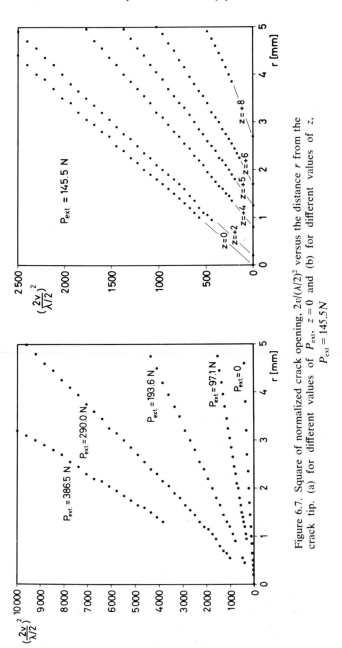

Figure 6.7. Square of normalized crack opening, $2v/(\lambda/2)^2$ versus the distance r from the crack tip. (a) for different values of P_{ext}, $z = 0$ and (b) for different values of z, $P_{ext} = 145.5N$

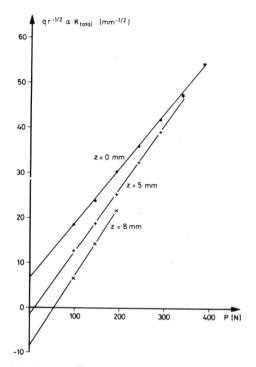

Figure 6.8. Ratio $q/r^{1/2}\alpha K_{\text{total}}$ as a function of external load P.

tip (the crack lies in the (x, z) plane). This showed that the residual stresses were distributed symmetrically about the center plane of the specimen $(z = 0)$, i.e., that they are constant for planes $z = $ const. (neglecting effects close to the free surfaces of the specimen). From the isochromatic fringe pattern [8] in the (y, z) plane which is shown in Figure 6.9, the stresses $\sigma_{y\,\text{res}}(z)$ can be evaluated when σ_z is assumed to be zero;

$$\sigma_{y\,\text{res}} - \sigma_{z\,\text{res}} = \frac{NS}{D} \, \text{N/mm}^2 \tag{6.12}$$

where $N = $ order of isochromatic fringes,
$S = $ fringe value 10.2 N/mm for Na-light [9]
$D = $ depth of the photoelastic specimen $= 39.8$ mm.

To eliminate edge effects the residual stresses where photoelastically

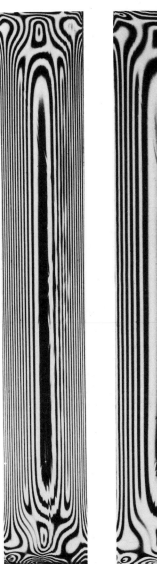

Figure 6.9. Isochromatic fringes showing the residual stress distribution in the (y, z) plane for depth of specimen $D = 38.9\,\text{mm}$ and $20.1\,\text{mm}$.

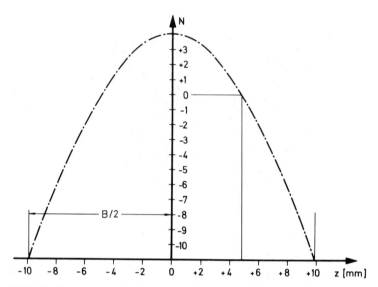

Figure 6.10. Distribution of residual stresses through the plate thickness B. Fringe order N
as a function of the thickness coordinate z.

evaluated for two different values of depth D. The resulting parabolic
distribution plotted in Figure 10 shows a maximum tensile stress $\sigma_{y\,res}(0) =$
1.2 N/mm² in the center of the plate and a maximum compressive stress
$\sigma_{y\,res}(\beta/2) = 2.8$ N/mm² at the plate surfaces $(z = \pm\beta/2)$. The neutral axes
are located at $z = 4.7$ mm.

In Figure 6.11a, a plot of the relative residual stress distribution is
compared with the ratio $K_{int}(z)/K_{int}(0)$. The variation of $K_{ext}(z)/K_{ext}(0)$
with thickness shown in Figure 6.11b may be attributed to the effect of
crack front curvature which can be seen in the variation of the relative
crack length $a(z)/a(0)$. Both $K_{int}(z)$ and $K_{ext}(z)$ were evaluated
from a least squares fit of the experimental data to equation (6.11) which is a
linear function of P_{ext}. The error bars shown in Figure 11 indicate the
resulting standard deviations.

Effect of time dependent behavior. In materials which behave viscoelastically, the crack opening displacement varies with time. Using the
same optical arrangement (Figure 6.2a), photographs of the interference
patterns between fracture surfaces in a specimen of a viscoelastic
material, VP 1527 (a polyester resin), under constant load were taken at
successive time intervals starting immediately after loading. The opening

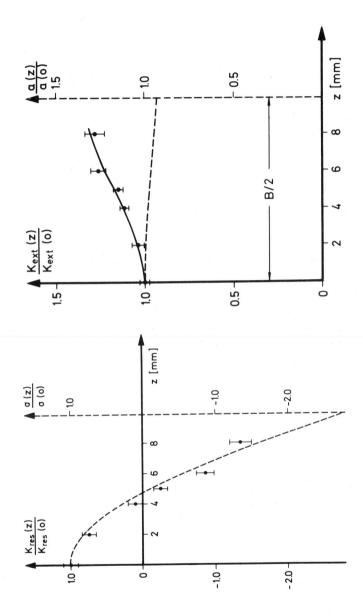

Figure 6.11. Relative stress intensity factors versus thickness coordinate z. (a) $K_{int}(z)/K_{int}(0)$ (.) in comparison to the residual stress distribution $\sigma(z)/\sigma(0)$ (—) and (b) $K_{ext}(z)/K_{ext}(0)$; relative crack length $a(z)/a(0)$ for comparison ($a(0) = 49.9$ mm)

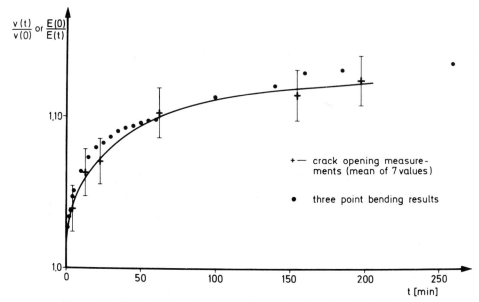

Figure 6.12. Ratio of Young's moduli $E(0)/E(t)$ as a function of loading time

displacement was determined at different times for seven values of the distance coordinate r. The values $v(t)/v(0)$ are plotted in Figure 6.12.

When time dependent behavior is important, consideration of equation (6.8) yields

$$\frac{v(t)}{v(0)} = \frac{E(0)}{E(t)} \tag{6.13}$$

In order to verify this result, three point bending tests using bars measuring $120 \times 10 \times 10 \text{ mm}^3$ of the same material were conducted to determine the ratio $E(0)/E(t)$. These results are also plotted in Figure 6.12. Both sets of data agree within experimental scatter.

6.4   Conclusion

The described experimental technique is a very precise tool for the determination of local stress intensity factors. Its application, however, is relatively difficult and is restricted to transparent materials. More use of the technique should be made to evaluate three dimensional crack problems.

References

[1] Sommer, E., An Optical Method for Determining the Crack-Tip Stress Intensity Factor, *Journal of Engineering Fracture Mechanics*, *I*, pp. 705–718 (1970).

[2] Sommer, E., Experimental Methods for the Determination of Stress Intensity Factors under Various Loading Conditions, in: *Proceedings on, Prospects of Fracture Mechanics*, edited by G. C. Sih, H. C. van Elst, D. Broek, Noordhoff International Publishing Leyden, pp. 593–607 (1974).

[3] Born, M., Optik, Springer/Berlin (1965).

[4] Stahn, D., Institut für Werkstoffmechanik, Freiburg, unpublished results.

[5] Paris, P. C., and Sih, G. C., Stress Analysis of Cracks, in: *Symposium on Fracture Toughness Testing and its Applications*, ASTM STP 381, pp. 30–83 (1965).

[6] Wiederhorn, S. M., Influence of Water Vapor on Crack Propagation in Soda-Lime Glass, *Journal of the American Ceramic Society*, 50, pp. 407–414 (1967).

[7] Brown, W. F. and Srawley, J. E., *Plane Strain Crack Toughness Testing of High Strength Metallic Materials*, ASTM STP 410 (1966).

[8] Föppl, L., Mönch, E., *Praktische Spannungsoptik*, Springer/Berlin (1959).

[9] Soltész, U., Experimentelle Untersuchungen zur stoßangeregten Wellenausbreitung in streifenförmigen Platten, Dissertation, Universität Karlsruhe (1973).

Author's index

Achenbach, J. D. 142, 161
Aleck, B. J. 79, 159
Andonian, A. A. 180, 187
Atluri, S. N. 180, 187

Baron, M. L. 134, 142, 159
Barriage, J. B. 102, 159
Bartholome, G. 180, 182, 187
Becker, E. B. 18, 19, 156
Beinert, J. 281, 283, 284, 297, 299, 303, 305, 319, 322, 324, 325, 326, 327, 328, 329, 330
Besumer, P. M. 180, 182, 187
Bhadra, P. 72, 157
Bickle, W. 23, 157
Blauel, J. G. 293, 329
Blonzou, C. H. 190, 249
Born, M. 331, 347
Bowie, O. L. 228, 252
Bradley, W. B. 175, 186
Broekhoven, M. J. G. 180, 182, 183, 184, 187
Brown, W. F. 337, 347
Buhler Vidal, J. O. 60, 132, 134, 158, 160
Bushnell, D. 19, 156

Carey, J. J. 138, 161
Carr, F. L. XXIII, LV
Cartwright, D. J. 155, 161, 162, 189, 248
Chandrashekhara, K. 107, 159
Chen, T. L. 47, 48, 49, 50, 80, 94, 107, 158, 159, 160
Chiang, F. 88, 114, 159, 160
Chiu, S. T. 175, 186

Chow, C. L. 176, 187
Clark, A. B. J. 310, 330
Clock, L. 39, 150, 157
Cohen, I. M. 180, 182, 187
Coker, E. G. 27, 157, 165, 186
Collatz, I. 17, 156
Cotterell, B. 296, 329
Cox, H. L. 150, 161
Crosley, P. B. 321, 330
Czoboly, E. XXII, LV

Dafermos, K. 79, 159
Dally, J. W. 136, 138, 159, 165, 175, 177, 186, 187
Daniel, I. M. 139, 161
del Rio, C. J. 67, 73, 157, 101, 114, 159, 160
DeSisto, T. S. XXIII, LV
Dowling, A. R. 180, 187
Dudderar, T. D. 245, 252
Duffy, J. 79, 157
Dunham, R. S. 19, 156
Durelli, A. J. 4, 17, 21, 22, 25, 27, 42, 44, 47, 48, 49, 50, 60, 67, 72, 73, 80, 82, 88, 94, 96, 101, 102, 106, 107, 110, 114, 132, 136, 138, 139, 151, 155, 156, 157, 158, 159, 160, 161

Erdogan, F. XVII, LV
Erf, R. K. 22, 23, 156, 157
Eringer, A. C., 153, 161
Etheridge, M. J. 175, 187

Fang, S. J. 142, 161
Feng, H. C. 19, 50, 101, 114, 157, 158, 159, 160

349

Fessler, H. 253, 279
Filon, L. N. G. 27, 157, 165, 186
Fischer, G. 21, 156
Fleischman, T. S. 182, 187
Föeppl, L. 40, 157, 286, 329, 342, 347
Forsyth, P. J. LI, LVI
Fourney, W. L. 175, 187
Freudenthal, A. M. 151, 161
Frocht, M. M. 262, 280

Gdoutos, E. 190, 243, 244, 248, 251, 297, 330
Gehlen, P. C. 321, 330
Gillemot, F. XXII, LV
Gillemot, L. F. XII, LV
Gilman, J. D. 180, 182, 187
Gorman, H. J. 245, 252
Griffel, W. 2, 41, 155, 157
Gross, B. E. 174, 176, 186, 187
Gross, D. 279, 330
Grostad, A. 180, 187
Gurney, C. 149, 161

Hahn, H. G. 42, 157
Hahn, G. T. 321, 330
Halbleib, W. F. 138, 159
Hardrath, W. H. 182, 187
Hartranft, R. J. XLIX, LVI, 154, 161, 253, 254, 255, 261, 273, 276, 279, 280
Havas, I. XXII, LV
Hellen, T. K. 180, 187
Hemann, J. H. 142, 161
Heywood, R. B. 42, 157
Hoagland, R. G. 321, 330
Hofer, K. E. 95, 159
Hogan, M. B. 42, 157
Hulbert, I. E. 185, 186

Inglis, C. E. 24, 157
Ioakimidis, N. I. 189, 190, 198, 199, 202, 208, 243, 244, 245, 248, 249, 250
Irwin, G. 155, 162, 175, 176, 187, 189, 248
Isida, M. 27, 157

Jimbo, Y. 134, 159
Joakimides, N. 190, 248
Jolles, M. 174, 175, 176, 186, 187
Jones, M. H., 176, 187

Kalthoff, J. F. 281, 283, 284, 296, 297, 299, 303, 305, 319, 322, 324, 325, 326, 327, 328, 329, 330
Kanninen, M. F. 321, 330
Kao, R. A. 19, 156
Katsamanis, F. 190, 243, 251
Kassir, M. 167, 186
Kerkhof, F. 293, 329
Key, S. 19, 156
Kim, B. S. 153, 161
Kipp, M. E. 254, 280
Kirsch, G. 24, 139, 157
Klemm, W. 283, 322, 325, 326, 327, 329
Knowles, J. K. 253, 279
Kobayashi, A. S. 102, 159, 175, 186
Kobayashi, T. 175, 187, 293, 310, 329
Kosolov, G. V. 27, 157
Krieg, R. 19, 156

Lake, G. J. LIV, LVI
Larson, F. R. XXIII, LV
Laura, P. A. 18, 156
Lee, Han-Chow, 82, 101, 159
Lehr, E. 28, 157
Lipson, C. 39, 150, 157
Lopardo, V. J. 44, 47, 158

Maiden, D. E. 175, 186
Maier, M. 281, 284, 297, 299, 303, 305, 329
Mall, S. 175, 187
Manogg, P. 281, 282, 284, 291, 293, 294, 305, 312, 328, 329
Mansell, D. O. 253, 279
Matthews, A. T. 134, 142, 159
McEvily, A. J. 3, 4, 155
McGowan, J. J. 174, 176, 186, 187
McLean, J. L. 180, 182, 187
Mendelson, A. 174, 186

Metcalf, J. T. 293, 310, 329
Miksch, M. 180, 187
Milios, J. 190, 243, 251
Mindlin, R. D. 153, 161
Mirabile, M. LV
Mönch, E. 286, 329, 342, 347
Mow, C. C. 136, 142, 159
Muki, R. 154, 161
Mullinx, B. R. 254, 280
Mulzet, A. P. 49, 157
Muskhelishvili, N. T. 18, 156, 189, 197, 200, 202, 231, 248

Neal, D. M. 228, 252
Neuber, H. 32, 42, 43, 146, 157, 158
Nishimura, G. 134, 159
Nisitani, H. 42, 157
Noll, G. 39, 150, 157

Oden, J. T. 19, 156
Ome, G. 132, 159
Oppel, G. 165, 186

Paipetis, S. A. 190, 243, 250
Pao, Y. H. 135, 136, 142, 159
Papadopoulos, G. A. 190, 219, 249, 251
Paris, P. C. 155, 162, 336, 347, 189, 201, 202, 217, 248
Parks, V. J. 21, 27, 44, 47, 49, 50, 60, 72, 73, 80, 82, 84, 88, 94, 95, 96, 101, 106, 107, 114, 151, 156, 157, 158, 159, 160, 161
Pavlin, V. 132, 159
Peters, W. H. 175, 180, 182, 183, 186, 187
Peterson, R. E. 2, 3, 41, 155, 157
Perrone, N. 19, 156
Philips, E. 4, 21, 22, 25, 155, 156, 157
Post, D. 136, 159, 165, 175, 186, 253, 265, 266, 279

Raftopoulos, D. 190, 243, 251
Rao, A. K. 18, 39, 156, 157
Rashid, Y. R. 180, 182, 187

Razem, C. 190, 243, 250
Reynen, J. 180, 182, 187
Rice, J. R. 297, 330
Richardson, M. K. 18, 156
Riley, W. F. 102, 197, 110, 138, 139, 142, 159, 160, 161, 165, 186
Ripling, E. J. 321, 330
Roark, R. J. 2, 155
Rooke, D. P. 155, 161, 162, 189, 248
Rosenfield, A. R. 321, 330
Rossmanith, H. P. 175, 187
Ryder, D. A. LI, LVI

Sanford, R. J. 310, 330
Savin, G. N. 4, 155, 40, 43, 157
Schmitt, W. 180, 182, 187
Schroedl, M. A. 174, 186
Seidelmann, U. 281, 297, 314, 330
Shea, R. 142, 161
Shiryaer, Ya M. 148, 154, 161
Sih, G. C. XVII, XXIII, XXVII, XXX, XXXIV, XXXVI, XLIV, XLV, XLVI, XLIX, LV, LVI, 154, 155, 161, 162, 167, 176, 186, 189, 201, 202, 223, 248, 253, 254, 255, 261, 273, 276, 279, 296, 320, 330, 336, 347
Smith, C. W. 174, 175, 176, 180, 182, 183, 186, 187, 253, 254, 279
Smith, D. G. 253, 254, 279
Sneddon, I. N. 288, 294, 329
Soltész, U. 281, 293, 297, 329, 330, 342, 347
Sommer, E. 331, 347
Sonntag, G. 40, 157
Stahn, D. 334, 347
Stassinakis, C. A. 190, 231, 234, 249, 251
Sternberg, E. 42, 154, 158, 161
Srawley, J. R. 176, 187

Tada, H. 155, 162, 189, 248
Tetelman, A. S. 3, 155
Theocaris, P. S. 79, 102, 159, 160, 190, 198, 199, 208, 218, 219, 223,

Theocaris, P. S. (*contd.*)
224, 226, 228, 229, 231, 234, 238, 240,
242, 243, 244, 245, 248, 249, 250,
251, 252, 287, 291, 297, 329, 330
Thireos, C. G. 190, 238, 240, 250
Thomas, A. G. LIV, LVI
Tsao, C. 4, 21, 22, 25, 84, 155, 156,
157, 159

Wade, B. G. 175, 186
Wang, N. M. 253, 279
Wei, R. P. XVII, LV

Wells, A. A. 136, 159, 165, 175, 186,
254, 280
Westergaard, H. M. 189, 248
Wiederhorn, S. M. 337, 347
Williams, M. L. 234, 236, 252, 288,
294, 329
Wilson, H. B. Jr. 18, 156
Winkler, S. 281, 318, 322, 323, 324,
325, 326, 327, 328, 329, 330

Young, W. C. 2, 39, 155, 157

Zienkiewicz, O. C. 19, 156

Subject index

Anisotropic materials 306, 307

Boundary layer 261, 276

Caustics
  basic equations 195
  branched crack 229
  general description 191, 282
  interface cracks 232, 233
  moving cracks 302, 303, 306, 323, 325
  notches 237
  plates and shells 239, 241
  through cracks 201, 205, 206, 213, 215, 216, 217, 220, 225, 227, 292, 295
Circular cylinder
  external crack XXXVII
  internal crack XXXVI
Cleavage XXIV
Coatings 22
Complex variable formulation 197
Concentrated loads XXXVIII, 12, 130, 131
Couple stress 153, 154
Crack growth XXII, XXVIII, XXXVIII, XL, XLIX
Crack opening displacement 335, 341
Crack tip stresses 168, 169, 172, 173, 259, 288, 296, 297
Cranz-Schardin camera 314
Critical ligament XXXI, XXXIII
Cyclic loading XLVII, XLVIII

Damage accumulation XLIX
Dilatation XXV
Distortion XXV

Dynamic stress concentration factors 135, 136, 144, 145
Dynamic stress intensity factors 301, 305, 315, 317, 320, 324, 326, 327

Effective strain XLIX
Effective stress XLIX

Failure criterion XXVII, XXVIII, XXXIII, XLIX
Fatigue XLVI
Finite strain 44, 47, 49, 50
Fracture toughness XXXV, XXXVI
Fringe pattern (crack) 334, 340
Frozen stress 175, 177, 264

Geometric nonlinearity 43
Grids 20

Hartranft-Sih theory 254
Heterogeneity 4, 9
Holography 22
Hysteresis XLVII, XLVIII

Interferometry 24
Isochromatics
  concrete (coating) 13
  circular holes 164
  circular ring 63
  cracks 90, 93, 166, 179, 268, 269, 270, 295
  elliptical holes 84
  inclusions 14
  polyurethane disk 55
  reinforcing ring 102
  residual stress 343
  rocket grain 128

Isochromatics (*contd.*)
  shrinkage 15
  slot 92, 108
  toroidal cavities 118
Isoclinics 125
Isopachics 246
Isothetics 119

Microcracks XXV
Micropolar 153
Microstructure LIII
Moiré 21
Multiple fringes 268, 269, 270

Optical coefficients 289

Photoelasticity 19, 163
Principle stresses 262

Residual stress 344

Shape change (see distortion)
Shear XXIV
Size effect XXII, XLIII
Speckle 23
Strain energy density factors
  critical   value XXIII,   XXXII,
    XXXIII, XLIV
  mixed mode XXXIV
Strain energy density function
  average value XLIX
  critical   value XXI,   XXIII,
    XXVIII, XXXII, XLI
  dilatational XX, XXVI, XXIX,
    XXX, XXXI, XXXII
  distortional XX, XXVI, XXIX,
    XXX, XXXI, XXXII
  general XX, XXVII, 48
  maximum XXVIII, XXX
  minimum XXVIII, XXIX
Strain gages 23
Strains
  circular bars 125
  embedded elbow 122, 123, 124
  rings 58, 59, 60, 61, 62
Stress concentration factors
  angular corners 89

bonded and free boundaries 71,
  72, 73
circular holes in plate 64, 65, 66,
  104, 105
circular holes in shell 103
circular rings 51, 52
cubic box 130, 131, 132, 133
definition 2
disk 56
elliptical holes in plate 29, 30,.31,
  38, 45, 46, 47, 48, 49, 85, 86
end of bar 109, 110
external notches (deep) 33, 35
external notches (shallow) 34, 36,
  37
hollow cylinder 115, 116, 117
hollow sphere 53, 54
internal flanges 96, 97, 98, 99,
  100, 101
notches in disk 81, 82
plate with fixed edges 68, 69, 75,
  77, 78, 83
plate with multiple holes 94
rocket grain 127
slot 111, 112, 113
Stress intensity factors
  across plate thickness 260, 273,
    274, 275, 278, 279
  branched crack 230
  caustic 291, 293
  complex 202
  corner crack 183, 184, 185
  circumferential crack XXXVII
  crack opening displacement 336,
    345
  edge crack 228
  line crack 154
  penny-shaped crack XXXIV
  plate with crack 242
  surface crack 179

Transmission polariscope 165, 267

Volume change (see dilatation)

Wave pattern 137, 140, 143

COLOPHON

letter: times 10/12, 9/11, 8/10
setter: European Printing Corporation
printer: Samsom Sijthoff Grafische Bedrijven
binder: Callenbach